Handbook of Spectroscopy

Handbook of Spectroscopy

Edited by **Jason Penn**

New York

Published by NY Research Press,
23 West, 55th Street, Suite 816,
New York, NY 10019, USA
www.nyresearchpress.com

Handbook of Spectroscopy
Edited by Jason Penn

International Standard Book Number: 978-1-63238-281-8 (Hardback)

Printed in the United States of America.

Contents

reface

It is often said that books are a boon to mankind. They document every progress and pass on the knowledge from one generation to the other. They play a crucial role in our lives. Thus I was both excited and nervous while editing this book. I was pleased by the thought of being able to make a mark but I was also nervous to do it right because the future of students depends upon it. Hence, I took a few months to research further into the discipline, revise my knowledge and also explore some more aspects. Post this process, I begun with the editing of this book.

This book is a compilation of substantial information on spectroscopy. Spectroscopy is the field pertaining to absorption and emission of electromagnetic radiation through the interface between matter and energy. It is based on the principle that energy varies on the particular wavelength of electromagnetic radiation. It has proven to be critical as a research instrument in multiple spheres like chemistry, physics, biology, medicine and ecology. The spheres of study are growing rapidly and scientists are delving into unchartered areas in this sphere through the introduction of novel procedures. The essential motive of this book is to emphasize on the current spectroscopic techniques like Magnetic Induction Spectroscopy, Laser-Induced Breakdown Spectroscopy, etc. The book covers topics from fundamental to advanced levels of spectroscopy which will offer a great source of knowledge for education and research functions.

I thank my publisher with all my heart for considering me worthy of this unparalleled opportunity and for showing unwavering faith in my skills. I would also like to thank the editorial team who worked closely with me at every step and contributed immensely towards the successful completion of this book. Last but not the least, I wish to thank my friends and colleagues for their support.

Editor

General Spectroscopy

Application of FTIR Spectroscopy in Environmental Studies

Claudia Maria Simonescu

Additional information is available at the end of the chapter

1. Introduction

FTIR Spectroscopy is a technique based on the determination of the interaction between an IR radiation and a sample that can be solid, liquid or gaseous. It measures the frequencies at which the sample absorbs, and also the intensities of these absorptions. The frequencies are helpful for the identification of the sample's chemical make-up due to the fact that chemical functional groups are responsible for the absorption of radiation at different frequencies. The concentration of component can be determined based on the intensity of the absorption. The spectrum is a two-dimensional plot in which the axes are represented by intensity and frequency of sample absorption.

The infrared region of the electromagnetic spectrum extends from the visible to the microwave (Figure 1).

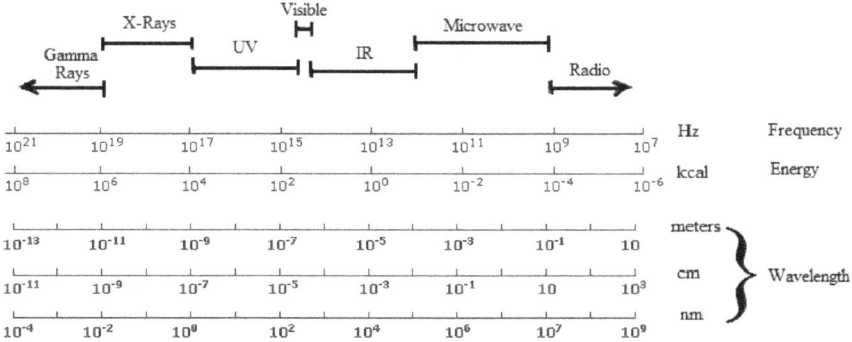

Figure 1. Schematic representation of the electromagnetic spectrum (adapted from http://www2.chemistry.msu.edu/faculty/reusch/VirtTxtJml/Spectrpy/UV-Vis/spectrum)?

Infrared radiation is divided into:

- near (NIR, ν = 10,000 – 4,000 cm^{-1});
- middle (MIR, ν = 4,000 – 200 cm^{-1}) and
- far (FIR, ν = 200 – 10 cm^{-1}).

Because all compounds show characteristic absorption/emission in the IR spectral region and based on this property they can be analyzed both quantitatively and qualitatively using FT-IR spectroscopy.

Today FT-IR instruments are digitalized and are faster and more sensitive than the older ones. FT-IR spectrometers can detect over a hundred volatile organic compounds (VOC) emitted from industrial and biogenic sources. Gas concentrations in stratosphere and troposphere were determined using FT-IR spectrometers (Puckrin et al., 1996).

In case of environmental studies FTIR Spectroscopy is used to analyze relevant amount of compositional and structural information concerning environmental samples (Grube et al., 2008). The analysis can be performed also to determine the nature of pollutants, but also to determine the bonding mechanism in case of pollutants removal by sorption processes. Techniques for measuring gas pollutants such as continuous air pollutants analyzer (SO$_2$, NO$_2$, O$_3$, NH$_3$), on-line gas chromatography (GC) used simple real-time instruments to quantify gas pollutants. They need to use several sensors in order to analyze multiple gas pollutants simultaneously.

FT-IR spectroscopy coupled with other spectroscopic techniques such as AAS (atomic absorption spectroscopy) have been used to assess the impact of industrial and natural activities on air quality (Kumar et al., 2005; Childers et al., 2001).

In addition to the traditional transmission FTIR (T-FTIR) methods (e.g. KBr-pellet or mull techniques), modern reflectance techniques are widely used today in environmental, agricultural, pharmaceuticals, and food studies. These modern techniques are attenuated total reflection FTIR (ATR-FTIR), and diffuse reflectance infrared Fourier transform spectroscopy (DRIFTS). The choice of the method to be used depends on many factors such as: the information needed (bulk versus surface analysis), the physical form of the sample, the time required for sample preparation (Majedová et al., 2003).

In the following there will be presented some of the most important research studies related to the involvement of FTIR spectroscopy in environmental studies.

2. Traditional transmission FT-IR (T-FTIR) spectroscopy in environmental studies

Transmission spectroscopy is the oldest and most commonly used method for identifying either organic or inorganic chemicals providing specific information on molecular structure, chemical bonding and molecular environment. It can be applied to study solids, liquids or gaseous samples being a powerful tool for qualitative and quantitative studies.

FTIR instrument's principle of function is the following: IR radiation from the source that hits the beam splitter is partly directed towards the two mirrors arranged as shown in Figure 2. One of the two mirrors is stationary, and the other is moved at a constant velocity during data acquisition. As it can be seen in Figure 2 at first the IR beams are reflected by mirrors, after that are recombined at the beam splitter, and then passed through the sample and reach the detector. This records all wavelengths in the IR range. After the two beams reflected by the mirrors recombine, they will travel different distances, and the recombination will lead to constructive and destructive interference. The result will be an interferogram. After the recombined beam has passed through the sample the detector will record the Fourier transform of the IR spectrum of the sample. The data obtained are then processed by a computer that performs an additional Fourier transform to back-transform the interferogram into an IR spectrum (Smith *et al.*, 2011; Blum and John, 2012).

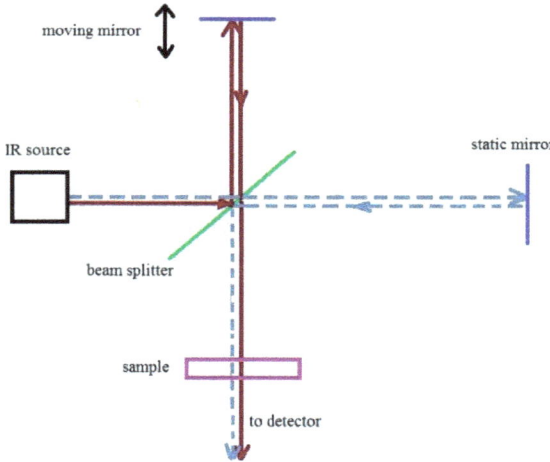

Figure 2. A schematic representation of an interferometer used in FTIR spectrometers (adapted from Blum and John, 2012 with permission (originally published in Drug Test. Analysis, DOI: 10.1002/dta.374))

The potential value of FTIR spectroscopy to a wide range of environmental applications has been demonstrated by numerous research studies. Some of them are presented below.

A review by McKelvy and coworkers containing 132 references at the chapter related to environmental applications of infrared spectroscopy (McKelvy *et al.*, 1998) covers the published literature about relevant applications of infrared spectroscopy for chemical analysis. The literature research was made for the period November 1995 to October 1997. The review contains aspects about infrared accessories and sampling techniques, infrared techniques, applications of infrared spectroscopy in environmental analysis, synthesis chemistry, food and agriculture, biochemistry and also the books and reviews appeared in that period for this subject (McKelvy *et al.*, 1998). An other review concerning the near-infrared and infrared spectroscopy was made by Workman Jr. This review covers the period 1993-1999 and presents the application of the near infrared spectral region to all types of analyses (Workman Jr, 1999).

The basic principle and methods of FTIR spectroscopy of the atmosphere are presented by Bacsik and coworkers in 2004 (Bacsik *et al.*, 2004). The same group of researchers published a review article related to the most significant and frequent applications of FTIR spectroscopy to the study of the atmosphere (Basick *et al.*, 2005). The authors summarized the basic literature in the field of special environmental applications of FTIR spectroscopy, such as power plants, petrochemical and natural gas plants, waste disposals, agricultural, and industrial sites, and the detection of gases produced in flames, in biomass burning, and in flares (Basick *et al.*, 2005).

Applications of FTIR spectroscopy to agricultural soils analysis were presented and discussed by Raphael in the book entitled "Fourier Transforms - New Analytical Approaches and FTIR Strategies" (Raphael, 2011). Chapter 19 of the same book presents the application of FTIR spectroscopy in waste management, and chapter 21 presents the study of trace atmospheric gases using Ground-Based Solar Fourier Transform Infrared Spectroscopy (Smidt *et al.*, 2011; Paton-Wals, 2011).

In case of air pollution the Fourier transform infrared (FTIR) instrument is used succesfully for measuring gas pollutants due to its many advantages such as: multiple gas pollutants will be monitored in real time, the IR spectra of sample can be analyzed and preserved for a long time, can be use to detect and measure directly both criteria and toxic pollutants in ambient air, measures also organic and inorganic compounds, can be also used to characterize and analyze microorganisms and monitor biotechnological processes, is generally installed at one location, but can be also portable and operated using battery for short-term survey, presents sensitivity from very low parts per million to high percent levels, can be applied to the analysis of solids, liquids and gases, no reagent is needed, and data acquisition is faster than with other physico-chemical techniques (Santos *et al.*, 2010).

The basic principle of FT-IR spectroscopy used in air pollutants detection and measuring is that every gas has its own „fingerprint" or absorption spectrum. The entire infrared spectrum will be monitored and FTIR sensor will read the different fingerprints of the gases present in the air sample. In case of determination of gas concentrations from stratosphere, the FT-IR spetrometers have to be designed with a fine resolution (0.01 cm^{-1}) due to the lower atmospheric pressure, and with a lower resolution between 0.05 cm^{-1} and 2 cm^{-1} for tropospheric gases determination. This is due to pressure broadening effects that result in broadened absorption lines. In troposphere water vapor concentrations are higher than those from stratosphere and they have a negative effect on the FT-IR spectrum measurements. The strong interference of water vapor in troposphere is overcome by detecting chemical substances in narrow bands of the IR spectrum where water absorption is very weak.

The total precipitable water vapour (PWV) from air which is responsible for the greenhouse effect being the most important trace gas can be measure using FT-IR spectroscopy. When it was compared with other instruments such as a Multifilter Rotating Shadow-band Radiometer (MFRSR), a Cimel sunphotometer, a Global Positioning System (GPS) receiver and daily radiosondes (Vaisala RS92) it was estimated that FTIR spectrometer provides very precise trophospheric water vapour data, but when area-wide coverage and real-time data availability is very important, the GPS and the RS92 data are more appropriate. FTIR spectroscopy can be use also as a reference when assessing the accuracy of the other techniques, but those who use this technique have to be aware of the FTIR's significant clear sky bias (Schneider, 2010).

Animal farms are major sources of air pollution with ammonia and greenhouse gases. Air concentration of these pollutants may be higher or lower depending on the systems used. In addition, these systems have to correspond both in terms of animal welfare, and in terms of environmental protection. If it is considered animal welfare, the straw based systems are considered animal friendly systems, and when it is considered the environmental protection, the slurry based systems are preferred, due to lower ammonia (NH_3) and greenhouse gas (GHG) emissions. For slurry based systems air pollutants emissions were intensively researched, and the specific emission factors for several slurry-based housing systems for pigs are mentioned in the "Guidance document on control techniques for preventing and abating emissions of ammonia" developed by the UN/ECE "Expert Group on Ammonia Abatement" of the "Executive Body for the Convention on Long-Range Transboundary Air Pollution" (EB.AIR/1999/2). The straw based systems have not been extensively studied in terms of emissions of air pollutants. There are few research studies regarding these systems. Thus, high resolution FTIR spectrometry was used in order to determine the emissions of ammonia (NH_3), nitrous oxide (N_2O), methane (CH_4), and volatile organic compounds (VOC) at a commercial pig farm in Upper Austria using a straw flow system by Amon and coworkers (Amon et al., 2007). The straw flow system is an animal friendly housing system for fattening pigs, being often equated with deep litter where there is no separation between the lying and the excretion areas. In deep litter systems most of the pigs welfare requirements are fulfilled. The main disadvatages of these systems are that there is a high straw consumption, the pigs are dirtier and the deep litter are characterized by high levels of NH_3 and greenhouse gases (GHG). Thus the level of NH_3 and greenhouse gases (GHG) has to be monitored in order to control and to avoid air pollution and to take appropriate measures for environment protection. For the pig farm monitored by Amon and coworkers it can be concluded that the straw flow system may combine recommendations of animal welfare and environmental protection (Amon et al., 2007).

Environmental problems are also due to the incorect application of manure. The main air pollutants associated with manure application are ammonia, and nitrous oxides. In order to develop new environmentaly friendly methods for manure applications all aspects have to be investigated. For this purpose Galle and coworkers made some area-integrated measurements of ammonia emissions after spreading of pig slurry on a wheat field, based on gradient measurements using FTIR spectroscopy. They concluded that the gradient method is valuable for measurement of ammonia emissions from wide area, although the detection limits of the system limits its use to the relatively high emissions (Galle et al., 2000).

In another study Jäger and coworkers reported that FTIR spectroscopy is capable of measuring low concentrations of CO_2, CH_4, N_2O and CO as well as isotope ratios (especially that of $^{13}CO_2$) in gas samples. The concentration levels of these gases are close to them in environmental air (Jäger *et al.* 2011). In the same paper the authors discussed also about the accuracy and stability of the FTIR instrument.

Volcanoes are considered important natural sources of air pollution. The most abundant gas typically released into the atmosphere by volcanoes is water vapor (H_2O), followed by carbon dioxide (CO_2) and sulfur dioxide (SO_2). Other gases such as hydrogen sulfide (H_2S), carbon monoxide (CO), hydrochloric acid (HCl), hydrofluoric acid (HF), hydrogen (H_2), helium (He), silicon tetrafluoride (SiF_4), carbon oxysulfide (COS) are released by volcanoes in small amounts. From the most dangerous to human, animals and agriculture are carbon dioxide, sulfur dioxide and hydrofluoric acid. Therefore it is important to monitor volcanic activities.

The first report about determination of HCl and SO_2 in volcanic gas dates from 1993 when Mori and coworkers used an FT-IR spectrometer during a stage of dome lava extrusion of the Unzen volcano (Mori *et al.*, 1993). Other gases including H_2O, CO_2, CO, COS, SO_2, HF were measured using a remote FT-IR spectral radiometer (Mori and Notsu, 1997; Francis *et al.*, 1996; Love *et al.*, 1998; Burton *et al.*, 2000; Mori and Notsu, 2008).

A telescope-attached FT-IR spectral radiometer was used to study the volcanic gases in seven active volcanoes from Japan. For one of the volcanoes monitored the authors have been used infrared radiation from hot lava domes, for three of them they used infrared radiation of the hot ground surface, and for the other three they used scattered solar light, as infrared sources. The observations over 15 years suggest that HCl/SO_2 and HF/HCl ratios are the most promising parameters reflecting volcanic activity among various parameters observable in remote FI-IR measurements (Notsu and Mori, 2010).

Oppenheimer and coworkers used thermal imaging and spectroscopic (FTIR) techniques to characterize phase-locked cycles of lava lake convection and gas plume composition of the Erebus volcano, Antarctica - a volcano continuously active for decades being now in steady-state. The authors identified a striking, cyclic correspondence between the surface motion of lava lake, and its heat and gas output. They concluded that this can be a reflection of unsteady, bi-directional magma flow in the conduit feeding the lake. It was also determined the ratio between gases emitted by volcanic lake, and the very tight correlation between CO_2 and CO was attributed to the redox equilibrium established in the lava lake. These results have a great contribution to the understanding of the laboratory models for magma convection degassing and volcanic gas geochemistry (Oppenheimer et al., 2009).

FTIR technique offers the potential for the non-destructive, simultaneous, real-time measurement of multiple gas phase compounds in complex mixtures such as cigarette smoke (Bacsik *et al.*, 2007a). Thus, in a study Bacsik and coworkers reported using of FTIR spectroscopy to study the mainstream cigarette smoke from cigarettes of different stated strengths (regular and various light cigarettes with different reported nicotine, tar and CO contents) (Bacsik, 2007b). The cigarette smoke is a very complex mixture that mainly consists of hydrocarbons and both carbon and nitrogen oxides. The results obtained by the authors reveal the fact that the strength of the cigarettes does not have a significant bearing on the quantity of the observed components obtained (Bacsik, 2007b).

An other anthropic source of air pollution is aircraft flight. The main pollutants released by aircrafts are unburnt hydrocarbons, carbon monoxide, and nitrogen oxides. The level of these pollutants is higher near the airport. For modern aircrafts the level of pollutants emissions is lower due to the using of more efficient turbine engine. Nevertheless the civil aviation authorities require the monitoring emissions from aircraft in airports and in the vicinity of airports. For this a non-intrusive Fourier Transform Infrared (FTIR) spectroscopy has been used to detect hydrocarbons in emissions from gas turbine engines (Arrigone and Hilton, 2005). The advantages of this mentioned techniques reported by Arrigone and Hilton are: it is non-intrusive—no sampling system is required and there is no physical interference with the exhaust plume while measurements are made; is useful for simultaneous monitoring of several species; the equipment is portable and can be simply set up and used outside the laboratory in engine test facilities, airfields (Arrigone and Hilton, 2005).

All these advantages encourage the use of FTIR spectroscopy as a valuable tool in monitoring emissions from aircraft in airports.

Quantitative information about air components and air pollutants is needed to study the impact of pollutants (gaseous, liquids or solids) on human health and atmospheric chemistry. To obtain these information an infrared spectral database was created. This database was completed with spectral information of gases emitted by biomass burning by Johnson and coworkers. The following classes of compounds: singly- and doubly-nitrogen-substituted aromatic, terpenes, hemi-terpenes, retenes and other pyrolysis biomarker compounds, carboxylic acids and dicarboxylic acids were identified in gases from biomass burning (Johnson et al., 2010).

Throughout, latest years, the significance of bioaerosols has been discussed in environmental and occupational hygiene. Identification of microorganisms using cultivation and microscopic examination is time consuming and alone does not provide sufficient information with respect to the evaluation of health hazards in connection with bioaerosol exposure. FT-IR spectroscopy has widely been used for the characterization and identification of bacteria and yeasts, due to the fact that they are hydrophilic microorganisms and can easily be suspended in water for sample preparation (Essendoubi et al., 2005; Duygu et al., 2009). The identification of airborne fungi using FT-IR spectroscopy was described by Fischer and coworkers. They found that the method was suited to reproducibly differentiate *Aspergillus* and *Penicillium* species. The results obtained can serve as a basis for the development of a database for species identification and strain characterization of microfungi (Fischer et al., 2006).

Studies on heavy metals and organic compounds removal from wastewaters using different natural and synthetic materials are many. The important role of FTIR spectroscopy in such studies is either related to the characterization of sorbents, chemical modified sorbents, or to establish the mechanism involved in sorption processes (Cheng *et al.*, 2012; Chen and Wang, 2012; Xu *et al.*, 2012; Ma *et al.*, 2012a; Wang *et al.*, 2011; Jordan *et al.*, 2011; Bardakçi and Bahçeli, 2010; Pokrovsky *et al.*, 2008; Parolo *et al.*, 2008).

Biosorption is considered as an alternative process for the removal of heavy metals, metalloid species, compounds and particles from aqueous solution by biological materials (Mungasavalli *et al.*, 2007). Biomaterials are adsorbent materials with high heavy metals adsorption capacity. They have many advantages such as reusability, low operating cost, improved selectivity for specific metals of interests, removal of heavy metals found in low concentrations in wastewaters, short operation time, and no production of secondary compounds which can be toxic (Mungasavalli *et al.*, 2007). FTIR spectroscopy can be used for characterization of biomaterials used in depolluting processes, but also to characterize materials obtained after chemical modification of them. Thus we used FTIR spectroscopy to characterize the material obtained after chemical modification of chitosan with glutardialdehyde in order to obtain a product with good sorption properties (Deleanu *et al.*, 2008), but also to characterize the materials obtained after alkaline treatment of bentonite to increase its capacity to retain ammonium ions from synthetic solutions (Simonescu *et al.*, 2005).

FT-IR spectroscopy has been used to identify the nature of possible sorbent (biosorbent) – pollutants (heavy metals, inorganic compounds, organic compounds) interactions.

For copper removal by fungal biomass to determine the characteristic functional groups that are responsible for biosorption of copper ions were made biomass's FTIR spectra before and after the bisorption process took place. The bonding mechanism between copper and biomass (fungal strain, cyanobacteria or other microorganism) (Yee *et al.*, 2004; Burnett *et al.*, 2006) can be determined by interpreting the infrared absorption spectrum.

We used in our studies fungal strains in order to remove heavy metals from synthetic waters which contain also copper in the form of copper sulfide nanoparticles, but also copper in dissolved state. In case of copper biosorption by *Aspergillus oryzae* ATCC 20423 the FTIR spectra registered are presented in Figure 3. The FTIR spectrum for *Aspergillus oryzae* ATCC 20423 before copper biosorption is presented in Figure 3a, the FTIR spectrum of *Aspergillus oryzae* ATCC 20423 after growth in the presence of copper solution with 25 mg copper/L is presented in Figure 3b, the FTIR spectrum of *Aspergillus oryzae* ATCC 20423 after growth in the presence of copper solution with 50 mg copper/L is presented in Figure 3c, the FTIR spectrum of *Aspergillus oryzae* ATCC 20423 after growth in the presence of copper solution with 75 mg copper/L is presented in Figure 3d, and the FTIR spectrum of *Aspergillus oryzae* ATCC 20423 after growth in the presence of copper solution with 100 mg copper/L is presented in Figure 3e.

From the Figure 3 it can be seen that all five FTIR spectra present distinct peaks in the following ranges: 3393 – 3418 cm^{-1}, 2926 – 2968 cm^{-1}, 1629 – 1638 cm^{-1}, 1404 – 1405 cm^{-1}, 1073 - 1077 cm^{-1}, and 529 – 533 cm^{-1}. The broad and strong band situated in the range 3393 – 3418 cm^{-1} can be attributed to overlapping of –OH and –NH stretching. The band from the range 2926 – 2968 cm^{-1} is attributed to the C-H stretching vibrations. The strong peak at 1629 – 1638 cm^{-1}

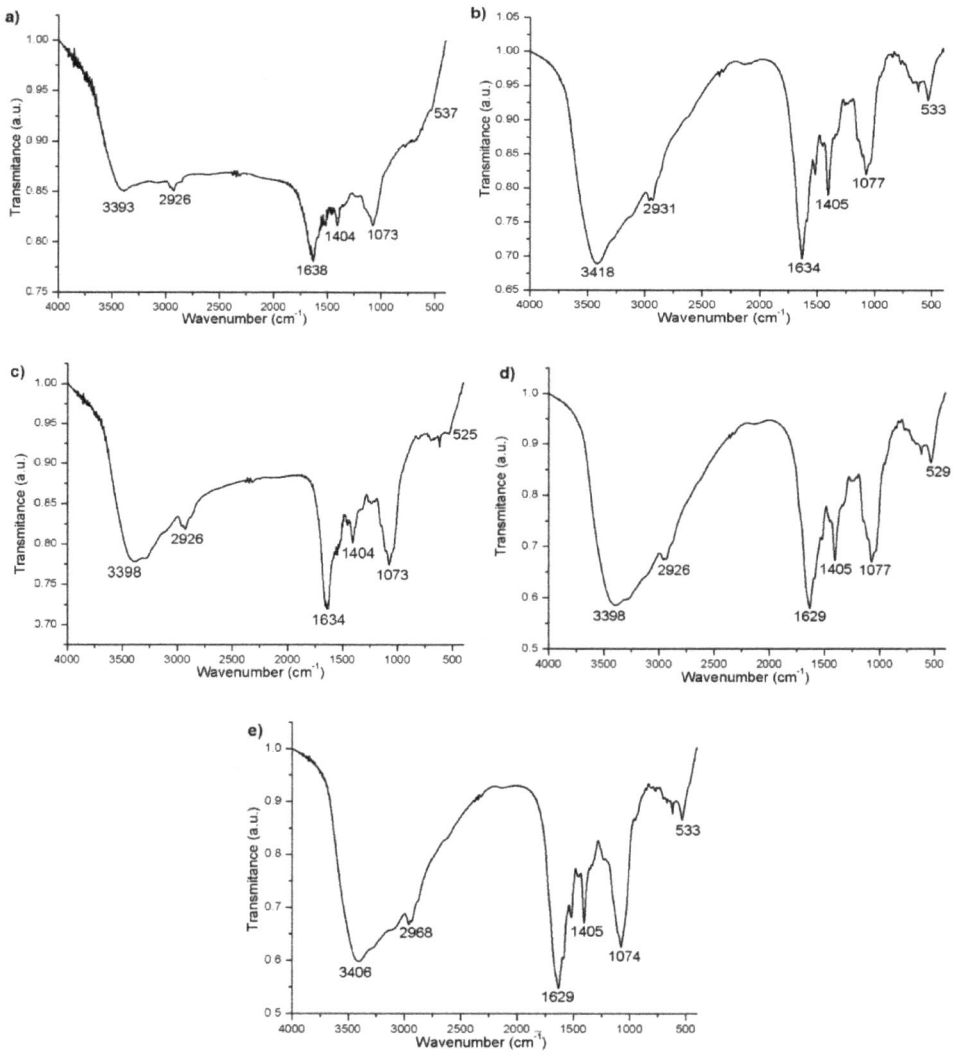

Figure 3. FT-IR spectra of *Aspergillus oryzae* ATCC 20423 unloaded (a) and loaded with Cu(II) ions (b-e)

can be due to a C=O stretching in carboxyl or amide groups. The peak at 1404 – 1405 cm^{-1} is attributed to N-H bending in amine group. The band observed at 1073 - 1077 cm^{-1} was assigned to the CO stretching of alcohols and carboxylic acids. Thus *Aspergillus oryzae* ATCC 20423 biomass contains hydroxyl, carboxyl and amine groups on surface.

From the Figures 3b-e it can be seen that the stretching vibration of OH group was shifted from 3393 cm^{-1} to 3418 cm^{-1} (3b), to 3398 cm^{-1} (3d), 3406 cm^{-1} (3e). These results revealed that chemical interactions between the copper ions and the hydroxyl groups occurred on the biomass surface. The carboxyl peak observed for unloaded biomass at 1638 cm^{-1} is shifted to 1634 cm^{-1} or 1629 cm^{-1}. This decrease in the wave number of the peak characteristic for C=O group from carboxylic acid revealed that interacts with carbonyl functional group are present between biomass and copper ions. These results indicated that the free carboxyl groups changed into carboxylate, which occurred during the reaction of the metal ions and carboxyl groups of the biosorbent.

No frequency changes were observed in the C-H and -NH$_2$ groups of biomass after copper biosorption. In addition, all FTIR spectrum of *Aspergillus oryzae* ATCC 20423 loaded with copper ions contain bands at 533, 529, 525 cm^{-1} which can be attributed to Cu-O stretching modes (Simonescu and Ferdes, *in press*).

The similar FT-IR results were reported for the biosorption Pb(II), Cd(II) and Cu(II) onto *Botrytis cinerea* fungal biomass (Akar *et al.*, 2005) and Pb(II) and Cd(II) from aqueous solution by macrofungus (*Lactarius scrobiculatus*) biomass (Anayurt *et al.*, 2009).

In our work we used also FT-IR spectroscopy in order to determine the characteristic functional groups which are responsible for biosorption of copper ions by *Polyporus squamosus*, *Aspergillus oryzae* NRRL 1989 (USA), *Aspergillus oryzae* 22343 (Simonescu *et al.*, 2012).

In case of biological degradation of pollutants a significant role can be attributed to biodegradation pathway due to the fact that different biodegradation pathways lead to different biodegradation products. Thus it is important to determine biodegradation pathways. For this purpose FTIR spectroscopy is a relevant tool for rapid determination of the resulting biotransformation product or mixtures. With this respect, Huang and coworkers investigated the ability of FT-IR to distinguish two different m-cresol metabolic pathways in *Pseudomonas putida* NCIMB 9869 after growth on 3,5-xylenol or m-cresol. From this study, it can be concluded that FT-IR spectral fingerprints were shown to differentiate metabolic pathways of m-cresol within the same bacterial strain and thus FTIR spectroscopy might provide a rapid, non-destructive, cost-effective approach for assessing of products resulted in biological degradation of pollutants (Huang *et al.*, 2006).

The main directions of use of FTIR spectroscopy in waste management are about getting information regarding the stage of organic matter for process and product control, and for monitoring of landfill remediation. For this purpose, Smidt and Meissl used FTIR spectroscopy to asses the stage of organic matter decomposition in waste materials (Smidt and Meissl, 2007). The results obtained confirm that FTIR spectroscopy represents an appropriate tool for process and quality control, for the assessment of abandoned landfills and for monitoring and checking of the successful landfill remediation (Smidt and Meissl, 2007).

The structural changing in biodegradation processes can be determined by FTIR analysis Thus Tomšič and coworkers studied structural changes of cellulose fabric modified by imidazolidinone biodegradation after different period using electron microscopic and spectroscopic analyses (Tomšič *et al.*, 2007). Also FT-IR spectroscopy is a quick and useful method to monitor the composting process (Grube *et al.*, 2006). The aim of them study was to elucidate the typical IR absorption bands and correlation of band growth rates with the compost maturity or degradation degree. The results of this study revealed that IR spectroscopy is a simple, quick and informative method that can be used instead of several time consuming chemical methods for monitoring of routine composting processes.

Soil is a complex medium with important ecological functions. Its functions depend on its characteristics. FTIR spectroscopy can be used to describe soil characteristics in the form of complex multivariate data sets. Thus FTIR spectroscopy has been used by Elliott and coworkers to investigate soils at different stages of recovery from degradation following opencast mining and from undisturbed land (Elliott et al., 2007). When a FT-IR spetrometer was used to determine gases from soils and rock formations no other gases than CO_2 have been detected except CO in the open-path compartment dedicated to atmosphere analysis (Pironon et al., 2009).

The use of living organisms to manage or remediate polluted soils named bioremediation represents an emerging technology. This technology is defined as the elimination, attenuation or transformation of polluting or contaminating substances by the use of biological processes. The results in situ bioremediation depend by microbial strains from contaminated site. The biodegradation process can be monitored by FTIR spectroscopy. For this purpose Bhat and coworkers performed a study about remediation of hydrocarbon contaminated soil through microbial degradation. The bacterial strains involved in bioremediation process were collected to be isolated from contaminated soil. FTIR spectra of untreated and treated soil samples revealed that the isolated bacterial strains have a substantial potential to remediate the hydrocarbon contaminated soils (Bhat et al., 2011).

Biomineralization has an important role for pollutants removal from environment. It has been known the mecanism involved in such processes to establish the nature of intemediates and final compound formed. FTIR spectroscopy is well-suited for such investigations, because it provides simultaneously molecular-scale information on both organic and inorganic constituents of a sample. Consequently FTIR spectroscopy was used in several complementary sample introduction modes as transmission (T-FTIR), attenuated total reflectance (ATR-FTIR), diffuse reflectance (DRIFTS) to analyze the processes of cell adhesion, biofilm growth, and biological Mn-oxidation by Pseudomonas putida strain GB-1 by Parikh and Chorover (Parikh and Chorover, 2005).

Fourier Transform Infrared (FT-IR) and Attenuated Total Reflectance (ATR) spectroscopy in the mid infrared (MIR) wavelength range (2500 – 16,000 nm) have been also developed for contaminant detection in water (Gowen et al., 2011a). The authors tested the near infrared spectroscopy (NIRS) for the detection and quantification of pesticides including Alachlor and Atrazine in aqueous solution. Calibration models were built to predict pesticide concentration using PLS regression (PLSR). The proposed method shows potential for direct measurement of low concentrations of pesticides in aqueous solution. The research was performed in the laboratory conditions, and it is well known that the NIR spectrum of aqueous samples is susceptible to changes in the environment (e.g. temperature, humidity) and sample (e.g. pH, turbidity). Thus further experiments are necessary to test the effect of such perturbations on predictive ability (Gowen et al., 2011b).

By joining FTIR spectroscopy with two dimensional correlation analysis (2DCORR) there will be obtained a device with improved performance in the study of complex environmental systems (Noda and Ozaki, 2005). The two dimensional correlation analysis (2DCORR) is a method to visualize the dynamic relationship between the variables in multivariate data set with application of the complex cross-correlation function. With the help of this analysis there will be identified the spectral features which change in phase (i.e. linearly correlated among them) and out of phase (partially or not at all correlated among them) (Mecozzi *et al.*, 2009). This technique can be applied to study the evolution of environmental complex systems. Mecozzi and coworkers applied FTIR spectroscopy joined with two dimensional correlation analysis (2DCORR) to identify the aggregation pathways of extractable humic substance from marine sediments, and to compare the molecular modifications determined by the actions of different pollutants on the marine algae *Dunaliella tertiolecta* that is a biomarker of environmental quality (Mecozzi *et al.*, 2009). From this study it can be concluded that FTIR spectroscopy joined with 2DCORR analysis can be an important tool for evaluating toxic effects on the marine life.

3. Attenuated Total Reflection – Fourier Transform Infrared (ATR-FTIR) spectroscopy in environmental studies

Attenuated Total Reflection – Fourier Transform Infrared (ATR-FTIR) Spectroscopy was introduced in 1960s (Harrick, 1967), and now is widely used in many areas.

The principle of this is FTIR technique is that light introduced into a suitable prism at an angle exceeding the critical angle for internal reflection develops an evanescent wave (a special type of electromagnetic radiation) at the reflecting surface. Interaction of this evanescent wave with the sample determines ATR spectrum recording. The main charactesistic of this techniques is the fact that particle samples are deposited on the surface of a horizontal ATR crystal for spectroscopic analysis (Figure 4). Zinc selenide (ZnSe) or Ge crystals are the most commonly used in ATR-FTIR spectroscopy.

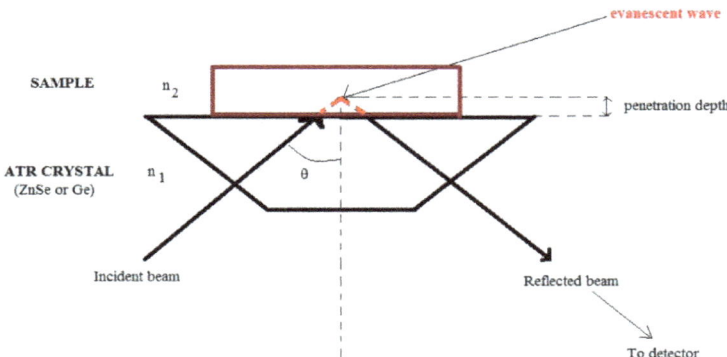

Figure 4. The principle of ATR-FTIR where n_1 and n_2 are the refractive indices of the crystal and the sample, respectively.

The main advantages of ATR-FTIR spectroscopy are: can be applied to a large variety of materials such as: powders, liquids, gels, pastes, pellets, slurries, fibers, soft solid materials, surface layers, polymer films, samples after evaporation of a solvent being a versatile and non-destructive technique; is useful for surface characterization, opaque samples; faster sampling being a non-destructive technique; is considered an extremely robust and reliable techique for quantitative studies involving liquids; excellent sample-to-sample reproducibility.

All these advantages make that ATR-FTIR spectroscopy to be used for: analysis of processes at surfaces (Freger and Ben-David, 2005), surface modification (Lehocký et al. 2003; Janorkar et al., 2004), surface degradation (Bokria et al., 2002), study of enzymatic degradation of a substrate film attached to a solid surface (Snabe et al. 2002), study of sunscreens on human skin (Rintoul et al., 1998), research of cereal, food and wood systems (D'Amico et al. 2012), detection of microbial metabolic products on carbonate mineral surfaces (Bullen et al., 2008), self-assembled thin films (Gershevitz et al. 2004), grafted polymer layers (Granville et al. 2004), adsorption processes (Sethuraman et Belfort 2005; Al-Hosney et Grassian 2005) of biological (Jiang et al. 2005; Mangoni et al. 2004) and synthetic (Freger et al. 2002) materials.

The followings are some examples of *in situ* ATR-FTIR spectroscopy's application in environmental studies.

In recent years adsorptive removal of heavy metals from aqueous effluents have received much attention because numerous materials such as: clays, zeolites, activated carbon can be used as adsorbents. The adsorption of inorganic ions on metal oxides and hydroxides was resolved using *in situ* ATR-FTIR spectroscopy. In a review Lefèvre describes and discusses *in situ* ATR-FTIR used in order to obtain information on the sorption mechanism of sulfate, carbonate, phosphate, perchlorate on hematite, goetite, alumina, silica, TiO_2 (Lefèvre, 2004). This is due to the fact that FTIR technique allows to analyze the sorption/desorption phenomena *in situ* being helpful in determining of the speciation of sorbed inorganic anions or ternary inorganic complexes formed. In addition this technique offers the possibility to distinguish outer-sphere and inner-sphere complexes. In this regard Yoon and coworkers used *in situ* ATR-FTIR spectroscopy and quantum chemical methods to determine the types and structures of the adsorption complexes formed by oxalate at boehmite (γ-AlOOH)/water and corundum (α-Al_2O_3)/water interfaces (Yoon *et al.*, 2004). They found that the adsorption mechanism of a aqueous HO_x^- species involves loss of protons from this species during the ligand-exchange reaction. The results obtained are useful in establishing the transport model of toxic species in natural waters, and remediation of liquid wastes.

Contamination of soils and groundwater by radioactive wastes containing uranium and other actinides is a significant problem. The fate and transport of these kind of pollutants in aquifers, design of cost-effective remediation techniques for radioactive-contaminated soils, and developing of materials proper for encapsulation and disposal of nuclear waste require knowledge of mechanism of radioactive pollutants – sorbent interactions. For radioactive waste depositories one of the most important factors which has to be considered is the long-term safety of them. For this, natural or anthropogenic barriers for sorption of radionuclides around the depositories are placed. Sorption data at the laboratory scale are useful to predict the behaviour of real systems. For this purpose Lefèvre and coworkers used ATR-IR spectroscopy to study the sorption of uranyl ions onto titanium oxide (mixture of rutile and anatase) and hematite. They found that the uranyl sorption on titanium oxide in the pH range 4-7 occurs by formation of one surface complex with uranium atoms bounded by two different chemical environments (Lefèvre *et al.*, 2008), and in case of sorption on hematite they concluded that the same surface species is responsible for the uranyl sorption in the pH range 5-8 (Lefèvre *et al.*, 2006). Due to the fact that experiments were reversible the authors concluded that reaction of hematite deposit with uranyl ions is the same with the reaction of it in dispersed suspensions (Lefèvre *et al.*, 2006). The sorption of U(V) on different forms of titanium dioxide was also studied using ATR-IR spectroscopy by Comarmond and coworkers. They showed the effect of different sources of sorbent and its surface properties on radionuclide sorption (Comarmond *et al.*, 2011). On the same subject Müller and coworkers used the high sensitivity of the *in situ* ATR-FTIR spectroscopy to establish the mechanism of sorption processes of U(VI) onto TiO_2 even at concentrations down to the low micromolar range. The Mid-IR spectra of U(VI) aqueous solutions and of U(VI) sorption onto different TiO_2 samples is presented in Figure 5.

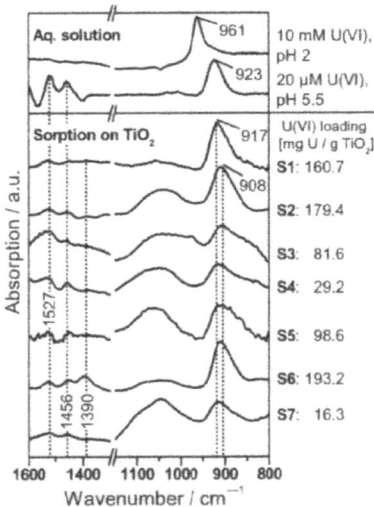

Figure 5. Mid-IR spectra of U(VI) aqueous solutions and of U(VI) sorption onto different TiO_2 samples (the values on the IR spectra are in cm^{-1}) (S1-S7 are different titania samples with different content of anatase and rutil, different particle size, and different origins) (from Müller *et al.*, 2012 used with permission (originally published in Geochimica et Cosmochimica Acta, http://dx.doi.org/10.1016/j.gca.2011.10.004))

By comparing the spectrum of the aqueous species spectra with the spectra of samples obtained after U(VI) sorption on TiO_2 it can be seen that the frequencies of the $v_3(UO_2)$ modes presented at 961 cm^{-1} for the aqueous species are significantly shifted (with 53-44 cm^{-1}) which suggests that uranyl surface complexes are formed at all titania samples.

This study is one complex due to the fact that authors performed researches to establish the influence of: stages of *in situ* sorption experiments (conditioning, sorption, and flushing), the contact time of U(VI) with the mineral, the initial U(VI) concentration, pH values, the origin and manufacturing procedure of TiO_2 samples and the absence of atmospheric-derived carbonate on the species formed in sorption processes of U(VI) on TiO_2. The results obtained by authors are relevant to the most environmental scenarios (Müller *et al.*, 2012).

Sorption of Np(V) onto TiO_2, SiO_2 and ZnO was investigated using ATR-FTIR spectroscopy. The results showed obtaining structurally similar bidentate surface complexes for all sorbents used (Müller *et al.*, 2009).

ATR-FTIR spectra confirmed formation of actinyl-carbonato complexes from interaction of actinide with hematite at a specific pH value. This can control the actinide transport in numerous subsurface receptors due to the abundance of carbonate in aquifers (Bargar *et al.*, 1999).

The influence of dissolved CO_2 on UO_2^{2+} sorption process was determined by Foerstendorf and Heim using ATR-FT-IR spectroscopy. They obtained a similar surface complex of the uranyl ion at the ferrihydrite-phase irrespective of the presence of atmospheric CO_2. Sorption of actinide ion on mineral phase determines a change of the carbonate ion from a monodentate to a bidentate ligand (Foerstendorf and Heim, 2008).

ATR-FTIR and FT-IR spectroscopy together with other techniques were used to determine the fate and transport of radionuclides in natural environments. The main mechanisms that are responsible for these are: sorption on organic (living matter and humic materials), sorption on inorganic materials (soil media and minerals), precipitation of them under oxic conditions, reduction in presence of microorganisms, and structural incorporation in different mineral host phases (Duff et *al.*, 2002).

Citric acid being a naturally-occuring acid commonly found in soils, and also a strong complexant of UO_2 is often found as a component of radioactive waste. Advantages such as: its biodegradability and complexing efficiency make from it a good candidate for remediation of uranium contaminated soils (Kantar and Honeyman, 2006). Factors with influence on the uranyl adsorption process to oxide minerals in presence of citric acid were determined by Logue and coworkers. Redden and coworkers have proposed formation of a ternary uranyl-citrate complexes on goethite (Redden *et al.*, 2001). Establishing the interactions between UO_2, citrate and mineral surfaces on a molecular level represents a key factor for modeling adsorption phenomena affecting transport in soils. For this purpose Pasilis and Pemberton used ATR-FTIR to elucidate the mechanism of UO_2 adsorption on aluminium oxide in the presence of citrate. They found that there is an enhanced citrate adsorption to Al_2O_3 in the presence of uranyl. This result suggests that uranyl may be the central link between two citrate ligands, and the uranyl is associated with the surface through a bridging citrate ligand. One other observation is that uranyl citrate complexes interact with citrate adsorbed to Al_2O_3 through outer shere interactions (Pasilis and Pemberton, 2008).

In recent years it ATR-FTIR spectroscopy has been used to investigate the atmospheric heteregenous reactions. Thus Al-Hosney and Grassian (2005) used this technique to investigate water adsorption on the surface of $CaCO_3$. They further used T-FTIR in order to investigate the role of surface adsorbed water in adsorption reactions of SO_2 and HNO_3 (Zhao and Chen, 2010). In other study Schuttlefield and coworkers (2007a) used ATR-FTIR spectroscopy to provide detailed information about water uptake and phase transitions for atmospherically relevant particles. To determine the factors involved in water uptake on the large fraction of dust present in the Earth's atmosphere, Schuttlefield and coworkers (2007b) used a variety of techniques, including ATR-FTIR. They concluded that water uptake on the clay minerals depends on the type and the source of the clay. These results are important because mineral dust aerosol provides a reactive surface in trophosphere being involved in reactions for atmosphere. The role of halogens in the aging process of organic aerosols was determined by Ofner and coworkers (2012) using long-path FTIR spectroscopy (LP-FTIR), attenuated-total reflection FTIR (ATR-FTIR), UV/VIS spectroscopy, and ultrahigh resolution mass spectroscopy (ICR-FT/MS). They concluded that the aerosol-halogen interaction might strongly contribute to the influence of organic aerosols on the climate system (Ofner *et al.*, 2012).

Khalizov and coworkers (2010) investigated the heterogeneous reaction of nitrogen dioxide (NO_2) on fresh and coated soot surfaces to assess its role in night-time formation of nitrous acid (HONO) in the atmosphere using ATR-FTIR (Khalizov *et al.*, 2010).

Segal-Rosenheimer and Dubowski (2007) combined two setups of FTIR for the parallel analysis of both condensed and gas phases of products resulted at the oxidation of cypermethrin (a synthetic pyrethroid being one of the most important insecticides in wide-scale use both indoors and outdoors) by gaseous ozone (Segal-Rosenheimer and Dubowski, 2007).

ATR-FTIR and T-FTIR methods provide detailed information on the composition of PM (particulate matter) samples. Both techniques can be used for qualitative and quantitative studies of particulate samples. Thus Veres (2005) used both methods to analyse particulate matter collected on Teflon Filters in Columbus – Ohio. He mentioned that ATR spectroscopy has limited applications in quantitative studies since it has a penetration depth of only a few microns, and this method can be replaced by transmission spectroscopy which penetrates into the bulk of substance (Veres, 2005).

Several groups of researchers used ATR-FTIR to particulate matter analysis. Thus Shaka and Saliba (2004) used ATR-FTIR spectroscopy in order to determine the concentration and the chemical composition of particulate matter at a coastal site in Beirut, Lebanon. Kouyoumdjian and Saliba (2006) determined the levels of the coarse (PM10-2.5) and fine (PM2.5) particles in the city of Beirut using ATR-FTIR spectroscopy. They also showed that nitrate, sulfate, carbonate and chloride were the main anionic constituents of the coarse particles, whereas sulfate was mostly predominant in the fine particles in the form of $(NH_4)_2SO_4$. Ghauch and coworkers (2006) used the same technique for the determination of small amounts of pollutants like the organic fraction of aerosols in the French cities of Grenoble and Clermont-Ferrand.

The applications of ATR-FTIR cover a wide range of subjects such as estimating of soil composition and fate of some soil components.

Monitoring of nitrate in soil is very important for managing fertilizer application and controlling nitrate leaching. This monitoring help to adjust nitrate level in soils in order to mantain the soil fertility, or to detect soil pollution. Due to the technological limitations, *in situ* or near real-time monitoring of soil nitrate is currently not feasible. In this purpose can be used the following methods: nitrate selective electrodes (Sibley, 2010), ion sensivitive field effect transistor (Birrell and Hummel, 2001), mid-infrared spectroscopy, and more particularly attenuated total reflectance (ATR) with Fourier transform infrared (FTIR) spectroscopy. Thus Raphael Linker submitted a report to the Grand Water Research Institute about simultaneous determination of $^{15}NO_3$-N and $^{14}NO_3$-N in aqueous solutions, soil extracts and soil pastes. The results obtained show that a combination of ATR-FTIR analysis with appropriate chemometrics can be successfully used to monitor $^{15}NO_3$-N and $^{14}NO_3$-N concentrations in soil during an incubation experiment (Linker, 2010). From the studies performed about measurement of nitrate concentration in soil pastes it can be concluded that ATR-FTIR appears to be a promising tool for direct and close to real-time determination of nitrate concentration in soils, with minimal treatment of the soil samples (Linker *et al.*, 2004; Linker *et al.* 2005; Linker *et al.*, 2006; Linker *et al.*, 2010). The same technique was used by Du and coworkers in order to evaluate net nitrification rate in *Terra Rosa* soil (Du *et al.*, 2009). ATR-FTIR spectroscopy was the technique preferred to mass spectrometry due to reduced cost, it is not time consuming, and doesn't require long and laborious preparation procedures. The results obtained have made major contributions for the estimation of the contribution of applied nitrogen and mineralized nitrogen to net nitrification rates ((Du *et al.*, 2009).

Soil paste was used by Choe and coworkers in order to improve the contact between sample and ATR crystal in case of using of the ATR-FTIR spectroscopy to determine the level of nitrate in soils. By comparing the nitrate peak intensity of soil pastes and their supernatant, it was shown that the nitrate dissolved in soil solution of the paste mainly responded to the FTIR signal. The results obtained are useful for the monitoring of nutrients in soils (Choe *et al.* 2010).

4. Diffuse Reflectance Infrared Fourier Transform (DRIFTS) spectroscopy in environmental studies

DRIFTS spectroscopy is considered a technique more sensitive to surface species than transmission measurements and is an excellent *in situ* technique. The principle is simple one: when incident light strikes a surface, the light that penetrates is reflected in all directions. This reflection is called diffuse reflectance. If the light that leaves the surface will pass through a thin layer of the reflecting materials, its wavelength content will have been modified by the optical properties of the matrix. The wavelength and intensity distribution of the reflected ligth will contain structural information on the substrate (Analytical Spectroscopy available at: http://www.analyticalspectroscopy.net/ap3-11.htm) (Figure 6).

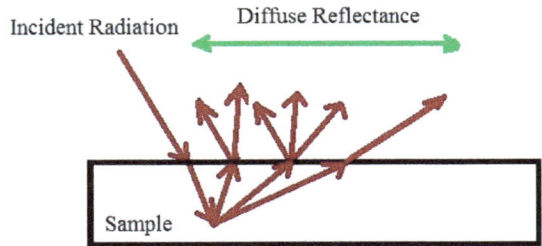

Figure 6. The principle of Diffuse Reflectance Infrared Fourier Transform Spectroscopy (adapted from Analytical Spectroscopy available at: http://www.analyticalspectroscopy.net/ap3-11.htm)

The main advantages of DRIFTS spectroscopy are: fast measurement of powdered samples, minimal or no sample preparation, ability to detect minor components, ability to analyze solid, liquid or gaseous samples, is one of the most suitable method for the examination of rough and opaque samples, high sensitivity, high versatility, capability of performing of the measurements under real life conditions.

In the environmental studies diffuse reflectance Fourier transform infrared (DRIFTS) spectroscopy is considered an alternative methodology for the quantitative analysis of nitrate in environmental samples (Verma and Deb, 2007a). It is considered a new, rapid and precise analytical method for the determination of the submicrogram levels of nitrate (NO_3^-) in environmental samples like soil, dry deposit samples, and coarse and fine aerosol particles. The DRIFTS method is a feasible nondestructive and time saving method for quantitative analyses of nitrate in soil, dry deposit and aerosol samples.

It is well known that soil can act as sinks as well as sources of carbon. A major fraction of carbon in soils is contained in the soil organic matter (SOM). It contributes to plant growth through its effect on the physical, chemical, and biological properties of the soil. Characterization of soil organic matter (SOM) is important for determining the overall quality of soils. For this DRIFTS spectroscopy can be used. This method only takes a few minutes, and is much faster than fractionating of soil samples using chemical and physical methods and determining the carbon contents of the fractions (Zimmermann *et al.*, 2007). In another study, Rumpel and coworkers tested diffuse reflectance infrared Fourier transform (DRIFT) spectroscopy in combination with multivariate data analysis [partial least squares (PLS)] as a rapid and inexpensive means of quantifying the lignite contribution to the total organic carbon (TOC) content of soil samples (Rumpel *et al.*, 2001). DRIFTS spectroscopy is also considered to be one of the most sensitive infrared technique to analyze humic substances (Ding *et al.*, 2000). Studies by Ding and coworkers demonstrate that both DRIFT and ^{13}C NMR are suitable for examining the effect of tillage on the distribution of light fraction in soil profile (Ding *et al*, 2002). More recently Ding and coworkers examined the effect of cover crops on the chemical and structural composition of SOM using chemical and DRIFT spectroscopic analysis. From this study it was concluded that both organic carbon (OC) and light fraction (LF) contents were higher in soils under cover crop treatments with and without fertilizer N than soils with no cover crop. Thus cover crops had a profound influence on the SOM and LF characteristics (Ding *et al.*, 2006). In other study Janik and coworkers (1995) showed that the use of diffuse reflectance infrared Fourier-transformed Spectroscopy (DRIFT) in combination with partial least squares algorithm (PLS) is a fast and low-cost method to predict carbon content and other soil properties such as clay content and pH. Zimmermann and coworkers evaluated the possibility of using of DRIFT-spectroscopy to estimate the soil organic matter content in soil samples from sites across Switzerland (Zimmermann *et al.*, 2004). It was concluded that DRIFT spectroscopy is a tool to predict changes in soil organic matter contents in agricultural soils resulting from changes in soil management. In other study Nault and coworkers used DRIFT spectroscopy to compare changes in organic chemistry of 10 species of foliar litter undergoing *in situ* decomposition for 1 to 12 years at four forested sites representing a range of climates in Canada (Nault *et al.*, 2009). This study demonstrated that DRIFT spectroscopy is a fast and simple analysis method for analyzing large numbers of samples to give good estimates of litter chemistry. Thus DRIFTS spectroscopy is considered a more faster technique to analyse the composition and the dynamics of organic matter in solis compared with FTIR spectroscopy (Tremblay and Gagné, 2002; Spaccini *et al.*, 2001).

Earth's atmosphere contains aerosols of various types and concentrations devided in: anthropogenic products, natural organic and inorganic products. The negative effects of these components refers to interaction with Earth's radiation budget and climate. In direct way aerosols scatter sunlight directly back into space, and indirect aerosols in the lower atmosphere can modify the size of cloud particles, and consequently changing the way in which clouds reflect and absorb sunlight. Aerosols act also as sites for chemical reactions to take place. As an exemple of these kind of reactions can be mentioned destruction of stratospheric ozone. The inorganic component of aerosols consist of inorganic salts (e.g. sulfate, nitrate, and ammonium). The most used method for analyzing these salts is ion chromatography (IC) (Chen *et al.*, 2003). The main disadvantages of this method are: time required for sample preparation and analysis that is up to 1 week, and the fact that this method is a destructive method of analysis. IR spectroscopy offers a simple and rapid alternative to IC for aerosols analysing, but it is imprecise and therefore only semi-quantitative. Advances in optics and detectors have allowed the development of more precise IR spectroscopy methods such as FTIR and DRIFT spectroscopy. FTIR spectroscopy was employed to determine on-site chemical composition of aerosol samples and to investigate the relationship between particle compositions and diameters (Tsai and Kuo, 2006). DRIFTS spectroscopy was used for quantitative analysis of atmospheric aerosols (Tsai and Kuo, 2006). The components of aerosols determined quantitative in area investigated were SO_4^{2-}, NO_3^- and NH_4^+. Compared with IC method, the DRIFT spectroscopy is a non-destructive, and quantitative method for aerosols analyzing.

Nitrogen dioxide, one of the key participants in atmospheric chemistry has been determined using DRIFT spectroscopy. Compared with other methods for nitrogen dioxide determination such as chemiluminiscence and fluorescence method that are multi-reagent procedure with the increased possibility of the experimental errors, the DRIFTS spectroscopy involves using NaOH–sodium arsenite solution as an absorbing reagent.

Another advantage of DRIFTS spectroscopy is that it can determine ambient nitrogen dioxide, in terms of nitrite, at submicrogram level (Verma *et al.*, 2008).

The feasibility of employing diffuse reflectance Fourier transform infrared (DRIFT) spectroscopy as a sensitive tool in the submicrogram level determination of sulphate (SO_4^{2-}) was checked by Verma and Deb in a study performed in 2007. The level of sulphate in environmental samples analysed like coarse and fine aerosol particles, dry deposits and soil was in range of ppb. The DRS-FTIR absorption spectrum of these real samples are presented in Figure 7.

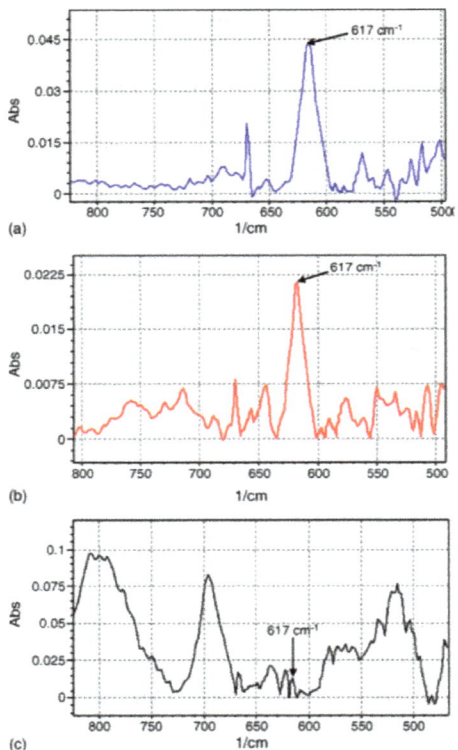

Figure 7. DRS-FTIR absorption spectrum of: (a) aerosol samples; (b) dry deposition sample; (c) soil sample (from Derma and Deb, 2007b used with permission (originally published in Talanta, doi:10.1016/j.talanta.2006.07.056))

For all real samples analyzed two-point baseline corrections were performed to obtain the quantitative absorption peak for sulphate at around 617 cm^{-1} (Verma and Deb, 2007b). The DRIFT method involved in this study did not require pretreatment of samples being reagent less, nondestructive, very fast, repeatable, and accurate and has high sample throughput value (Verma and Deb, 2007b). On the same topic Ma and coworkers have published paper entitled, "A case study of Asian dust storm particles: Chemical composition, reactivity to

SO_2 and hygroscopic properties". This paper presents a study about characterization of Asian dust storm particles using multiple analysis methods such as SEM-EDS, XPS, FT-IR, BET, TPD/mass and Knudsen cell/mass. The atmospheric dust particles are responsible by absorption and scattering of solar radiation and indirect acting as cloud condensation nucleus. The composition, source and size distribution of dust storm are important in predicting them impacts on climate and atmospheric environment. The dust particles can react with gaseous components or pollutants from the atmosphere such as sulfur dioxide. Thus numerous studies were performed to determine the role of dust in SO_2 chemistry (Prince et al., 2007; Ullerstam et al., 2002, 2003; Zhang et al., 2006; Ma et al., 2012b). The morphology, elemental fraction, source distribution, true uptake coefficient for SO_2 and higroscopic behavior were studied. The major components of Asian dust storm particles were aluminosilicate, SiO_2 and $CaCO_3$ mixed with some organic and nitrate compounds. The particles analyzed by Ma and coworkers are coming from anthropogenic sources and local sources after long transportation. Between SO_2 uptake coefficient and mass was established a linear dependence. Consequently DRIFTS and FTIR spectroscopy combined with other analitical methods will provide important information about the effects of dust storm particle on the atmosphere (Ma et al., 2012b).

One of the most important application of DRIFTS spectroscopy is to investigate sorption-uptake processes on different materials in order to reduce the impact of pollutants. Thus Valyon and coworkers studied N_2 and O_2 sorption on synthetic and natural mordenites, and on molecular sieves 4A, 5A and 13X using DRIFT spectroscopy (Valyon et al., 2003). Kazansky and coworkers used DRIFTS spectroscopy to study sorption of N_2, both pure and in mixtures with oxygen, O_2, by zeolites NaLSX and NaZSM-5 (Kazansky et al., 2004). Llewellyn and Theocharis studied carbon dioxide adsorption on silicate using DRIFTS spectroscopy (Llewellyn and Theocharis, 1991). Heterogeneous oxidation of gas-phase SO_2 on different iron oxides was investigated in situ using a White cell coupled with Fourier transform infrared spectroscopy (FTIR) and diffuse reflectance infrared Fourier transform spectroscopy (DRIFTS) by Fu and coworkers (Fu et al., 2007). From this study it can be concluded that adsorbed SO_2 could be oxidized on the surface of most iron oxides to form a surface sulfate species at ambient temperature, and the surface hydroxyl species on the iron oxides was the key reactant for the heterogeneous oxidation (Fu et al., 2007). Heterogeneous reaction of NO_2 with carbonaceous materials (commercial carbon black, spark generator soot, Diesel soot from passenger car and high-purity graphite) at elevated temperature (400°C) was studied using DRIFT spectroscopy. Different infrared signals appear when NO_2 is adsorbed either on aliphatic or graphitic domains of soot (Muckenhuber and Grothe, 2007).

Gas sensors are playing an important role in the detection of toxic pollutants such as CO, H_2S, NO_x, SO_2, and inflammable gases such as hydrocarbons, H_2, CH_4. Diffuse Reflectance Infrared (DRIFT) spectroscopy has been used to characterize them. Thus, the studies performed by Harbeck in him Dissertation have shown that thick film sensors can easily be characterised in different working conditions (at elevated temperatures, in the presence of humidity) using Diffuse Reflectance Infrared (DRIFT) spectroscopy. He characterized undoped and Pd-doped SnO₂ sensor surfaces at different temperatures using two different methods in parallel: DRIFT spectroscopy and electrical measurements. Simultaneous recording of the DRIFT spectra and the sensor resistance helped him to clarify the role of the individual surface species in the sensing mechanism. The results of his work show that several reactions take place in the presence of CO depending both on temperature and humidity. It was found that all surface species are involved in the reactions and it is supposed that parallel and consecutive CO reactions take place on the surface (Harbeck, 2005).

DRIFT spectroscopy is also suitable for application to studies of surface phenomena and large specific surface materials such as the sensing layers. In this purpose Bârsan and Weimar investigated the effect of water vapour in CO sensing by using Pd doped SnO₂ sensors obtained using thick film technology as an example of the basic understanding of sensing mechanisms applied to sensors. The results obtained show that all parts of the sensor (sensing layer, electrodes, substrate) have influence to the gas detection and their role has to be taken into consideration when one attempts to understand how a sensor works (Bârsan and Weimar, 2003).

All the examples mentioned above show the importance of DRIFT spectroscopy in analyzing of environmental samples either liquid, solid or gaseous.

5. Open Path FT-IR spectroscopy in environmental studies

The open-path FT-IR Spectroscopy is conventionally used for monitoring gaseous air pollutants, but can also be used for monitoring both the gaseous or particulate air pollutants. The principle of function is the same with classical FTIR Spectroscopy, except the cell into the sample will be injected which it is extended to open atmosphere (Minnich and Scotto, 1999). In this technique the infrared light sources can be either natural solar light, or light coming from a heated filament situated behind the target gas. The infrared signal passes through a sample and chemical vapors present in sample will absorb the infrared energy at different wavelengths. All compounds in the vapor will give unique fingerprints of absorbance features which will be compared to a library of spectra on the computer. This comparison will be useful to identify and quatify in real time.

The advantages of open-path FT-IR Spectroscopy include: no sample collection, handling ar preparation is necessary; good sensivity for certain species; real time data collection and reporting; ability to simultaneously and continuously analyze many compounds; remote, long-path measurements; *in situ* application; stored data can be used and re-analyzed for a divers range of volatile or non-volatile compounds; cost effectiveness (Marshall *et al.*, 1994).

The main disadvantage of OP-FTIR is considered to be the fact that it can be applied only to the cases with high concentrations of gases such as stack measurement, landfill measurement, and fence-line monitoring (Hong *et al.*, 2004). Thus Perry *et al.* (1995) and Tso and Chang applied OP-FTIR to determine the VOC and ammonia concentrations in industrial areas, the concentration of pollutants being in this area in the level of 0.1 ppm

(Perry *et al.*, 1995; Tso and Chang, 1996). Childers *et al.* applied OP-FTIR spectroscopy for the measurement of ammonia, methane, carbon dioxide, and nitrous oxide in a concentrated swine production facility. The pollutants concentration was in the reanges 0.1 – 100 ppm. The results have led authors to conclude that the confinement barns was the significant source of ammonia emission, and the waste treatment lagoon was the major source of methane (Childers *et al.*, 2001). A similar research was performed by Hedge et al. in oder to monitor methane and carbon dioxide emitted form a landfill in northern of Taiwan (Hedge *et al.*, 2003), and Thorn et al. used OP-FTIR to measure phosphine concentrations in the air surrounding the large fumigated structures of a tobacco warehouse (Thorn *et al.*, 2001). OP-FITR was used by Harris and coworkers to monitor ammonia and methane emissions from animal housing and waste lagoons due to the ability to detect multiple compounds simultaneously (Harris *et al.*, 2007).

Levine and Russwurm described in an article the use of the open-path FT-IR Spectroscopy in remote sensing of aiborne gas and vapor contaminants (Levine and Russwurm, 1994). Applying open-path Fourier transform spectroscopy for measuring aerosols was described by Wu and coworkers (Wu *et al*, 2007).

Air monitoring during site remediation using open-path FTIR Spectroscopy was reported by Minnich and Scotto (Minnich and Scotto, 1999), and monitoring trace gases from aircraft emissions using the same technique was reported by Haschberger (Haschberger, 1994).

The use of OP-FTIR spectroscopy for identification of fugitive organic compound (VOC) emission sources and to estimate emission rates at an Air Force base in United States was described by Hall (Hall, 2004). Galle *et al.* have demonstrated advantages of FTIR over traditional point-measurement methods by providing detection over large sampling areas (Galle *et al.*, 2001).

OP-FTIR was successfully applied by Walter et al., and Kagann *et al.* for the measurements of air quality criteria pollutants such as ozone, carbon dioxide, sulfur dioxide, and nitrogen dioxide in ambient air (Walter *et al.*, 1999; Kagan *et al*, 1999). Grutter and coworkers used OP-FTIR spectroscopy to measure trace gases over Mexico City. This was the first report on the concentration profiles of acetylene, ethylene, ethane, propane, and methane in this region. Specific correlation between the profiles and wind direction were made in order to determine the main sources that contribute to these profiles (Grutter *et al.*, 2003).

A comparison between different analysis techniques applied to ozone and carbon monoxide detection was made by Briz and coworkers. They compared classical least-squares (CLS) procedures with line-by-line method (SFIT) to analyze OP-FTIR spectra and concluded that discrepancies observed in CLS-based methods were induced by the experimental background reference spectrum, and SFIT results agreement well with the standard extractive methods (Briz *et al.*, 2007). The same author together with other coworkers proposed a new method for calculating emission rates from livestock buildings applying Open-Path FTIR spectroscopy (Briz *et al.*, 2009). The method was applied in a cow shed in the surroundings of La Laguna, Tenerife Island (Spain), and results obtained revealed that the livestock building behaves such as an accumulation chamber, and methane emission factor was lower than the proposed by Emission Inventory (Briz *et al.*, 2009).

As was described by Lin and coworkers an open-path Fourier transform infrared spectroscopy system can be used for monitoring of VOCs in industrial medium. They used this system to monitor VOCs emissions from a paint manufacturing plant, and they determined seven VOCs in ambient environment. The same system was also used to determine the VOCs in a petrochemical complex. The results obtained were correlated with meteorological data and were effective in the depiction of spatial variations in indentifying sources of VOC emissions. They also mentioned another important advantage of OP-FTIR spectroscopy such as the ability to obtain more comprehensive data than by using the traditional multiple, single-point monitoring methods. It can be concluded that OP-FTIR can be useful in both industrial hygiene and environmental air pollutat regulatory enforcement (Lin *et al.*, 2008).

Ammonia, CO, methane, ethane, ethylene, acetylene, propylene, cyclohexane, and O-xylene were identified as major emissions in a coke processing area from Taiwan using OP-FTIR system by Lin and coworkers (Lin *et al.*, 2007). Main gaseous byproducts (CO, CO_2, CH_4 and NH_3) of thermal degradation (pyrolysis) of biomass in forest fires were determined accurately using OP-FTIR. The results obtained in this study can help to improve the modelling of the pyrolysis processes in physical-based models for predicting forest fire behaviour (de Castro *et al.*, 2007). An other reasearch in this field was performed by Burling and coworkers who measured trace gas emissions from biomass burning of fuel types from the southeastern and southwestern United States (Burling *et al.*, 2010) with the help of OP-FTIR. The authors detected and quantified 19 gas-phase species in these fires: CO_2, CO, CH_4, C_2H_2, C_2H_4, C_3H_6, HCHO, HCOOH, CH_3OH, CH_3COOH, furan, H_2O, NO, NO_2, HCNO, NH_3, HCN, HCl, and SO_2. The emission factors depend on the fuel composition and fuel types.

All the advantages of OP-FTIR spectroscopy and all the studies mentioned above demonstrate the utility of OP-FTIR in measuring and monitoring of atmospheric gases. This technique has increasingly been accepted by different environmental agencies as a tool in the measurement and the monitoring of the atmospheric gases (Russwurm and Childers, 1996; Russwurm, 1999).

6. Conclusion

All these presented above show the importance of FTIR spectroscopy in environmental studies. The major advantages of this technique are: real time data collection and reporting, excellent sample-to-sample reproductibility, enhanced frequency accuracy, high signal-to-noise ratios, superior sensitivity, analytical performance. In addition, the measurement is very rapid so that a large number of samples can be analyzed. Consequently FTIR spectroscopy coupled with other techniques is widely used to determine the nature of pollutants (gaseous, liquid or solid), to monitor environment, to asses the impact of pollution on health and environment, to determine the level of decontamination processes.

The modern techniques such as attenuated total reflection FTIR (ATR-FTIR), and diffuse reflectance infrared Fourier transform spectroscopy (DRIFTS), but also traditional transmission FTIR can be used for such studies according to the information needed, the physical form of the sample, and the time required for the sample preparation.

Author details

Claudia Maria Simonescu

Department of Analytical Chemistry and Environmental Engineering, Faculty of Applied Chemistry and Materials Science, „Politehnica" University of Bucharest, Romania

Acknowledgement

The author wants to thank all the authors who gave her the permission to cite them works, and to the publishers for reusing of some figures from the papers published by them.

7. References

Akar T, Tunali S, Kiran I (2005) *Botrytis cinerea* as a new fungal biosorbent for removal of Pb(II) from aqueous solutions. *Biochemical Engineering Journal* 25(30): 227-235.

Al-Hosney HA, Grassian VH (2005) Water, sulfur dioxide and nitric acid adsorption on calcium carbonate: A transmission and ATR-FTIR study. *Phys. Chem. Chem. Phys.* 7: 1266-1276.

Amon B, Kryvoruchko V, Fröhlich M, Amon T, Pöllinger A, Mösenbacher I, Hausleitner A (2007) Ammonia and greenhouse gas emissions from a straw flow system for fattening pigs: Housing and manure storage. *Livestock Science* 112: 199–207.

Anayurt RA, Sari A, Tuzen M (2009) Equilibrium, thermodynamic and kinetic studies on biosorption of Pb(II) and Cd(II) from aqueous solution by macrofungus (*Lactarius scrobiculatus*) biomass. *Chemical Engineering Journal* 151(1-3): 255-261.

Arrigone GM, Hilton M (2005) Theory and practice in using Fourier transform infrared spectroscopy to detect hydrocarbons in emissions from gas turbine engines. *Fuel* 84: 1052–1058.

Bacsik Z, Mink J (2007a) Photolysis-assisted, long-path FT-IR detection of air pollutants in the presence of water and carbon dioxide. *Talanta* 71: 149–154.

Bacsik Z, McGregor J, Mink J (2007b) FTIR analysis of gaseous compounds in the mainstream smoke of regular and light cigarettes. *Food and Chemical Toxicology* 45: 266–271.

Bacsik Z, Mink J, Keresztury G (2005) FTIR spectroscopy of the atmosphere part 2. applications. *Appl. Spectrosc. Rev.* 40: 327–390.

Bacsik Z, Mink J, Keresztur G (2004) FTIR spectroscopy of the atmosphere I. principles and methods. *Appl. Spectrosc. Rev.* 39: 295–363.

Bardakçi B, Bahçeli S (2010) FTIR study on modification of transition metal on zeolites for adsorption. *Indian Journal of Pure & Applied Physics* 48: 615-620.

Bargar JR, Reitmeyer R, Davis JA (1999) Spectroscopic Confirmation of Uranium (VI)-Carbonato Adsorption Complexes on hematite. *Environ. Sci. Technol.* 33: 2481–2484.

Bârsan N, Weimar U (2003) Understanding the fundamental principles of metal oxide based gas sensors; the example of CO sensing with SnO_2 sensors in the presence of humidity. *J. Phys.: Condens. Matter* 15: R813–R839.

Bhat MM, Shankar S, Shikha, Yunus M, Shukla RN (2011) Remediation of Hydrocarbon Contaminated Soil through Microbial Degradation- FTIR based prediction. *Advances in Applied Science Research* 2(2): 321-326.

Birrell S.J, and Hummel JW (2001) Real-time multi-ISFET/FIA soil analysis system with automatic sample extraction. *Comp. and Elect. in Agric.* 32(1): 45-67.

Blum M-M, John H (2012) Historical perspective and modern applications of Attenuated Total Reflectance – Fourier Transform Infrared Spectroscopy (ATR-FTIR). *Drug Test. Analysis* 4: 298-302.

Bokria JG, Schlick S (2002) Spatial Effects in the Photodegradation of Poly(acrylonitrile-butadiene-styrene): A Study by ATR-FTIR. *Polymer* 43: 3239-3246.

Briz S, Barrancos J, Nolasco D, Melián G, Padrón E, Pérez N (2009) New Method for Estimating Greenhouse Gas Emissions from Livestock Buildings Using Open-Path FTIR Spectroscopy, *Remote Sensing of Clouds and the Atmosphere XIV*, edited by Richard H. Picard, Klaus Schäfer, Adolfo Comeron, Evgueni Kassianov, Christopher J. Mertens, Proc. of SPIE 7475: 747510-1-747510-9.

Briz S, de Castro AJ, Diez S, Lopez F, Schaefer K (2007) Remote Sensing by Open-Path FTIR Spectroscopy: Comparison of Different Analysis Techniques Applied to Ozone and Carbon Monoxide Detection. *Journal of Quantitative Spectroscopy and Radiative Transfer* 103(2): 314-330.

Bullen HA, Oehrle SA, Bennett AF, Taylor NM, Barton HA (2008) Use of Attenuated Total Reflectance Fourier Transform Infrared Spectroscopy to Identify Microbial Metabolic Products on Carbonate Mineral Surfaces. *Applied and Environmental Microbiology* 74(14): 4553-4559.

Burling IR, Yokelson RJ, Griffith DW, Johnson TJ, Veres P, Roberts J, Warneke C, Urbanski SP, Reardon J, Weise DR, Hao W & De Gouw JA (2010) Laboratory measurements of trace gas emissions from biomass burning of fuel types from the southeastern and southwestern United States. *Atmospheric Chemistry and Physics* 10(22): 11115-11130.

Burnett P-GG, Daughney JC, Peak D (2006) Cd adsorption onto *Anoxybacillus flavithermus*: Surface complexation modeling and spectroscopic investigations. *Geochimica et Cosmochimica Acta* 70: 5253–5269.

Burton MR, Oppenheimer C, Horrocks LA, Francis PW (2000) Remote sensing of CO_2 and H_2O emission rates from Masaya volcano, Nicaragua. *Geology* 28: 915–918.

Chen Y, Wang J (2012) Removal of radionuclide Sr^{2+} ions from aqueous solution using synthesized magnetic chitosan beads. *Nuclear Engineering and Design* 242: 445-451.

Cheng C, Wang JN, Xu L, Li AM (2012) Preparation of new hyper cross-linked chelating resin for adsorption of Cu^{2+} and Ni^{2+} from water. *Chinese Chemical Letters* 23: 245–248.

Childers JW, Thompson Jr EL, Harris DB, Kirchgessner DA, Clayton M, Natschke DF, Phillips WJ (2001) Multi-pollutant concentration measurements around a concentrated swine production facility using open-path FTIR spectrometry. *Atmospheric Environment* 35(11): 1923-1936.

Choe E, van der Meer F, Rossiter D, van der Salm C, Kim K-W (2010) An Alternate Method for Fourier Transform Infrared (FTIR) Spectroscopic Determination of Soil Nitrate Using Derivative Analysis and Sample Treatments. *Water Air Soil Pollut.* 206:129–137.

Comarmond MJ, Payne TE, Harrison JJ, Thiruvoth S, Wong HK, Augtherson RD, Lumpkin GR, Müller K, Foerstendorf H (2011) Uranium Sorption on Various Forms of Titanium Dioxide – Influence of Surface Area, Surface Charge, and Impurities. *Environ. Sci. Technol.* 45(13): 5536-5542.

D'Amico S, Hrabalova M, Müller U, Berghofer E (2012) Influence of ageing on mechanical properties of wood to wood bonding with wheat flour glue. *European Journal of Wood and Wood Products*, DOI: 10.1007/s00107-012-0595-x

Deleanu C, Simonescu CM, Căpăţînă C (2008) Comparative study on the adsorption of Cu(II) ions onto chistosan and chemical modified chitosan. *Proceedings of 12th Conference on Environment and Mineral Processing* – Part III – 5.-7.6. Ostrava, Czech Republic, 201-207.

Ding G, Liu X, Herbert S, Novak J, Amarasiriwardena D, Xing B (2006) Effect of cover crop management on soil organic matter. *Geoderma* 130: 229–239, doi:10.1016/j.geoderma.2005.01.019

Ding G, Novak JM, Amarasiriwardena D, Hunt PG, and Xing B (2002) Soil Organic Matter Characteristics as Affected by Tillage Management. *Soil Sci. Soc. Am. J.* 66:421–429.

Ding, G, Amarasiriwardena D, Herbert S, Novak J, and Xing B (2000) Effect of cover crop systems on the characteristics of soil humic substances. p. 53–61. *In* E.A. Ghabbour and G. Davis (ed.) *Humic substances: Versatile components of plants, soil and water*. Press, Orlando, FL. The Royal Society of Chemistry, Cambridge.

Du CW, Linker R, Shaviv A & Zhou JM (2009) *In Situ* Evaluation of Net Nitrification Rate in *Terra Rossa* Soil Using a Fourier Transform Infrared Attenuated Total Reflection N-15 Tracing Technique. *Applied Spectroscopy* 63: 1168-1173.

Duff MC, Coughlin JU, Hunter DB (2002) Uranium co-precipitation with iron oxide minerals. *Geochimica et Cosmochimica Acta.* 66(20): 3533–3547.

Duygu D (Yalcin), Baykal T, Açikgöz İ, Yildiz K (2009) Fourier Transform Infrared (FT-IR) Spectroscopy for Biological Studies. G.U. *Journal of Science* 22(3): 117-121.

Elliott GN, Worgan H, Broadhurst D, Draper J, Scullion J (2007) Soil differentiation using fingerprint Fourier transform infrared spectroscopy, chemometrics and genetic algorithm-based feature selection. Soil Biology & Biochemistry 39: 2888–2896.

Essendoubi M, Toubas D, Bouzaggou M, Pinon J.-M, Manfait M, Sockalingum GD (2005) Rapid identification of *Candida* species by FT-IR microspectroscopy. Biochim Biophys Acta. 1724(3):239-47, http://dx.doi.org/10.1016/j.bbagen.2005.04.019

Fischer G, Braun S, Thissen R, Dott W. (2006) FT-IR spectroscopy as a tool for rapid identification and intra-species characterization of airborne filamentous fungi. Journal of Microbiological Methods 64: 63– 77, doi:10.1016/j.mimet.2005.04.005

Foerstendorf H, Heim K (2008) Sorption of uranium(VI) on ferrihydrite – Infleunce of atmospheric CO_2 on surface complex formation invetsigated by ATR-FT-IR spectroscopy. NRC7- Seventh International Conference on Nuclear and Radiochemistry, Budapest, Hungary

Freger V and Ben-David A (2005) Use of Attenuated Total Reflection Infrared Spectroscopy for Analysis of Partitioning of Solutes between Thin Films and Solution. Analytical Chemistry B. 77(18): 6019–6025.

Freger V, Gilron J, Belfer S (2002) TFC polyamide membranes modified by grafting of hydrophilic polymers: an FT-IR/AFM/TEM study. J. Membr. Sci. 209: 283-292.

Fu H, Wang X, Wu H, Yin Y, Chen J (2007) Heterogeneous Uptake and Oxidation of SO_2 on Iron Oxides. J. Phys. Chem. C 111: 6077-6085.

Galle B, Klemedtsson L, Bergqvist B, Ferm M, Törnqvist K, Griffith DWT, Jensen N-O, Hansen F (2000) Measurements of ammonia emissions from spreading of manure using gradient FTIR techniques. Atmospheric Environment 34: 4907-4915.

Gershevitz O, Sukenik CN (2004) In Situ FTIR-ATR Analysis and Titration of Carboxylic Acid-Terminated SAMs. J. Am. Chem. Soc. 126: 482-483.

Ghauch A, Deveau P-A, Jacob V, Baussand P (2006) Use of FTIR spectroscopy coupled with ATR for the determination of atmospheric compounds. Talanta. 68(15): 1294–1302.

Gowen A, Tsenkova R, Bruen M and O'Donnell C (2011a) Vibrational Spectroscopy for Analysis of Water for Human Use and in Aquatic Ecosystems. *Critical Reviews in Environmental Science and Technology.*

Gowen A, Tsuchisaka Y, O'Donnell C, Tsenkova R (2011b) Investigation of the Potential of Near Infrared Spectroscopy for the Detection and Quantification of Pesticides in Aqueous Solution. *American Journal of Analytical Chemistry* 2: 53-62. doi:10.4236/ajac.2011.228124

Granville AM, Boyes SG, Akgun B, Foster MD, Brittain WJ (2004). Synthesis and Characterization of Stimuli-Responsive Semi-Fluorinated Polymer Brushes by Atom Transfer Radical Polymerization. *Macromolecules* 37: 2790-2796.

Grube M, Muter O, Strikauska S, Gavare M, Limane B (2008) Application of FT-IR spectroscopy for control of the medium composition during the biodegradation of nitro aromatic compounds. *J. Ind. Microbiol. Biotechnol.* 35: 1545-1549.

Grube M, Lin JG, Lee PH, Kokorevicha S (2006) Evaluation of sewage sludge-based compost by FT-IR spectroscopy. *Geoderma* 130: 324– 333.

Grutter M, Flores E, Basaldud R, Ruiz-Suárez LG (2003) Open-path FTIR spectroscopic studies of the trace gases over Mexixo City. Atmos. Oceanic Opt. 16(3): 232-236.

Hall FE Jr (2004) Case Study: Environmental Monitoring Using Remote Optical Sensing (OP-FTIR) Technology at the Oklahoma City Air Logistics Center Industrial Wastewater Treatment Facility. *Fed. Facilities Environ. J.* 15: 21-37.

Harbeck S (2005) Characterisation and Functionality of SnO_2 Gas Sensors Using Vibrational Spectroscopy. Dissertation. available at: http://tobias-lib.uni-tuebingen.de/volltexte/2005/1693/pdf/Serpil_Harbeck_thesis_final_druck.pdf Accessed 2012 March 11.

Harrick NJ (1967) *Internal Reflection Spectroscopy*, Wiley-Interscience, New York.

Harris DB, Shores RC, Thoma ED (2007) Using Tnable Diode Lasers to Measure Emissions form Animal Housing and Waste Lagoons, U.S. EPA, NRMRL, Research Traingle Park,

NC. *16th Annual International Emission Inventory Conference – Emission Inventories: Integration, Analysis, and Commnications*. 14-17 May 2007, NC. 16 pp.

Haschberger P (1994) Remote measurement of trace gases from aircraft emissions using infrared spectroscopy, Mitt.- Dtsch. Forschungssanst. Luft-Raumfahrt (94-06, *Impact of Emisssions from Aircraft and Spacecraft Unpon the Atmosphere*): 100-105.

Hegde U, Chang TC, Yang SS (2003) Methane and Carbon Dioxide Emissions from Shanchu-ku Landfill Site in Northern Taiwan. *Chemosphere* 52: 1275-1285.

Hong DW, Heo GS, Han JS, Cho SY (2004) Application of the open path FTIR with COL1SB to measurements of ozone and VOCs in the urban area. *Atmospheric Environment* 38: 5567-5576.

Huang WE, Hopper D, Goodacre R, Beckmann M, Singer A, Draper J (2006) Rapid characterization of microbial biodegradation pathways by FT-IR spectroscopy. *Journal of Microbiological Methods* 67: 273–280.

Janorkar AV, Metters AT, Hirt DE (2004) Modification of Poly(lactic acid) Films: Enhanced Wettability from Surface-Confined Photografting and Increased Degradation Rate Due to an Artifact of the Photografting Process, *Macromolecules* 37: 9151-9159.

Jäger F, Gluschke O, Doll R (2011) FTIR- and NDIR-Spectroscopy measurements on environmental air. How accurate are $^{13}CO_2$ isotope ratio and trace gas measurements with an outdoor instrument? *Geophysical Research Abstracts* Vol. 13, EGU2011-8715-1.

Jiang CH, Gamarnik A, Tripp CP (2005) Identification of lipid aggregate structures on TiO_2 surface using headgroup IR bands. *J. Phys. Chem.* 109: 4539-4544.

Johnson TJ, Profeta LTM, Sams RL, Griffith DWT, Yokelson RL (2010) An infrared spectral database for detection of gases emitted by biomass burning. *Vibrational Spectroscopy* 53: 97–102.

Jordan N, Foerstendorf H, Weiß, Heim K, Schild D, Brendler V (2011) Sorption of selenium(VI) onto anatase: Macroscopic and microscopic characterization. *Geochimica et Cosmochimica Acta* 75(6): 1519-1530.

Kagann RH, Wang CD, Chang KL, Lu CH (1999) Open-Path FTIR Measurement of Criteria Pollutants and Other Ambient Species in an Industrial City. *Proc. SPIE* 3534: 140-149.

Kantar C. and Honeyman B. D. (2006) Citric acid enhanced remediation of soils contaminated with uranium by soil flushing and soil washing. *J. Environ. Eng.* 132: 247–255.

Kazansky VB, Sokolova NA, Bülov M (2004) DRIFT spectroscopy study of nitrogen sorption and nitrogen–oxygen transport co-diffusion and counter-diffusion in NaLSX and NaZSM-5 zeolites. *Microporous and Mesoporous Materials* 67: 283–289.

Khalizov AF, Cruz-Quinones M, and Zhang R (2010) Heterogeneous Reaction of NO_2 on Fresh and Coated Soot Surfaces. *J. Phys. Chem.* A 114: 7516–7524.

Kouyoumdjian H and Saliba NA (2006) Mass concentration and ion composition of coarse and fine particles in an urban area in Beirut: effect of calcium carbonate on the absorption of nitric and sulfuric acids and the depletion of chloride. *Atmos. Chem. Phys.* 6: 1865–1877.

Kumar R, Singh G, Pal AK (2005) Assessment of impact of coal and minerals related Industrial activities in Korba industrial belt of Chhattisgarh through spectroscopic techniques. *Mineral Processing Technology* 605-612.

Lefèvre G, Kneppers J, Fédoroff M (2008) Sorption of uranyl ions on titanium oxide studied by ATR-IR spectroscopy. *J. Colloid. Interface Sci.* 327(1): 15-20.

Lefèvre G, Noinville S, Fédoroff M (2006) Study of uranyl sorption onto hematite by in situ attenuated total reflection-infrared spectroscopy. *J. Colloid. Interface Sci.* 296(2): 608-613.

Lefèvre G (2004) In situ Fourier-transform infrared soectroscopy studies of inorganic ions adsorption on metal oxides and hydroxides. *Advances in Colloid and Interface Science* 107: 109-123.

Lehocký M, Drnovská H, Lapíková B, Barros-Timmons AM, Trindade T, Zembala M, Lapík L (2003) Plasma Surface Modification of Polyethylene. *Colloids Surf. A* 222: 125-131.

Levine SP, Russwurm GM (1994) Fourier-transform infrared optical remote sensing for monitoring airbone gas and vapor contaminants in the field, *TrAC. Trends Anal. Chem.* (Pers. Ed), 13(7): 258-262.

Lin C, Liou N, Sun E (2008) Applications of Open-Path Fourier Transform Infrared for Identification of Volatile Organic Compound Pollution Sources and Characterization of Source Emission Behaviors. *Journal of the Air & Waste management Association* 58(6): 821-828.

Lin C, Liou N, Chang PE, Yang JC, Sun E (2007) Fugitive coke oven gas emission profile by continuous line averaged open-path Fourier transform infrared monitoring. *J Air Waste Manag Assoc.* 57(4): 472-479.

Linker R (2010) Development of chemometric tools for FTIR determination of N-species in environmental systems, Final Report Submitted to the Grand Water Research Institute available at: http://gwri-ic.technion.ac.il/pdf/gwri_abstracts/2110/1.pdf. Accessed 2012 March10.

Linker R, Weiner M, Shmulevich I, Shaviv A (2006). Nitrate Determination in Soil Pastes using Attenuated Total Reflectance Mid-infrared Spectroscopy: Improved Accuracy via Soil Identification. *Biosystems Engineering* 94 (1): 111–118, doi:10.1016/j.biosystemseng.2006.01.014

Linker R, Shmulevich I, Kenny A, Shaviv A (2005), Soil identification and chemometrics for direct determination of nitrate in soils using FTIR-ATR mid-infrared spectroscopy. *Chemosphere* 61: 652–658.

Linker R, Kenny A, Shaviv A, Singher L, Shmulevich I (2004). FTIR/ATR nitrate determination of soil pastes using PCR, PLS and cross-correlation. *Applied Spectroscopy*, 58: 516–520.

Llewellyn PL, and Theocharis CR (1991) A diffuse reflectance fourier transform infra-red study of carbon dioxide adsorption on silicalite-I. *J. Chem. Technol. Biotechnol.* 52: 473–480. doi: 10.1002/jctb.280520405

Love SP, Goff F, Counce D, Siebe C, Delgado H (1998) Passive infrared spectroscopy of the eruption plume at Popocatépetl volcano, Mexico. *Nature* 396: 563–567; doi:10.1038/25109.

Ma X, Li L, Yang L, Su C, Wang K, Jiang K (2012a) Preparation of hybrid CaCO$_3$–pepsin hemisphere with ordered hierarchical structure and the application for removal of heavy metal ions *Journal of Crystal Growth* 338(1): 272-279

Ma Q, Liu Y, Liu C, Ma J, He H (2012b) A case study of Asian dust storm particles: Chemical composition, reactivity to SO$_2$ and hygroscopic properties. *Journal of Environmental Sciences* 24(1): 62–71.

Majedová J (2003) FTIR techniques in clay mineral studies – Review. *Vibrational Spectroscopy* 31: 1–10.

Mangoni ML, Papo N, Mignogna G, Andreu D, Shai Y, Barra D, Simmaco M (2003) Ranacyclins, a new family of short cyclic antimicrobial peptides: biological function, mode of action and parameters involved in target specificity. *Biochemistry* 42: 14023-14035.

Marshall TL, Chaffin CT, Hammaker RM, Fateley WG (1994) An intorduction to open-path FT-IR, Atmospheric monitoring. *Environ. Sci. Technol.* 28(5): 224A-232A.

McKelvy ML, Britt TR, Davis BL, Gillie JK, Graves FB, Lentz LA (1998) Infrared Spectroscopy. *Anal. Chem.* 70: 119R-177R.

Mecozzi M, Moscato F, Pietroletti M, Quarto F, Oteri F, Cicero AM (2009) Applications of FTIR spectroscopy in environmental studies supported by two dimensional correlation analysis. *Global NEST Journal*, 11(4): 593-600.

Minnich TR, Scotto RL (1999) Use of Open-Path FTIR Spectroscopy to Adrees Air Monitoring Needs During Site Remediations. Invited Article Published in „Remediati", Summer 1999, 1-16.

Mori T., Notsu K (2008) Temporal variation in chemical composition of the volcanic plume from Aso volcano, Japan, measured by remote FT-IR spectroscopy. *Geochem. J.* 42: 133–140.

Mori T, Notsu K (1997) Remote CO, COS, CO$_2$, SO$_2$, HCl detection and temperature estimation of volcanic gas. *Geophys. Res. Lett.* 24: 2047–2050.

Mori T, Notsu K, Tohjima Y, Wakita H (1993) Remote detection of HCl and SO$_2$ in volcanic gas from Unzen volcano, Japan. *Geophys. Res. Lett.* 20: 1355–1358.

Muckenhuber H, and Grothe H (2007) A DRIFTS study of the heterogeneous reaction of NO$_2$ with carbonaceous materials at elevated temperature. *Carbon.* 45 (2): 321–329.

Müller K, Foerstendorf H, Meusel T, Brendler V, Lefèvre G, Comarmond MJ, Payne TE (2012), Sorption of U(VI) at the TiO$_2$–water interface: An in situ vibrational spectroscopic study. *Geochimica et Cosmochimica Acta.* 76: 191–205. http://dx.doi.org/10.1016/j.gca.2011.10.004

Müller K, Foerstendorf H, Brendler V, Bernhard G (2009) Sorption of Np(V) onto TiO$_2$, SiO$_2$, and ZnO: An in Situ ATR FT-IR Spectroscopic Study. *Environ. Sci. Technol.* 43(20): 7665–7670.

Mungasavalli DP, Viraraghavan T, Chunglin Y (2007) Biosorption of chromium from aqueous solutions by pretreated Aspergillus niger: batch and column studies. *Colloids. Surf. A Physicochem. Eng. Aspects* 301: 214-223.

Nault JR, Preston CM, Trofymow JAT, Fyles J, Kozak L, Siltanen M, Titus B (2009) Applicability of Diffuse Reflectance Fourier Transform Infrared Spectroscopy to the

Chemical Analysis of Decomposing Foliar Litter in Canadian Forests. *Soil Science.* 174(3): 130-142, DOI: 10.1097/SS.0b013e318198699a.

Noda I. and Ozaki I. (2005) Two-dimensional correlation spectroscopy. *Application in vibrational and optical spectroscopy,* John Wiley & Sons, UK

Notsu K, Mori T (2010) Chemical monitoring of volcanic gas using remote FT-IR spectroscopy at several active volcanoes in Japan. *Applied Geochemistry* 25: 505–512.

Ofner JN, Balzer N, Buxmann J, Grothe H, Schmitt-Kopplin P, Platt U and Zetzsch C (2012) Halogenation processes of secondary organic aerosol and implications on halogen release mechanisms. *Atmos. Chem. Phys. Discuss.*12: 2975–3017.

Oppenheimer C, Lomakina AS, Kyle PR, Kingsbury NG, Boichu M (2009) Pulsatory magma supply to a phonolite lava lake. *Earth and Planetary Science Letters* 284: 392–398, doi:10.1016/j.epsl.2009.04.04

Parikh SJ, Chorover J (2005) FTIR Spectroscopic Study of Biogenic Mn-Oxide Formation by *Pseudomonas putida* GB-1. *Geomicrobiology Journal,* 22:207–218.

Parolo ME, Savini MC, Vallés JM, Baschini MT, Avena MJ (2008) Tetracycline adsorption on montmorillonite: pH and ionic strength effects. *Applied Clay Science* 40: 179–186.

Pasilis SP, Pemberton JE (2008) Spectroscopic investigation of uranyl(VI) and citrate coadsorption to Al_2O_3. *Geochimica et Cosmochimica Acta* 72: 277–287.

Paton-Walsh C (2011). Remote Sensing of Atmospheric Trace Gases by Ground-Based Solar Fourier Transform Infrared Spectroscopy, *Fourier Transforms - New Analytical Approaches and FTIR Strategies,* Prof. Goran Nikolic (Ed.), ISBN: 978-953-307-232-6, InTech, Available from: http://www.intechopen.com/books/fourier-transforms-new-analytical-approaches-and-ftir strategies/remote-sensing-of-atmospheric-trace-gases-by-ground-based-solar-fourier-transform-infrared-spectroscopy Accessed 2012 March 12.

Perry SH, McKane PL, Pescatore DE, DuBois AE, Kricks RJ (1995) Maximizing the use of open-path FTIR for 24-h monitoring around the process area of an industrial chemical facility. *AWMA Conference on Optical Remote Sensing for Environmental and Process Monitoring SPIE* 2883: 333–344.

Pironon J, de Donato Ph, Barrès O, Garnier Ch (2009) On-line greenhouse gas detection from soils and rock formations. *Energy Procedia* 1: 2375–2382.

Pokrovsky OS, Martinez RE, Golubev SV, Kompantseva EI, Shirokova LS (2008) Adsorption of metals and protons on Gloeocapsa sp. cyanobacteria: A surface speciation approach. *Applied Geochemistry* 23: 2574–2588.

Prince A P, Kleiber P, Grassian V H, Young M A (2007) Heterogeneous interactions of calcite aerosol with sulfur dioxide and sulfur dioxide-nitric acid mixtures. *Physical Chemistry & Chemical Physics* 9(26): 3432–3439.

Puckrin E, Evans WFJ, Adamson TAB (1996) Measurement of tropospheric ozone by thermal emission spectroscopy. *Atmospheric Environment* 30(4): 563-568.

Raphael L (2011). Application of FTIR Spectroscopy to Agricultural Soils Analysis, *Fourier Transforms - New Analytical Approaches and FTIR Strategies,* Prof. Goran Nikolic (Ed.), ISBN: 978-953-307-232-6, InTech, Available from:

http://www.intechopen.com/books/fourier-transforms-new-analytical-approaches-and-ftir-strategies/application-of-ftir-spectroscopy-to-agricultural-soils-analysis. accessed 2012 March 12.

Redden G, Bargar J, and Bencheikh-Latmani R (2001) Citrate enhanced uranyl adsorption on goethite: an EXAFS analysis. *J. Colloid Interface Sci.* 244: 211–219.

Rintoul L, Panayiotou H, Kokot S, George G, Cash G, Frost R, Bui T, Fredericks P (1998) Fourier transform infrared spectrometry: a versatile technique for real world samples. *Analyst* 123: 571–577.

Rumpel C, Janik LJ, Skjemstad JO, Kögel-Knabner I (2001) Quantification of carbon derived from lignite in soils using mid-infrared spectroscopy and partial least squares. *Organic Geochemistry* 32(6): 831–839.

Russwurm GM (1999) Compendium of Methods for the Determination of Toxic Organic Compounds in Ambient Air – second edition. Long-Path Open-Path Fourier Transform Infrared Monitoring Of Atmospheric Gases – Method TO-16: 16-1-16-41. available at http://www.epa.gov/ttnamti1/files/ambient/airtox/tocomp99.pdf accessed 2012 March 12.

Russwurm GM, Childers JW (1996) FT-IR Open-Path Monitoring Guidance Document. U. S. Environmental Protection Agency, Research Triangle Park, NC, EPA/600/R-96/040, April 1996.

Santos C, Fraga ME, Kozakiewicz Z, Lima N (2010) Fourier transform infrared as a powerful technique for the identification and characterization of filamentous fungi and yeast. *Research in Microbiology* 161: 168-175.

Schneider M, Romero PM, Hase F, Blumenstock T, Cuevas E, Ramos R (2010) Continuous quality assessment of atmospheric water vapour measurement techniques: FTIR, Cimel, MFRSR, GPS, and Vaisala RS92. *Atmos. Meas. Tech.* 3: 323-338, doi:10.5194/amt-3-323-2010.

Schuttlefield J, Al-Hosney H, Zachariah A, and Grassian VH (2007a) Attenuated Total Reflection Fourier Transform Infrared Spectroscopy to Investigate Water Uptake and Phase Transitions in Atmospherically Relevant Particles. *Appl. Spectrosc.* 61: 283-292.

Schuttlefield JD, Cox D, and Grassian VH (2007b) An investigation of water uptake on clays minerals using ATR-FTIR spectroscopy coupled with quartz crystal microbalance measurements. *J. Geophys. Res.* 112. D21303, doi:10.1029/2007JD008973.

Scott RPW *Analytical Spectroscopy* available at: http://www.analyticalspectroscopy.net/ap3-11.htm Accessed 2012 March 11.

Segal-Rosenheimer M and Dubowski Y (2007) Heterogeneous Ozonolysis of Cypermethrin Using Real-Time Monitoring FTIR Techniques. *J. Phys. Chem.* C 111: 11682-11691.

Sethuraman A, Belfort G (2005) Protein structural perturbation and aggregation on homogeneus surfaces. *Biophys. J.* 88: 1322-1333.

Shaka H and Saliba N (2004) Concentration measurements and chemical composition of PM10-2.5 and PM2.5 at a coastal site in Beirut, Lebanon. *Atmospheric Environment* 38: 523 – 531.

Sibley KJ,. Brewster GR, Astatkie T, Adsett JF, and Struik PC (2010). In-Field Measurement of Soil Nitrate Using an Ion-Selective Electrode, *Advances in Measurement Systems*,

Milind Kr Sharma (Ed.), ISBN: 978-953-307-061-2, InTech, Available from: http://www.intechopen.com/books/advances-in-measurement-systems/in-field-measurement-of-soil-nitrate-using-an-ion-selective-electrode Accessed 2012 April 30.

Simonescu CM, Dima R, Ferdeş M, Meghea A (2012) Equilibrium and Kinetic Studies on the Biosorption of Cu(II) onto *Aspergillus niger* Biomass. Rev. Chim. (Bucharest). ISSN 0034-7752 63(2): 224-228.

Simonescu CM, Deleanu C, Bobirică L, Melinescu A, Giurginca M (2005) Bentonite and Na-bentonite used in ammonium removal from wastewaters. *Proceedings of 14th Romanian International Conference of Chemistry and Chemical Engineering* (RICCCE XIV). ISBN: 973-718-284-7, ISBN: 973-718-288-X, Ed. Printech, Bucharest 22 – 24 September 2005, 6: S06-262.

Simonescu CM, Ferdeş M Fungal biomass for Cu(II) uptake from aqueous systems. Polish Journal of of Environmental Studies. *in press*

Smidt E, Böhm K and Schwanninger M (2011) The Application of FT-IR Spectroscopy in Waste Management, Fourier Transforms - *New Analytical Approaches and FTIR Strategies*, Prof. Goran Nikolic (Ed.), ISBN: 978-953-307-232-6, InTech, Available from: http://www.intechopen.com/books/fourier-transforms-new-analytical-approaches-and-ftir-strategies/the-application-of-ft-ir-spectroscopy-in-waste-management   Accessed 2012 March 12.

Smidt E, Meissl K (2007) The applicability of Fourier transform infrared (FT-IR) spectroscopy in waste management. *Waste Management* 27: 268–276.

Smith BC (2011) *Fundamentals of Fourier Transform Infrared Spectroscopy*, 2nd Edn, CRC Press, Boca Raton, FL, USA.

Snabe T, Petersen BS (2002) Application of infrared spectroscopy (attenuated total reflection) for monitoring enzymatic activity on substrate films. *Journal of Biotechnology* 95:145-155.

Spaccini R, Piccolo A, Haberhauer G, Stemmer M, and Gerzabek MH (2001) Decomposition of maize straw in three European soils as revealed by DRIFT spectra of soil particle fractions. *Geoderma* 99:245–260.

Thorn TG, Marshall TL, Chaffin CT (2001) Open-Path FTIR Air Monitoring of Phosphine around Large Fumigated Structures; *Field Anal. Chem. Technol.* 5: 116-120.

Tomšič B, Simončič B, Orel B, Vilčnik A, Spreizer H (2007) Biodegradability of cellulose fabric modified by imidazolidinone. *Carbohydrate Polymers* 69: 478–488.

Tremblay L, and Gagné J-P (2002) Fast quantification of humic substances and organic matter by direct analysis of sediments using DRIFT spectroscopy. *Anal. Chem.* 74:2985–2993.

Tsai YI, Kuo S-C (2006) Development of diffuse reflectance infrared Fourier transform spectroscopy for the rapid characterization of aerosols. *Atmospheric Environment* 40: 1781–1793.

Tso TL, Chang SY (1996) Unambiguous identification of fugitive pollutants and the determining of annual emission flux as a diurnal monitoring mode using oper.-path Fourier transform infrared spectroscopy. *Analytical Sciences* 12: 311–319.

Ullerstam M, Johnson M S, Vogt R, Ljungström E (2003) DRIFTS and Knudsen cell study of the heterogeneous reactivity of SO_2 and NO_2 on mineral dust. *Atmospheric Chemistry and Physics* 3(6): 2043–2051.

Ullerstam M, Vogt R, Langer S, Ljungström E (2002) The kinetics and mechanism of SO_2 oxidation by O_3 on mineral dust. *Physical Chemistry & Chemical Physics* 4(19): 4694–4699.

Valyon J, Lónyi F, Onyestyák G, Papp J (2003) DRIFT and FR spectroscopic investigation of N_2 and O_2 adsorption on zeolites. *Microporous and Mesoporous Materials* 61(1–3): 147–158, ZEOLITE '02 (Proceedings of the 6th Internationanl Conference on the Occurrence, Properties and Utilization of Natural Zeolites), http://dx.doi.org/10.1016/S1387-1811(03)00362-7.

Veres P (2005) FTIR Analysis of Particulate Matter Collected on Teflon Filters in Columbus, OH - A Senior Honors Thesis - The Ohio State University June 2005.

Verma SK, Deb MK, Verma D (2008) Determination of nitrogen dioxide in ambient air employing diffuse reflectance Fourier transform infrared spectroscopy. *Atmospheric Research* 90: 33–40.

Verma SK, Deb MK (2007a) Nondestructive and rapid determination of nitrate in soil, dry deposits and aerosol samples using KBr-matrix with diffuse reflectance Fourier transform infrared spectroscopy (DRIFTS). *Analytica Chimica Acta* 582: 382–389.

Verma SK, Deb MK (2007b) Direct and rapid determination of sulphate in environmental samples with diffuse reflectance Fourier transform infrared spectroscopy using KBr substrate. *Talanta* 71: 1546–1552. doi:10.1016/j.talanta.2006.07.056

Walter WT, Perry SH, Han JS, Park CJ (1999) Open-Path FTIR Ozone Measurements in Korea; *Proc. SPIE* 3534: 133-139.

Wang S-L, Lee J-F (2011) Reaction mechanism of hexavalent chromium with cellulose. *Chemical Engineering Journal* 174: 289– 295.

Workman Jr. JJ (1999) Review of Process and Non-invasive Near-Infrared and Infrared Spectroscopy: 1993–1999. *Applied Spectroscopy Reviews* 34(1&2): 1–89.

Wu C-F, Chen Y-L, Chen C-C, Yang T-T, Chang PE (2007) Applying open-path Fourier transform infrared spectroscopy for measuring aerosols, *Journal of Environmental Science and Health*, Part A. 42(8): 1131-1140.

Xu L, Wang JN, Meng Y, Li AM (2012) Fast removal of heavy metal ions and phytic acids from water using new modified chelating fiber. *Chinese Chemical Letters* 23: 105–108.

Yee N, Benning LG, Phoenix VR, Ferris FG (2004) Characterization of Metal-Cyanobacteria Sorption Reactions: A Combined Macroscopic and Infrared Spectroscopic Investigation. *Environ. Sci. Technol.* 38: 775-782.

Yoon TH, Johnson SB, Musgrave CB, Jr Brown GE (2004) Adsorption of organic matter at mineral/water interfaces: I. ATR-FTIR spectroscopic and quantum chemical study of oxalate adsorbed at boehmite/water and corundum/water interfaces. *Geochimica et Cosmochimica Acta* 68(22): 4505-4518.

Zhang X Y, Zhuang G S, Chen J M, Wang Y, Wang X, An Z Set al. (2006) Heterogeneous reactions of sulfur dioxide on typical mineral particles. *Journal of Physical Chemistry* B,110(25): 12588–12596.

Zhao Y and Chen Z (2010) Application of Fourier Transform Infrared Spectroscopy in the Study of Atmospheric Heterogeneous Processes. *Applied Spectroscopy Reviews* 45: 63-91.

Zimmermann M, Leifeld J, Fuhrer J (2007) Quantifying soil organic carbon fractions by infrared-spectroscopy. *Soil Biology & Biochemistry* 39: 224–231.

Zimmermann M, Leifeld J, Schmidt MWI, Fuhrer J (2004) Characterization of Soil Properties by DRIFT-Spectroscopy. Eurosoil Congress, Freiburg, D.Poster and Proceedings. available at

http://www.bodenkunde2.uni-freiburg.de/eurosoil/abstracts/id101_Zimmermann_full.pdf Accessed 2012 March11.

Chapter 2

Electronic (Absorption) Spectra of 3d Transition Metal Complexes

S. Lakshmi Reddy, Tamio Endo and G. Siva Reddy

Additional information is available at the end of the chapter

1. Introduction

1.1. Types of spectra

Spectra are broadly classified into two groups (i) emission spectra and (ii) absorption spectra

i. *Emission spectra* Emission spectra are of three kinds (a) continuous spectra,(b) band spectra and (c) line spectra.

Continuous spectra: Solids like iron or carbon emit continuous spectra when they are heated until they glow. Continuous spectrum is due to the thermal excitation of the molecules of the substance.

Band spectra: The band spectrum consists of a number of bands of different colours separated by dark regions. The bands are sharply defined at one edge called the head of the band and shade off gradually at the other edge. Band spectrum is emitted by substances in the molecular state when the thermal excitement of the substance is not quite sufficient to break the molecules into continuous atoms.

Line spectra: A line spectrum consists of bright lines in different regions of the visible spectrum against a dark background. All the lines do not have the same intensity. The number of lines, their nature and arrangement depends on the nature of the substance excited. Line spectra are emitted by vapours of elements. No two elements do ever produce similar line spectra.

ii. *Absorption spectra:* When a substance is placed between a light source and a spectrometer, the substance absorbs certain part of the spectrum. This spectrum is called the absorption spectrum of the substance.

Electronic absorption spectrum is of two types. d-d spectrum and charge transfer spectrum. d-d spectrum deals with the electronic transitions within the d-orbitals. In the charge – transfer spectrum, electronic transitions occur from metal to ligand or vice-versa.

2. Electronic spectra of transitions metal complexes

Electronic absorption spectroscopy requires consideration of the following principles:

a. *Franck-Condon Principle:* Electronic transitions occur in a very short time (about 10^{-15} sec.) and hence the atoms in a molecule do not have time to change position appreciably during electronic transition .So the molecule will find itself with the same molecular configuration and hence the vibrational kinetic energy in the exited state remains the same as it had in the ground state at the moment of absorption.

b. *Electronic transitions between vibrational states:* Frequently, transitions occur from the ground vibrational level of the ground electronic state to many different vibrational levels of particular excited electronic states. Such transitions may give rise to vibrational fine structure in the main peak of the electronic transition. Since all the molecules are present in the ground vibrational level, nearly all transitions that give rise to a peak in the absorption spectrum will arise from the ground electronic state. If the different excited vibrational levels are represented as υ_1, υ_2, etc., and the ground state as υ_0, the fine structure in the main peak of the spectrum is assigned to $\upsilon_0 \rightarrow \upsilon_0$, $\upsilon_0 \rightarrow \upsilon_1$, $\upsilon_0 \rightarrow \upsilon_2$ etc., vibrational states. The $\upsilon_0 \rightarrow \upsilon_0$ transition is the lowest energy (longest wave length) transition.

c. *Symmetry requirement:* This requirement is to be satisfied for the transitions discussed above.

Electronic transitions occur between split 'd' levels of the central atom giving rise to so called d-d or ligand field spectra. The spectral region where these occur spans the near infrared, visible and U.V. region.

Ultraviolet (UV)	Visible (Vis)	Near infrared (NIR)	
50,000 - 26300	26300 - 12800	12800 - 5000	cm^{-1}
200 - 380	380 - 780	780 - 2000	nm

3. Russel-Saunders or L-S coupling scheme

An orbiting electronic charge produces magnetic field perpendicular to the plane of the orbit. Hence the orbital angular momentum and spin angular momentum have corresponding magnetic vectors. As a result, both of these momenta couple magnetically to give rise to total orbital angular momentum. There are two schemes of coupling: Russel-Saunders or L-S coupling and j-j coupling.

a. The individual spin angular momenta of the electrons, s_i, each of which has a value of ± ½, combine to give a resultant spin angular momentum (individual spin angular momentum is represented by a lower case symbol whereas the total resultant value is given by a upper case symbol).

$$\sum s_i = S$$

Two spins of each $\pm \frac{1}{2}$ could give a resultant value of S =1 or S= 0; similarly a resultant of three electrons is $1\frac{1}{2}$ or $\frac{1}{2}$ .The resultant is expressed in units of $h/2\pi$. The spin multiplicity is given by (2S+1). Hence, If n is the number of unpaired electrons, spin multiplicity is given by n + 1.

b. The individual orbital angular momenta of electrons, l_i, each of which may be 0, 1 ,2, 3 , 4 ..... in units of $h/2\pi$ for s, p, d, f, g, ...orbitals respectively, combine to give a resultant orbital angular momentum, L in units of $h/2\pi$. $\sum l_i = L$

The resultant L may be once again 0, 1, 2, 3, 4.... which are referred to as S, P, D, F G,... respectively in units of $h/2\pi$.The orbital multiplicity is given by (2L+1).

0	1	2	3	4	5
S	P	D	F	G	H

c. Now the resultant S and L couple to give a total angular momentum, J. Hence, it is not surprising that J is also quantized in units of $h/2\pi$.The possible values of J quantum number are given as

$$J = \left(L + S \right), \left(L + S - 1 \right), \left(L + S - 2 \right), \left(L + S - 3 \right), ... \left| L - S \right|,$$

The symbol | | indicates that the absolute value (L − S) is employed, i.e., no regard is paid to ± sign. Thus for L = 2 and S = 1, the possible J states are 3, 2 and 1 in units of $h/2\pi$.

The individual spin angular momentum, s_i and the individual orbital angular momentum, l_i, couple to give total individual angular momentum, j_i. This scheme of coupling is known as spin-orbit coupling or j -j coupling.

4. Term symbols

4.1. Spectroscopic terms for free ion ground states

The rules governing the term symbol for the ground state according to L-S coupling scheme are given below:

a. The spin multiplicity is maximized i.e., the electrons occupy degenerate orbitals so as to retain parallel spins as long as possible (Hund's rule).
b. The orbital angular momentum is also maximized i.e., the orbitals are filled with highest positive m values first.
c. If the sub-shell is less than half-filled, J = L− S and if the sub-shell is more than half − filled, J = L +S.

The term symbol is given by $^{2S+1}L_J$. The left-hand superscript of the term is the spin multiplicity, given by 2S+1 and the right- hand subscript is given by J. It should be noted that S is used to represent two things- (a) total spin angular momentum and (b) and total angular momentum when L = 0. The above rules are illustrated with examples.

For d^4 configuration:

↑	↑	↑	↑	

m_l +2   +1   0   -1   -2

Hence, L = 3 -1 = 2 i.e., D; S = 2; 2S+1 = 5; and J = L- S = 0; Term symbol = $^5D_0$

For d^9 configuration:

↓↑	↓↑	↓↑	↓↑	↑

m_l +2   +1   0   -1   -2

Hence, L = +2+1+0-1 = 2 i.e., D ; S = 1 /2 ; 2S+1 = 2 ; and J = L+ S = 3/2 ; Term symbol = $^2D_{5/2}$

Spin multiplicity indicates the number of orientations in the external field. If the spin multiplicity is three, there will be three orientations in the magnetic field.- parallel, perpendicular and opposed. There are similar orientations in the angular momentum in an external field.

The spectroscopic term symbols for d^n configurations are given in the Table-1. The terms are read as follows: The left-hand superscript of the term symbol is read as singlet, doublet, triplet, quartet, quintet, sextet, septet, octet, etc., for spin multiplicity values of 1, 2, 3, 4. 5, 6, 7, 8, etc., respectively.$^1S_0$ (singlet S nought); $^2S_{1/2}$ (doublet S one–half); $^3P_2$ (triplet P twc); $^5I_8$ (quintet I eight). It is seen from the Table-1 that d^n and d^{10-n} have same term symbols, if we ignore J values. Here n stands for the number of electrons in d^n configuration.

d^n	Term	d^n	Term
d^0	$^1S_0$	d^{10}	$^1S_0$
d^1	$^2D_{3/2}$	d^9	$^2D_{5/2}$
d^2	$^3F_2$	d^8	$^3F_4$
d^4	$^5D_0$	d^6	$^5D_4$
d^5	$^6S_{5/2}$		

Table 1. Term symbols

It is also found that empty sub -shell configurations such as p^0, d^0, f^0, etc., and full filled sub-shell configurations such as p^6, d^{10}, f^{14}, etc., have always the term symbol $^1S_0$ since the resultant spin and angular momenta are equal to zero. All the inert gases have term symbols for their ground state $^1S_0$ .Similarly all alkali metals reduce to one electron problems since closed shell core contributes nothing to L , S and J; their ground state term symbol is given by $^2S_{1/2}$. Hence d electrons are only of importance in deciding term symbols of transition metals.

5. Total degeneracy

We have seen that the degeneracy with regard to spin is its multiplicity which is given by (2S+1). The total spin multiplicity is denoted by M_s running from S to -S. Similarly orbital degeneracy, M_L, is given by $(2L+1)$ running from L to -L. For example, L= 2 for D state and so the orbital degeneracy is $(2 \times 2+1) = 5$ fold. Similarly, for F state, the orbital degeneracy is seven fold. Since there are $(2L+1)$ values of M_L, and $(2S+1)$ values of M_s in each term, the total degeneracy of the term is given by: $2(L+1)(2S+1)$.

Each value of M_L occurs $(2S+1)$ times and each value of M_s occurs $(2L+1)$ times in the term. For ^3F state, the total degeneracy is $3 \times 7 = 21$ fold and for the terms $^3P, {}^1G, {}^1D, {}^1S$, the total degeneracy is 9,9,5,1 fold respectively. Each fold of degeneracy represents one microstate.

6. Number of microstates

The electrons may be filled in orbitals by different arrangements since the orbitals have different m_l values and electrons may also occupy singly or get paired. Each different type of electronic arrangement gives rise to a microstate. Thus each electronic configuration will have a fixed number of microstates. The numbers of microstates for p^2 configuration are given in Table-2 (for both excited and ground states).

m_l -1		↑	↑		↓	↓		↓	↓			↑			↑↓
0	↑		↑	↓		↓	↓		↑	↑	↑	↓		↑↓	
+1	↑	↑		↓	↓		↑	↑		↓	↓		↑↓		
m_L	+1	0	-1	+1	0	-1	+1	0	-1	+1	+1	-1	+2	0	

Table 2. Number of microstates for p^2 configuration

Each vertical column is one micro state. Thus for p^2 configuration, there are 15 microstates. In the above diagram, the arrangement of singlet states of paired configurations given in A (see below) is not different from that given in B and hence only one arrangement for each m_l value.

↑↓	↑↑		↑↓	↑↑
A	B		A	B

The number of microstates possible for any electronic configuration may be calculated from the formula,

Number of microstates = n! / r! (n - r)!

Where n is the twice the number of orbitals, r is the number of electrons and ! is the factorial.

For p^2 configuration, n= 3x2 =6; r = 2; n – r = 4

6! = 6 x 5 x 4 x 3 x 2 x 1 = 720; 2! = 2 x 1 =2; 4! = 4 x 3 x 2 x 1 = 24

Substituting in the formula, the number of microstates is 15.

Similarly for a d^2 configuration, the number of microstates is given by 10! / 2! (10 − 2)!

$$\frac{10\times9\times8\times7\times6\times5\times4\times3\times2\times1}{2\times1\left(8\times7\times6\times5\times4\times3\times2\times1\right)}=45$$

Thus a d^2 configuration will have 45 microstates. Microstates of different d^n configuration are given in Table-3.

d^n configuration	d^1,d^9	d^2,d^8	d^3,d^7	d^4,d^6	d^5	d^{10}
No.of microstates	10	45	120	210	252	1

Table 3. Microstates of different d^n configuration

7. Multiple term symbols of excited states

The terms arising from d^n configuration for 3d metal ions are given Table-4.

Configuration	Ion	Term symbol
d^1	Ti^{3+},V^{4+}	
d^9	Cu^{2+}	^2D
d^2	Ti^{2+},V^{3+},Cr^{4+}	
d^8	Ni^{2+}	$^3F, ^3P, ^1G, ^1D, ^1S$
d^3	Cr^{3+},V^{2+},Mn^{4+}	
d^7	Ni^{3+},Co^{2+}	$^4F, ^4P, ^2(H, G, F, D, D, P)$
d^4	Cr^{2+},Mn^{3+}	
d^6	Fe^{2+},Co^{3+}	$^5D , ^3(H, G, F, F, , D, P, P), ^1(I, ,G, G, F, D, D, S,S)$
d^5	Mn^{2+}, Fe^{3+}	
d^{10}	Zn^{2+}	$^6S, ^4(G, F, D, P), ^2(I, H, G, G, F, F), ^2(D, D, D, P, S)$
		^6S

Table 4. Terms arising from d^n configuration for 3d ions (n=1 to10)

8. Selection rules

8.1. La Porte selection rule

This rule says that transitions between the orbitals of the same sub shell are forbidden. In other words, the for total orbital angular momentum is $\Delta L = \pm 1$. This is La Porte allowed transitions. Thus transition such as $^1S \longrightarrow {}^1P$ and $^2D \longrightarrow {}^2P$ are allowed but transition such as $^3D \longrightarrow {}^3S$ is forbidden since $\Delta L = -2$. That is, transition should involve a change of one unit of angular momentum. Hence transitions from *gerade* to *ungerade* (g to u) or vice versa are allowed, i.e., $u \longrightarrow g$ or $g \longrightarrow u$ but not $u \longrightarrow u$ or $g \longrightarrow g$. In the case of p sub shell, both ground and excited states are odd and in the case of d sub shell both ground and excited states are even. As a rule transition should be from even to odd or vice versa.

The same rule is also stated in the form of a statement instead of an equation:

Electronic transitions within the same p or d sub-shell are forbidden, if the molecule has centre of symmetry.

8.2. Spin selection rule

The selection Rule for Spin Angular Momentum is

$$\Delta S = 0$$

Thus transitions such as $^2S \longrightarrow {}^2P$ and $^3D \longrightarrow {}^3P$ are allowed, but transition such as $^1S \longrightarrow {}^3P$ is forbidden. The same rule is also stated in the form of a statement,

Electronic Transitions between the different states of spin multiplicity are forbidden.

The selection Rule for total angular momentum, J, is

$$\Delta J = 0 \text{ or } \pm 1$$

The transitions such as $^2P_{1/2} \longrightarrow {}^2D_{3/2}$ and $^2P_{3/2} \longrightarrow {}^2D_{3/2}$ are allowed, but transition such as $^2P_{1/2} \longrightarrow {}^2D_{5/2}$ is forbidden since $\Delta J = 2$.

There is no selection rule governing the change in the value of n, the principal quantum number. Thus in hydrogen, transitions such as $1s \longrightarrow 2p$, $1s \longrightarrow 3p$, $1s \longrightarrow 4p$ are allowed.

Usually, electronic absorption is indicated by reverse arrow, $\longleftarrow$, and emission is indicated by the forward arrow, $\longrightarrow$, though this rule is not strictly obeyed.

8.3. Mechanism of breakdown of selection rules

8.3.1. Spin-orbit coupling

For electronic transition to take place, $\Delta S = 0$ and $\Delta L = \pm 1$ in the absence of spin-orbit coupling. However, spin and orbital motions are coupled. Even, if they are coupled very weakly, a little of each spin state mixes with the other in the ground and excited states by an amount dependent upon the energy difference in the orbital states and magnitude of spin –orbit coupling constant. Therefore electronic transitions occur between different states of spin multiplicity and also between states in which ΔL is not equal to ± 1. For example, if the ground state were 99% singlet and 1% triplet (due to spin– orbit coupling) and the excited state were 1% singlet and 99 % triplet, then the intensity would derive from the triplet –triplet and singlet-singlet interactions. Spin-orbit coupling provides small energy differences between degenerate state.

This coupling is of two types. The single electron spin orbit coupling parameter ζ, gives the strength of the interaction between the spin and orbital angular momenta of a single electron for a particular configuration. The other parameter, λ, is the property of the term. For high spin complexes,

$$\lambda = \pm \frac{\xi}{2S}$$

Here positive sign holds for shells less than half field and negative sign holds for more than half filled shells. S is the same as the one given for the free ion. The λ values in crystals are close to their free ion values. Λ decreases in crystal with decreasing Racah parameters B and C. For high spin d^5 configuration, there is no spin orbit coupling because ^6S state is unaffected by the ligand fields. The λ and ζ values for 3d series are given in Table-5.

Ion	Ti(II)	V(II)	Cr(II)	Mn(II)	Fe(II)	Co(II)	Ni(II)
Ξ (cm^{-1})	121	167	230	347	410	533	649
λ(cm^{-1})	60	56	57	0	-102	-177	-325

Table 5. λ and ζ values for 3d series

8.3.2. La Porte selection rule

Physically 3d (even) and 4p (odd) wave functions may be mixed, if centre of inversion (i) is removed. There are two processes by which i is removed.

a. The central metal ion is placed in a distorted field (tetrahedral field, Tetragonal distortions, etc.,) The most important case of distorted or asymmetric field is the case of a tetrahedral complex. Tetrahedron has no inversion centre and so d-p mixing takes place. So electronic transitions in tetrahedral complexes are much more intense, often by a factor 100, than in a analogous octahedral complexes. *Trans* isomer of [Co(en)$_2$Cl$_2$] $^+$ in aqueous solution is three to four times less intense than the *cis* isomer because the former is centro-symmetric. Other types of distortion include Jahn –Teller distortions.

b. Odd vibrations of the surrounding ligands create the distorted field for a time that is long enough compared to the time necessary for the electronic transition to occur (Franck Condon Principle).Certain vibrations will remove the centre of symmetry. Mathematically this implies coupling of vibrational and electronic wave functions. Breaking down of La Porte rule by vibrionic coupling has been termed as "Intensity Stealing". If the forbidden excited term lies energetically nearby a fully allowed transition, it would produce a very intense band. Intensity Stealing by this mechanism decreases in magnitude with increasing energy separation between the excited term and the allowed level.

9. Splitting of energy states

The symbols **A**(or **a**) and **B** (or **b**) with any suffixes indicate wave functions which are singly degenerate. Similarly **E** (or **e**) indicates double degeneracy and **T** (or **t**) indicates triple degeneracy. Lower case symbols, a_{1g}, a_{2g}, e_g, etc., are used to indicate electron wave functions(orbitals) and upper case symbols are used to describe electronic energy levels. Thus $^2T_{2g}$ means an energy level which is triply degenerate with respect to orbital state and also doubly degenerate with respect to its spin state. Upper case symbols are also used without any spin multiplicity term and they then refer to symmetry (ex., A_{1g} symmetry). The subscripts **g** and **u** indicate *gerade* (even) and *ungerade* (odd).

d orbitals split into two sets - t_{2g} orbitals and e_g orbitals under the influence crystal field. These have T_{2g} and E_g symmetry respectively. Similarly **f** orbitals split into three sets - a_{2u} (f_{xyz}) , t_{2u} ($f_{x(y^2-z^2)}$, $f_{y(z^2-x^2)}$, $f_{z(x^2-y^2)}$) and t_{1u} ($f_x{}^3$, $f_y{}^3$, $f_z{}^3$). These have symmetries A_{2u}, T_{2u} and T_{1u} respectively.

Splitting of **D** state parallels the splitting of **d** orbitals and splitting of **F** state splits parallels splitting of **f** orbitals. For example, **F** state splits into either T_{1u}, T_{2u} and A_{2u} or T_{1g}, T_{2g} and A_{2g} sub-sets. Which of these is correct is determined by **g** or **u** nature of the configuration from which **F** state is derived. Since **f** orbitals are **u** in character ^2F state corresponding to f^1 configuration splits into $^2T_{1u}$, $^2T_{2u}$, and $^2A_{2u}$ components; similarly ^3F state derived from d^2 configuration splits into $^3T_{2g}$, $^3T_{1g}$ and $^3A_{2g}$ components because **d** orbitals are **g** in character.

9.1. Splitting of energy states corresponding to d^n terms

These are given in Table-6.

Energy	Sub- states
S	A_1
P	T_1
D	$E + T_2$
F	$A_2 + T_1 + T_2$
G	$A_1 + E + T_1 + T_2$
H	$E + T_1 + T_1 + T_2$
I	$A_1 + A_2 + E + T_1 + T_2 + T_2$

Table 6. Splitting of energy states corresponding to d^n terms

The d-d spectra is concerned with d^n configuration and hence the crystal field sub-states are given for all the d^n configuration in Table -7.

Configuration	Free ion ground state	Crystal field substates	Important excited states	Crystal field state
d^1, d^9	^2D	$^2T_{2g}$, ^2E_g		
d^2, d^8	^3F	$^3T_{1g}$, $^3T_{2g}$, $^3A_{2g}$	^3P	$^3T_{1g}$
d^3, d^7	^4F	$^4T_{1g}$, $^4T_{2g}$, $^4A_{2g}$	^4P	$^4T_{1g}$
d^4, d^6	^5D	$^5T_{2g}$, ^5E_g		
d^5	^6S	$^6A_{1g}$		

Table 7. Crystal field components of the ground and some excited states of d^n (n=1 to 9) configuration

10. Energy level diagram

Energy Level Diagrams are described by two independent schemes - Orgel Diagrams which are applicable to weak field complexes and Tanabe –Sugano (or simply T-S) Diagrams which are applicable to both weak field and strong field complexes.

11. Inter-electronic repulsion parameters

The inter-electronic repulsions within a configuration are linear combinations of Coulombic and exchange integrals above the ground term. They are expressed by either of the two ways: Condon - Shortley parameters, F_0, F_2 and F_4 and Racah parameters, A, B and C. The magnitude of these parameters varies with the nature of metal ion.

11.1. Racah parameters

The Racah parameters are A, B and C. The Racah parameter A corresponds to the partial shift of all terms of a given electronic configuration. Hence in the optical transition considerations, it is not taken into account. The parameter, B measures the inter electronic repulsion among the electrons in the d-orbitals. The decrease in the value of the interelectronic repulsion parameter, B leads to formation of partially covalent bonding. The ratio between the crystal B^1 parameter and the free ion B parameter is known as nephelauxetic rato and it is denoted by β. The value of β is a measure of covalency. The smaller the value, the greater is the covalency between the metal ion and the ligands. The B and C values are a measure of spatial arrangement of the orbitals of the ligand and the metal ion.

Racah redefined the empirical Condon –Shortley parameters so that the separation between states having the maximum multiplicity (for example, difference between is a function of ^3F and ^3P or ^4F and ^4P is a function of a single parameter, B. However, separations between terms of different multiplicity involve both B and C

12. Tanabe –Sugano diagrams

Exact solutions for the excited sate energy levels in terms of Dq, B and C are obtained from Tanabe-Sugano matrices. However, these are very large (10 x 10) matrices and hand calculations are not feasible. For this reason Tanabe-Sugano have drawn energy level diagrams known as T-S diagrams or energy level diagrams. The T-S diagrams are valid only if the value of B, C and Dq ae lower for a complex than for the free ion value.

Quantitative interpretation of electronic absorption spectra is possible by using Tanabe – Sugano diagrams or simply T-S diagrams. These diagrams are widely employed to correlate and interpret spectra for ions of all types, from d^2 to d^8. Orgel diagrams are useful only qualitatively for high spin complexes whereas T-S diagrams are useful both for high spin and low spin complexes. The x-axis in T-S diagrams represent the ground state term. Further, in T-S diagrams, the axes are divided by B, the interelectronic repulsion parameter or Racah Parameter. The x-axis represents the crystal field strength in terms of Dq/ B or Δ / B and the Y-axis represents the energy in terms of E/B.

The energies of the various electronic states are given in the T-S diagrams on the vertical axis and the ligand field strength increases from left to right on the horizontal axis. The symbols in the diagram omit the subscript, g, with the understanding that all states are *gerade* states. Also, in T.S. diagrams, the zero of energy for any particular d^n ion is taken to be the energy of the ground state. Regardless of the ligand field strength, then, the horizontal axis represents the energy of the ground state because the vertical axis is in units of E/B and x-axis is also in units of Δ /B. Thus, the unit of energy in T-S diagram is B, Racah Parameter.

The values of B are different for different ions of the same d^n (or different d^n configuration) which is shown on the top of each diagram. One T-S diagram is used for all members of an isoelectronic group. Also some assumption is made about the relative value of C/B.

13. Electron spin resonance

Electron Spin Resonance (ESR) is a branch of spectroscopy in which radiation of microwave frequency is absorbed by molecules possessing electrons with unpaired spins. It is known by different names such as Electron Paramagnetic Resonance (EPR), Electron Spin Resonance (ESR) and Electron Magnetic Resonance (EMR). This method is an essential tool for the analysis of the structure of molecular systems or ions containing unpaired electrons, which have spin-degenerate ground states in the absence of magnetic field. In the study of solid state materials, EPR method is employed to understand the symmetry of surroundings of the paramagnetic ion and the nature of its bonding to the nearest neighbouring ligands.

When a paramagnetic substance is placed in a steady magnetic field (H), the unpaired electron in the outer shell tends to align with the field. So the two fold spin degeneracy is removed. Thus the two energy levels, $E_{1/2}$ and $E_{-1/2}$ are separated by $g\beta H$, where g is spectroscopic splitting factor and is called gyro magnetic ratio and β is the Bohr magneton. Since there is a finite probability for a transition between these two energy levels, a change in the energy state can be stimulated by an external radio frequency. When microwave frequency (υ) is applied perpendicular to the direction of the field, resonance absorption will occur between the two split spin levels. The resonance condition is given by, $h\nu = g\beta H$, where h is Planck's constant.

The resonance condition can be satisfied by varying υ or H. However, EPR studies are carried out at a constant frequency (ν), by varying magnetic field (H). For a free electron, the g value is 2.0023. Since h and β are constants, one can calculate the g factor. This factor determines the divergence of the Zeeman levels of the unpaired electron in a magnetic field and is characteristic of the spin system.

In the crystal systems, the electron spins couple with the orbital motions and the g value is a measure of the spin and orbital contributions to the total magnetic moment of the unpaired electron and any deviation of magnetic moment from the free spin value is due to the spin-orbit interaction. It is known that the crystal field removes only the orbital degeneracy of the ground terms of the central metal ion either partially or completely. The strong electrical fields of the surrounding ligands results in "Stark Splitting" of the energy levels of the paramagnetic ion. The nature and amount of splitting strongly depends on the symmetry of the crystalline electric field. The Stark splitting of the free ion levels in the crystal field determines the magnetic behaviour of the paramagnetic ion in a crystal. Whenever there is a contribution from the unquenched orbital angular momentum, the measured g values are isotropic as a result of the asymmetric crystal field since the contribution from the orbital motion is anisotropic. To decide the ultimate ground state of a paramagnetic ion in the crystal, the two important theorems, Kramers and Jahn-Teller, are useful. Using group theory, one can know the nature of the splitting of the free ion levels in the crystal fields of various symmetries.

Jahn-Teller theorem states that any nonlinear molecule in an electronically degenerate ground state is unstable and tends to distort in order to remove this degeneracy. The direction of distortion which results in greatest stabilization can often be deduced from EPR and other spectroscopic data.

Kramers' theorem deals with restrictions to the amount of spin degeneracy which can be removed by a purely electrostatic field. If the system contains an odd number of electrons, such an electrostatic field cannot reduce the degeneracy of any level below two. Each pair forms what is known as a Kramers' doublet, which can be separated only by a magnetic field. For example, Fe(III) and Mn(II) belonging to d^5 configuration, exhibit three Kramers' doublets labeled as $|\pm 5/2>$, $|\pm 3/2>$ and $|\pm 1/2>$.

If the central metal ion also possesses a non-zero nuclear spin, I, then hyperfine splitting occurs as a result of the interaction between the nuclear magnetic moment and the electronic magnetic moment. The measurement of g value and hyperfine splitting factor provides information about the electronic states of the unpaired electrons and also about the nature of the bonding between the paramagnetic ion and its surrounded ligands. If the ligands also contain non-zero nuclear spin, then the electron spin interacts with the magnetic moment of the ligands. Then one could expect super hyperfine EPR spectrum.

The g value also depends on the orientation of the molecules having the unpaired electron with respect to the applied magnetic field. In the case of perfect cubic symmetry, the g value does not depend on the orientation of the crystal. But in the case of low symmetry crystal fields, g varies with orientation. Therefore we get three values g_{xx}, g_{yy}, and g_{zz} corresponding to a, b and c directions of the crystal. In the case of tetragonal site $g_{xx} = g_{yy}$ which is referred to as $g_\perp$ and corresponds to the external magnetic field perpendicular to the Z-axis. When it is parallel, the value is denoted as $g_\parallel$. Hence one can deduce the symmetry of a complex by EPR spectrum i.e., cubic, tetragonal, trigonal or orthorhombic. Anyhow, it is not possible to distinguish between orthorhombic and other lower symmetries by EPR.

13.1. EPR signals of first group transition metal ions

Transition metal ions of 3d group exhibit different patterns of EPR signals depending on their electron spin and the crystalline environment. For example, $3d^1$ ions, VO^{2+} and Ti^{3+} have s = 1/2 and hence are expected to exhibit a single line whose g value is slightly below 2.0. In the case of most abundant ^{51}V, s = 1/2 and I = 7/2, an eight line pattern with hyperfine structure of almost equal intensity can be expected as shown in Fig-1. In the case of most abundant Ti, (s = 1/2 and I = 0), no hyperfine structure exists. However, the presence of less abundant isotopes (^{47}Ti with I = 5/2 and ^{49}Ti with I = 7/2) give rise to weak hyperfine structure with six and eight components respectively. This weak structure is also shown in Fig-1.

Cr(III), a d^3 ion, with s = 3/2 exhibits three fine line structure. The most abundant ^{52}Cr has I = 0 and does not exhibit hyperfine structure. However, ^{53}Cr with I = 3/2 gives rise to hyperfine structure with four components. This structure will be weak because of the low abundance of ^{53}Cr. Thus each one of the three fine structure lines of ^{53}Cr is split into four weak hyperfine lines. Of these, two are overlapped by the intense central line due to the most abundant ^{52}Cr and the other two lines are seen in the form of weak satellites.

Mn(II) and Fe(III) with d^5 configuration have s = 5/2 and exhibit five lines which correspond to a $|\pm5/2> \rightarrow |\pm3/2>$, $|\pm3/2> \rightarrow |\pm1/2>$ and $|+1/2> \rightarrow |-1/2>$ transitions. In the case of ^{55}Mn, which has I = 5/2, each of the five transitions will give rise to a six line hyperfine structure. But in powders, usually one observes the six-hyperfine lines corresponding to $|+1/2> \rightarrow |-1/2>$ transition only. The remaining four transition sets will be broadened due to the high anisotropy. Fe^{3+} yields no hyperfine structure as seen in Fig -1.

Co^{2+}, a d^7 configuration, with s value of 3/2 exhibits three fine structure lines. In the case of ^{59}Co (I = 7/2), eight line hyperfine pattern can be observed as shown in the Fig-1.

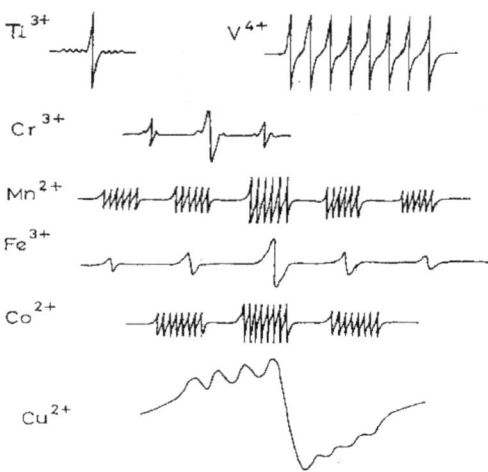

Figure 1. EPR signal of 3d ions

14. Survey of experimental results

14.1. Titanium

Titanium is the ninth most abundant element in the Earth's crust (0.6%). There are 13 known isotopes of titanium. Among them five are natural isotopes with atomic masses 46 to 50 and the others are artificial isotopes. The synthetic isotopes are all radioactive. Titanium alloys are used in spacecraft, jewelry, clocks, armored vehicles, and in the construction of buildings. The compounds of titanium are used in the preparation of paints, rubber, plastics, paper, smoke screens ($TiCl_4$ is used), sunscreens. The main sources of Ti are ilmenite and rutile.

Titanium exhibits +1 to +4 ionic states. Among them Ti^{4+} has d^0 configuration and hence has no unpaired electron in its outermost orbit. Thus Ti^{4+} exhibits diamagnetism. Hence no d-d transitions are possible. The ionic radius of Ti^{3+} is the same as that of Fe(II) (0.76 A.U). Ti(I) and Ti(III) have unpaired electrons in their outermost orbits and exhibit para magnetism

14.2. Electronic spectra of titanium compounds

The electronic configuration of Ti^3 is [Ar] $3d^1 4s^2$. It has five fold degeneracy and its ground state term symbol is ^2D. In an octahedral crystal field, the five fold degeneracy is split into $^2T_{2g}$ and ^2E_g states. Thus only one single electron transition, $^2T_{2g} \rightarrow {}^2E_g$, is expected in an octahedral crystal field. The separation between these energies is 10Dq, which is crystal field energy. Normally, the ground $^2T_{2g}$ state is split due to Jahn-Teller effect and hence lowering of symmetry is expected for Ti(III) ion. This state splits into $^2B_{2g}$ and ^2E_g states in tetragonal symmetry and the excited term ^2E_g also splits into $^2B_{1g}$ and $^2A_{1g}$ levels. Thus, *three bands* are expected for *tetragonal (C_{4v})* symmetry. Energy level diagram in tetragonal environment is shown in Fig -2.

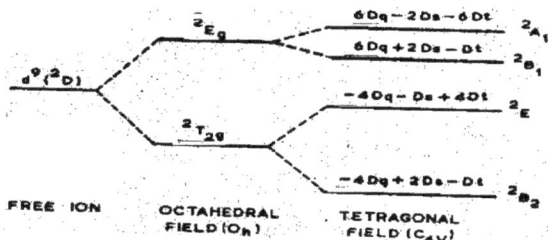

Figure 2. Energy level diagram of Ti³⁺ in octahedral and tetragonal fields

The transitions in the tetragonal field are described by the following equations.

$$^2B_{2g} \rightarrow {}^2E_g : \left[-4Dq - Ds + 4Dt - \left(-4Dq + 2Ds - Dt \right) \right] = -3Ds + 5Dt \tag{1}$$

$$B_{2g} \rightarrow {}^2B_{1g} : \left[6Dq + 2Ds - Dt - \left(-4Dq + 2Ds - Dt \right) \right] = 10Dq \tag{2}$$

$$^2B_{2g} \rightarrow {}^2A_{1g} : \left[6Dq - 2Ds - 6Dt - \left(-4Dq + 2Ds - Dt \right) \right] = 10Dq - 4Ds + 5Dt \tag{3}$$

In the above formulae, Dq is octahedral crystal field and Ds and Dt are tetragonal field parameters. The same sign of Dq and Dt indicates an axial elongation and opposite sign indicates an axial compression

14.2.1. EPR spectra of titanium compounds

When any Ti(III) compound in the form of powder is placed in a magnetic field, it gives a resonance signal. The single d-electron of Ti³⁺ has spin, s = 1/2. The abundance of isotopes is reported as ⁴⁶Ti ≈ 87%, ⁴⁸Ti ≈7.7% and ⁵⁰Ti ≈5.5% and have nuclear spin I = 0, 5/2 and 7/2 respectively. Electron spin and nuclear spin interactions give rise to (2I+1) hyperfine lines (0,6 and 8) and appear as satellite. Since ⁴⁶Ti abundance is more, the EPR signal contains only one resonance line which is similar to the one shown in Fig-3. The g value for this resonance is slightly less than 2.0.

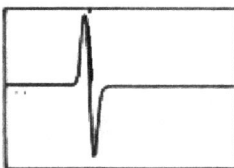

Figure 3. RT powered EPR spectrum of Ti(III).

14.2.2. Relation between EPR and optical absorption spectra

EPR studies for Ti³⁺ can be correlated with optical data to obtain the orbital reduction parameter.

$$\alpha = K\frac{\lambda_{(Covalency)}}{\lambda_{(Ionic)}} = \frac{(g_e - g_{11})\Delta E}{n\lambda_{(ionic)}} \tag{4}$$

where n is 8 for C$_{4v}$, ΔE is the energy of appropriate transition, λ is the spin-orbit coupling constant for Ti^{3+}, i.e., 154 cm^{-1} and k is the orbital reduction parameter.

14.2.3. Typical examples

EPR and optical absorption spectral data of selected samples are discussed as examples. The data chosen from the literature are typical for each sample and hence should be considered as representative only. For more complete information on specific example, the original references are to be consulted. X-band spectra and optical absorption spectra of the powdered samples are recorded at room temperature (RT).

14.2.4. Optical absorption studies

Ti(III) ion in solids is characterized by three broad bands around 7000, 12000 and 18000 cm^{-1}. These are due to the transitions from $^2B_{2g} \rightarrow ^2E_g$, $^2B_{2g} \rightarrow ^2B_{1g}$, and $^2B_{2g} \rightarrow ^2A_{1g}$ respectively. Three bands of titanite at 7140, 13700 and 16130 cm^{-1} and of anatase at 6945, 12050 and 18180 cm^{-1} are attributed to the above transitions. The optical absorption spectrum of lamprophyllite is also similar. The optical absorption spectrum of benitoite sample displays three bands at 8260, 10525 and 15880 cm^{-1}. From the observed band positions, the crystal field parameter in octahedral field, Dq and tetragonal field parameters, Ds and Dt, are given in Table-8.

Sample	Dq cm^{-1}	Ds cm^{-1}	Dt cm^{-1}
Titanite	1370	-1367	608
Anatase	1205	-1867	268
Lamprophyllite	877	-1426	1525
Benitoite	1050	-1945	485

Table 8. Crystal field parameters of Ti(III)

The magnitude of Dt indicates the strength of the tetragonal distortion. This is more in lamprophyllite when compared to the other samples.

i. X-band EPR spectra of the powdered sample of titanite shows a broad resonance line in the centre (335.9 mT). The measured g value is 1.957. Another resonance line is noticed at 341.4 mT with g =1.926. The central eight line transition is superimposed on the spectrum and the components are attributed to VO(II) impurity. The g value of Ti^{3+} is 1.957 and other g value is due to VO(II). The g value of 1.95 indicates that Ti^{3+} is in tetragonally distorted octahedral site.

ii. The EPR spectrum of anatase shows a large number of resonances centered around g value of 2 which is attributed to Ti^{3+}. The additional structures between g values of 2 and 4 are attributed to Fe(III) impurity in the compound. Both the ions are in tetragonally distorted environment.

iii. X band EPR of polycrystalline lamprophyllite sample indicates a broad resonance line with line width 56.6 mT and a g value of 2.0. This is due to the presence of Ti(III) in the compound. The broad line is due to the dipolar-dipolar interaction of Ti(III) ions. Even at liquid nitrogen temperature, only the line intensity increases indicating that Curie law is obeyed.

Using EPR and optical absorption spectral results of titanite, the covalency parameter is calculated using equation (4), $\alpha = \dfrac{(g_e - g_{11})\Delta E}{n\lambda_{(ionic)}}$. The α value obtained is 0.51, which indicates higher covalent character between ligand and metal ion.

15. Vanadium

Vanadium abundance in earth's crust is 120 parts per million by weight. Vanadium's ground state electron configuration is [Ar] $3d^3 4s^2$. Vanadium exhibits four common oxidation states −5, +4, +3, and +2 each of which can be distinguished by its color. Vanadium(V) compounds are yellow in color whereas +4 compounds are blue, +3 compounds are green and +2 compounds are violet in colour. Vanadium is used in making specialty steels like rust resistant and high speed tools. The element occurs naturally in about 65 different minerals and in fossil fuel deposits. Vanadium is used by some life forms as an active center of enzymes. Vanadium oxides exhibit intriguing electrochemical, photochemical, catalytical, spectroscopic and optical properties. Vanadium has 18 isotopes with mass numbers varying from 43 to 60. Of these, ^{51}V, natural isotope is stable:

15.1. Electronic spectra of vanadium compounds

Vanadium in its tetravalent state invariably exists as oxo-cation, VO^{2+} (vanadyl). The VO^{2+} ion has a single d electron which gives rise to the free ion term ^2D. In a crystal field of octahedral symmetry, this electron occupies the t_{2g} orbital and gives rise to ground state term $^2T_{2g}$. When the electron absorbs energy, it is excited to the e_g orbital and accordingly in octahedral geometry only one band corresponding to the transition, $^2T_{2g} \rightarrow {}^2E_g$, is expected. Because of the non-symmetrical alignment of the V=O bond along the axis, the site symmetry, in general, is lowered to tetragonal (C_{4v}) or rhombic (C_{2v}) symmetry. In C_{4v} site symmetry, $^2T_{2g}$ splits into $^2B_{2g}$ and ^2E_g, whereas ^2E_g splits into $^2B_{1g}$, $^2A_{1g}$. Hence three bands are expected in C_{4v} symmetry in the range of 11000 −14000, 14500 − 19000 and 20000 − 31250 cm^{-1}. The degeneracy of ^2E_g is also removed in C_{2v} symmetry resulting four bands. Energy level diagram of VO^{2+} in octahedral C_{4v} and C_{2v} symmetries are shown in Fig- 4. In the tetragonal C_{4v} symmetry transitions are described by the following equations.

$$^2B_{2g} \rightarrow {}^2E_g : \left[-4Dq - Ds + 4Dt - \left(-4Dq + 2Ds - Dt\right)\right] = -3Ds + 5Dt \tag{5}$$

$$B_{2g} \rightarrow {}^2B_{1g} : \left[6Dq + 2Ds - Dt - \left(-4Dq + 2Ds - Dt\right)\right] = 10Dq \tag{6}$$

$$^2B_{2g} \rightarrow {}^2A_{1g} : \left[6Dq - 2Ds - 6Dt - \left(-4Dq + 2Ds - Dt\right)\right] = 10Dq - 4Ds + 5Dt \tag{7}$$

In the above formulae, Dq is octahedral crystal field parameter and Ds, Dt are tetragonal field parameters. The same sign of Dq and Dt indicates an axial elongation and opposite sign indicates an axial compression.

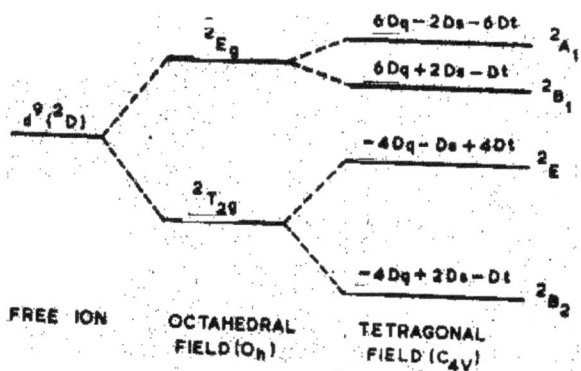

Figure 4. Energy level diagram indicating the assignment of the transitions in octahedral C₄ᵥ symmetry.

15.2. EPR spectra of vanadium compounds

The EPR signal is of three types. (i) is due to high concentration of vanadium. If the vanadium content in the compound is high, it gives a broad resonance line. Therefore the hyperfine line from ^{51}V cannot be resolved. The g value for this resonance is less than 2. (ii) VO^{2+} ion has s= ½ and I = 7/2. The EPR spectrum shows hyperfine pattern of eight equidistant lines. In C₄ᵥ symmetry two sets of eight lines are expected (sixteen-line pattern) whereas in C₂ᵥ symmetry three sets of eight lines are expected. Further in tetragonal distortion, $g_{11} < g_{\perp} < g_e$ which shows the presence of an unpaired electron in the d_{xy} orbital. This is characteristic feature of a tetragonally compressed complex.

Further lowering of symmetry gives rise to EPR spectrum which is similar to the one shown in g$_{yy}$ and g$_{zz}$ respectively. The hyperfine constants are designated as A₁, A₂ and A₃ respectively.

Using the EPR data, the value of dipolar term P and k term are calculated,

$$A_{11} = P\left[-\frac{4}{7} - k - \left(g_{11} - g_e\right) + \frac{3}{7}\left(g_{\perp} - g_e\right)\right] \tag{8}$$

$$A_{\perp} = P\left[\frac{2}{7} - k + \frac{11}{14}\left[g_{\perp} - g_e\right]\right] \tag{9}$$

$$g = \frac{1}{3}\left(g_{11} + 2g_{\perp}\right) \text{ and } A = \frac{1}{3}\left(A_{11} + 2A_{\perp}\right) \tag{10}$$

Using the EPR data, the admixture coefficients are calculated from the following formulae,

$$g_{11} = 2\left(3C_1^2 - C_2^2 - 2C_3^2\right) \tag{11}$$

$$g_\perp = 4C_1\left(C_2 - C_3\right) \text{ and } C_1^2 + C_2^2 + C_3^2 = 1 \tag{12}$$

$$A_{11} = P\left[g_{11} - \left(k + \frac{15}{7}\right)\left(1 - 2C_3^2\right) - \frac{3}{7}\left(1 + C_1C_2C_3\right)\right] \tag{13}$$

$$A_\perp = p\left[\frac{11}{14}g_\perp - 2C_1C_2\left(k + \frac{9}{7}\right)\right] \tag{14}$$

15.3. Relation between EPR and optical absorption spectra

The optical absorption results and EPR results are related as follows. EPR studies can be correlated with optical data to obtain the orbital coefficients β^{*2} and ε_π^{*2}.

$$g_{11} = g_e - \frac{8\lambda\beta^{*2}}{\Delta E_{xy}} \tag{15}$$

$$g_1 = g_e - \frac{2\lambda\varepsilon_\pi^{*2}}{\Delta E_{xz}} \tag{16}$$

Here g₁₁ and g⊥ are the spectroscopic splitting factors parallel and perpendicular to the magnetic field direction of g_e (i.e., 2.0023 for a free electron).

Δ E₁ is the energy of $^2B_{2g} \to {}^2B_{1g}$ and Δ E₂ is the energy of $^2B_{2g} \to {}^2E_g$.

λ is the spin-orbit coupling constant(160 cm⁻¹) for the free vanadium(VO²⁺).

15.4. Typical examples

EPR and optical absorption spectral data of certain selected samples are discussed. The data chosen from the literature are typical for each sample. The data should be considered as representative only. For more complete information on specific example, original references are to be consulted. X-band spectra of the powdered samples and optical absorption spectra are recorded at room temperature (RT).

X-band EPR spectra of the vanadium(IV) complex with DMF recorded in solutions reveal a well-resolved axial anisotropy with 16-line hyperfine structure. This is characteristic of an interaction of vanadium nuclear spin (^{51}V, I = 7/2) with S. The observed EPR parameters are g₁₁ =1.947, A₁₁ = 161.3 x 10⁻⁴ cm⁻¹ and g⊥ =1.978, A⊥ = 49.0 x 10⁻⁴ cm⁻¹. EPR parameters of several samples are available in literature and some of them are given in Table -9.

Mineral name	g_{11}	$g_\perp$	A_{11} mT	$A_\perp$ mT
Kainite	1.932	1.983	17.7	6.9
Apophyllite	1.933	1.982	18.02	6.02
Pascoite site I siteII	1.933 1.946	1.988 1.976	18.50 20.00	7.6 8.2
CAPH	1.933	1.993		

Table 9. Various EPR parameters of VO(II) in minerals

Using the EPR data, the admixture coefficients are calculated for apophyllite and pascoite minerals and are given in the Table -10.

Sample	C_1	C_2	C_3	K	P (x 10^{-4} cm^{-1})
Apophyllite	0.7083	0.7124	0.0028	0.86	122.7
Pascoite	0.7010	0.7116	0.0035	0.36	118.4
	0.7090	0.7285	0.03174	0.34	143

Table 10. Admixture coefficients of VO^{2+} ion

EPR spectrum of polycrystalline sample of wavellite with sixteen line pattern indicates the presence of VO^{2+} ion as an impurity. The EPR parameters calculated are $g_{zz}= 1.933$ and $g_{yy}=g_{xx} = 1.970$ and the corresponding A values are 19.0 and 6.2 mT.

15.5. Typical examples

a. (i) Divalent vanadium (V^{2+}) of d^3 configuration, containing halide and other ions in aqueous solutions, gives three transitions, i.e., $^4A_{2g} \rightarrow {}^4T_{2g}$, $^4A_{2g} \rightarrow {}^4T_{1g}(F)$ and $^4A_{2g} \rightarrow {}^4T_{1g}(P)$ in an octahedral geometry. In $\left[V\left(H_2O\right)_6^{2+} \right]$, the three bands are observed at 11400, 17100 and 24000 cm^{-1} along with some weak shoulders at about 20000 and 22000 cm^{-1}. The bands observed at 11400, 17100 and 24000 cm^{-1} are assigned to the transitions $^4A_{2g} \rightarrow {}^4T_{2g}$, $^4T_{1g}(F)$ and $^4T_{1g}(P)$ respectively. 10Dq is 11400 cm^{-1}. For divalent vanadium ion, Racah parameters are B = 860 and C = 4165 cm^{-1}. Calculated Racah parameters are expected to be less than the one in the free ion value. Accordingly, the weak shoulders observed at 20000 and 22000 cm^{-1} are assigned to $^4A_{2g} \rightarrow {}^2T_{2g}$, and $^4A_{2g} \rightarrow {}^2T_{1g}$, ^2E transitions.

(ii) The optical absorption spectrum of vanadium carboxylate tetrahydrate sample displays three bands at 11400, 17360 and 23920 cm^{-1}. These are assigned to the transitions, $^4A_{2g} \rightarrow {}^4T_{2g}$, $^4T_{1g}(F)$ and $^4T_{1g}(P)$ in an octahedral geometry.

b. Trivalent Vanadium (V^{3+}) (d^2) in aqueous solutions shows two stronger bands at about 15000 and 23000 cm^{-1} and some weaker bands at 11500, 18000 cm^{-1}. The stronger bands are assigned to the transitions, $^3T_{1g}(F) \longrightarrow ^3T_{2g}(F)$ and $^3T_{1g}(F) \longrightarrow ^3T_{1g}(P)$ in an octahedral environment. Since this ion contains two d electrons, it is not so easy to attribute to the other bands. Therefore T-S diagrams are used to identify the other bands. 10Dq is 16400 cm^{-1} and B = 623 (free ion B= 886 $cm^{-1}$and C =765 cm^{-1}). Third band could be expected at 32000 cm^{-1} due to $^3T_{1g}(F) \longrightarrow ^2A_{1g}$. This band corresponds to double electron transition and hence the intensity is expected to be lower than that of the first two bands. The weaker bands observed at 11500, 18000 cm^{-1} are attributed to the spin forbidden transitions, $^3T_{1g}(F) \longrightarrow ^1E_g$, $^1T_{2g}$ and $^1A_{1g}$.

c. (i) *Tetravalent vanadium* (V^{4+}) (d^1). The absorption spectrum of tetravalent vanadium compounds shows three transitions, $^2B_{2g} \longrightarrow ^2E_g$, $^2B_{2g} \longrightarrow ^2B_{1g}$ and $^2B_{2g} \longrightarrow ^2A_{1g}$. The $^2B_{2g} \longrightarrow ^2E_g$ is the most intense and $^2B_{2g} \longrightarrow ^2B_{1g}$ is the weakest. Accordingly, the bands observed in vanadium doped zinc hydrogen maleate tetrahydrate (ZHMT) at 13982, 16125 and 21047 cm^{-1} are assigned to the above three transitions respectively. The octahedral crystal field parameter, Dq (1613 cm^{-1}), and tetragonal field parameters, Ds (-2700 cm^{-1}) and Dt (1178 cm^{-1}), are evaluated.

(ii) The electronic absorption spectrum of the VO^{2+} in $CdSO_4.8H_2O$ recorded at room temperature shows bands at 12800, 13245, 14815, 18345 cm^{-1}. These bands are assigned to $^2B_{2g} \longrightarrow ^2E_g$, $^2B_{2g} \longrightarrow ^2B_{1g}$ and $^2B_{2g} \longrightarrow ^2A_{1g}$ transitions. The band observed at 12500 cm^{-1} is the split component of the band at 13245 cm^{-1}. The crystal field octahedral parameter, Dq (1465 cm^{-1}) and tetragonal field parameters, Ds (-2290 cm^{-1}) and Dt (1126 cm^{-1}) are evaluated.

Several examples are found in the literature. Some of them are given in the Table-11.

Sample	Transition from $^2B_2$ cm^{-1}			Dq cm^{-1}	Ds cm^{-1}	Dt cm^{-1}
	^2E,	$^2B_1$	$^2A_1$			
Cadmium ammonium phosphate hexahydradate(CAPH)	12270	16000	26625	1600	-3275	488
Aphophyllite	12500	15335	24385	1538	-2080	653
Pascoite site I	12255	14450	21415	1445	-2765	803
Site II	12255	16000	21415	1600	-2524	937

Table 11.

d. Pentavalent vanadium has no d electron and hence d-d transitions are not possible. Therefore, the observed bands in electronic absorption spectrum are ascribed to charge transfer bands. These appear around 37000, 45000 cm^{-1}. These are assigned to transitions from ligand orbitals to metal d-orbitals: $A_1 \longrightarrow T_2$ ($t_1 \longrightarrow 2e$) and $A_1 \longrightarrow T_2$ ($3t_2 \longrightarrow 2e$) in tetrahedral configuration for the ion VO_4^{3-}.

Vanadium doped silica gel also shows sharp band at 41520 cm^{-1} and shoulders at 45450 and 34480 cm^{-1}. These are also assigned to charge transfer transitions in tetrahedral environment of VO_4^{3-}. The minimum value of 10Dq for VO_4^{3-} is expected at about 16000 cm^{-1} in octahedral geometry. This is expected because the two bands at 34480 and 45450 cm^{-1} are from the ligand orbitals to two vacant d orbitals which are 10Dq apart. This would be about twice the energy separation (8000 cm^{-1}) observed for tetrahedral VO_4^{3-}. Hence the evidence does not satisfy the assignment of bands to d-d transitions. Therefore the bands are due to charge transfer transitions.

16. Chromium

Chromium is the 6th most abundant transition metal. Chromium is used in the manufacture of stainless steel and alloys. The ground state electronic configuration is [Ar] 3d^{4}4s^2. It exhibits +2 to +6 oxidation states. Most stable oxidation state are +2 (CrO), +3 (Cr$_2$O$_3$) and +6 (K$_2$Cr$_2$O$_7$).

16.1. Optical spectra

a.   Divalent chromium(d^2)

Cr^{2+} has a d^4 configuration and forms high spin complexes only for crystal fields less than 2000 cm^{-1}. The ground state term in an octahedral crystal field is ^{5}E$_g$ belonging to the $t_{2g}^3 e_g^1$ configuration. The excited state ^{5}T$_{2g}$ corresponds to promotion of one single electron to give $t_{2g}^2 e_g^2$ configuration. The d^4 electron is susceptible to Jahn-Teller distortion and hence Cr^{2+} compounds usually are of low symmetry. In lower symmetry, the excited quintet state of Cr^{2+} splits into three levels and the ground level quintet state splits into two levels. In the case of Cr^{2+}(H$_2$O)$_6$, the value of Dq is 1400 cm^{-1}. In spinels, Cr^{2+}is in the tetrahedral environment and Dq is about 667 cm^{-1}only.

b.   Trivalent chromium(d^3):

In octahedral symmetry, the three unpaired electrons are in t_{2g}^3 orbitals which give rise to ^{4}A$_{2g}$, ^{2}E$_g$, ^{2}T$_{1g}$ and ^{2}T$_{2g}$ states. Of these ^{4}A$_{2g}$ is the ground state. If one electron is excited, the configuration is $t_{2g}^2 e_g^1$ which gives two quartet states ^{4}T$_{1g}$ and ^{4}T$_{2g}$ and a number of doublet states. When the next electron is also excited, the configuration is $t_{2g}^1 e_g^2$ which gives rise to one quartet state ^{4}T$_{1g}$ and some doublet states.

$$^4F \rightarrow {}^4A_{2g}\left(F\right), {}^4T_{1g}\left(F\right), {}^4T_{2g}\left(F\right)$$

$$^4P \rightarrow {}^4T_{1g}\left(P\right)$$

$$^2G \rightarrow {}^2A_{1g}(G), {}^2T_{1g}(G), {}^2T_{2g}(G), {}^2E_g(G)$$

$$^2H \rightarrow {}^2E_g(H), 2\,{}^2T_{1g}(H), {}^2T_{2g}(H)$$

In both fields, $^4A_{2g}$(F) represents the ground state. Hence, three spin allowed transitions are observed in high spin state $^4A_{2g}$(F) $\rightarrow$ $^4T_{2g}$(F) (υ_1), $^4A_{2g}$(F)$\rightarrow$ $^4T_{1g}$(F) (υ_2) and $^4A_{2g}$(F) $\rightarrow$ $^4T_{1g}$(P) (υ_3). These spin allowed bands split into two components when the symmetry of Cr^{3+} ion is lowered from octahedral to C_{4v} or C_{3v}. Generally, $^4A_{2g}$(F) $\rightarrow$ $^4T_{1g}$(P) occurs in the UV-Vis region.

The strong field electronic configurations for the ground state and their terms are given as follows:

$$\left(t_{2g}\right)^3 \left(e_g\right)^0 : {}^4A_{2g}(F), {}^2E_g(G), {}^2T_{1g}(G), {}^2T_{2g}(G)$$

$$\left(t_{2g}\right)^2 \left(e_g\right)^1 : {}^4T_{1g}(F), {}^4T_{2g}(F), {}^2T_{2g}(H)$$

$$\left(t_{2g}\right)^1 \left(e_g\right)^2 : {}^4T_{1g}(P)$$

Racah parameter, B, is calculated with spin allowed transitions using equation (17)

$$B = \left(2v_1^2 + v_2^2 - 3v_1 v_2\right) \Big/ \left(15v_2 - 27v_1\right) \tag{17}$$

The octahedral crystal field parameter Dq is characteristic of the metal ion and the ligands. The Racah parameter, B depends on the size of the 3d orbital; B is inversely proportional to covalency in the crystal.

16.2. EPR spectra of chromium compounds

Cr^{3+} ion, splits into $|\pm 1/2\rangle$ and $|\pm 3/2\rangle$ Kramers' doublets in the absence of magnetic field, separated by 2D, D being the zero-field splitting parameter. This degeneracy can be lifted only by an external magnetic field. In such a case, three resonances are observed corresponding to the transitions, $|-3/2\rangle \leftrightarrow |-1/2\rangle$, $|-1/2\rangle \leftrightarrow |1/2\rangle$ and $|1/2\rangle \leftrightarrow |3/2\rangle$ at $g\beta B - 2D$, $g\beta B$ and $g\beta B + 2D$ respectively. In a powder spectrum, mainly the perpendicular component is visible. If all the three transitions are observed, the separation between the extreme sets of lines is 4D [$g\beta B + 2D - (g\beta B - 2D) = 4D$]. If D is equal to zero, a single resonance line appears with g ~ 1.98. If D is very large compared to microwave frequency, a single line is seen around g = 4.0.

16.3. Relation between EPR and optical absorption spectra

A comparison is made between the observed g_{eff} from EPR results and the calculated one from the optical spectrum. For Cr^{3+}, EPR and optical results are related by,

$$g_{11} = g_e - \frac{8\lambda}{\Delta E\left({}^4T_{1g}(F)\right)} \tag{18}$$

$$g_1 = g_e - \frac{8\lambda}{\Delta E\left({}^4T_{2g}(F)\right)} \tag{19}$$

Here g_{11} and $g_{\perp}$ are the spectroscopic splitting factors parallel and perpendicular to the magnetic field direction, g , the free electron value g_e, is 2.0023. These values give,

$$g_{eff} = \frac{1}{3}(g_{11} + g_1). \tag{20}$$

The value of D can also be estimated from the optical absorption spectrum. The ${}^4A_{2g}(F) \longrightarrow {}^4T_{2g}(F)$ component in the optical spectrum is due to the lowering of symmetry which also includes the D term.

$$D = \left(\frac{2\lambda}{10Dq}\right)^2 (\Delta_z - \Delta_x). \tag{21}$$

The spin-orbit splitting parameter, λ [for free ion, Cr^{3+} is 92 cm^{-1}] is related to Racah parameter (B) by the equation,

$$\lambda = 0.11(B + 1.08)^2 + 0.0062 \tag{22}$$

16.4. Typical examples

The data chosen from the literature are typical for each sample. The data should be considered as representative only. For more complete information on specific examples, the original references are to be consulted. X-band spectra and optical absorption spectra of the powdered sample are recorded at room temperature (RT).

1. Trivalent chromium [d^3]: The optical absorption spectrum of fuchsite recorded in the mull form at room temperature shows bands at 14925, 15070, 15715, 16400, 17730 and 21740 cm^{-1}. The two broad bands at 16400 and 21740 cm^{-1} are due to spin-allowed transitions, ${}^4A_{2g}(F) \rightarrow {}^4T_{2g}(F)$ and ${}^4T_{1g}(F)$ respectively. The band at 17730 cm^{-1} is the split component of the ${}^4T_{2g}(F)$ band. This indicates that the site symmetry of Cr^{3+} is C_{4v} or C_{3v}. The bands at 16400 and 21700 cm^{-1} are responsible for the green color of the mineral. The additional weak features observed for the υ_1 band at 15715 and 15070 cm^{-1} are attributed to the spin-forbidden transitions, ${}^4A_{2g} \rightarrow {}^2T_{1g}(G)$ and ${}^4A_{2g} \rightarrow {}^2E_g(G)$. Using equation (17), Racah parameter, B, is calculated (507 cm^{-1}). Substituting Dq and B values and using T-S diagrams for d^3 configuration and solving the cubic field energy matrices ,another Racah parameter, C is evaluated (2155 cm^{-1}) which is less than the free ion value [C =3850 cm^{-1}].

Several examples are available in the literature. Some of them are given in the Table-12.

Compound	^2E_g(G) cm^{-1}	$^2T_{1g}$(G) cm^{-1}	$^4T_{2g}$(F) (ν_1) cm^{-1}	$^2T_{2g}$(G) cm^{-1}	$^4T_{1g}$(F) (ν_2) cm^{-1}	$^2T_{1g}$(H) cm^{-1}	$^4T_{1g}$(P) (ν_3) cm^{-1}	Dq cm^{-1}	B cm^{-1}	C cm^{-1}	β	CFSE
Fuchsite quartz	15500	15995	19995	22720	27020	35700	43465	2000	677	3400	0.66	24000
Dickite	14690	15500	16260		23800		37000	1626	803		0.78	19512
Fuchite	15070	15715	16400 17730	14925	21740			1640	507	2155	0.49	19680
Chromate			17390		23810			1739				
Natural Ruby	14262 14296	16725 16919 17042	18170 17245	21012 21058 21389	24993			1830	732	2155	0.71	21960
Variscite			16660 18180	15380	21735	30295		1666	475	2200	0.46	19992
Synthetic Uvarovite			16670	18000	22730		28000					
Sr3Ga2Ge4O14 Garnet			16299		433.6			1629.9	712.3		0.69	19559
Ureyite			15600		22000				664		0.65	
Alexandrite	14000	-	16600	21000	25000							
Uvarovite			16600		23100							
Clinoclore								1834	668		0.728	63x350
Amesite								1782	737		0.899	58.0x
Muscovite								1610	737		0.89	55.6
Phlogopite								1690				58.0

1. The EPR spectrum of fuchsite recorded at room temperature (RT) clearly indicates a strong resonance line with a few weak resonances on either side of it. The g value for this centrally located strong line is 1.98. This is due to the main transition |-1/2> ↔ |1/2> of Cr^{3+}. The calculated value of D is around 270 G. For weak lines, D is around 160 G. Since the lines are equally spaced on either side of the strong resonance, E is zero. The strong line at g (1.98) value is observed indicating a high concentration of chromium.

2. The EPR spectrum of chromate shows a broad EPR signal with g value of 1.903 which may be due to Cr^{3+} which is in high concentration in the mineral. The chromium ion is in octahedral coordination.

3. EPR spectrum of zoisite at LNT gives a g and D values of 1.99 and 42.5 mT respectively which are due to Cr^{3+} in octahedral environment.

4. EPR spectrum of chromium containing fuchsite quartz shows a g value of 1.996 which may due to Cr^{3+} which is in octahedral environment.

5. EPR spectrum of blue sapphire shows four Cr^{3+} sites with the same g value of 1.98 having different D values (130,105,65 and 34 mT) . Green sapphire also has the same g value but different D values (132,114, 94 and 35 mT). The results suggest that chromium content is slightly different in different sapphires.

Table 12. Assignment of bands for Cr(III) with $^4A_{2g}$(F) ground state. All values are given in cm^{-1}

Several examples are given in the literature. Some of them are presented in the Table-13.

Compound	Observed			$^4T_{1g}$(F) (ν_2) cm^{-1}	$^4T_{2g}$(F) (ν_1) cm^{-1}	Calculated	
	$g_\perp$	g_{11}	g_{eff}			g_{eff}	λ (cm^{-1})
Varscite	1.958	1.9684	1.994	16660	21735	1.9615	75
Chromate			1.903				

Table 13. EPR parameters of Cr^{3+} compounds.

2. Tetravalent chromium (d^2):

Absorption spectra of Cr^{4+} in forsterite and garnet show the absorption band at 9460 cm^{-1} which is the typical of Cr^{4+} ions. It is attributed to the $^3A_{2g} \rightarrow {}^3T_{2g}$ transition. The absorption band at 19590 cm^{-1} is also attributed to $^3A_{2g} \rightarrow {}^3T_{1g}$ transition. The absorption band at 19590 cm^{-1} overlaps with the bands at 16130 and 23065 cm^{-1}.

17. Manganese

The atomic number of manganese is 25 and its outermost electronic configuration is [Ar] $3d^54s^2$. It exhibits several oxidation states, +2, +3, +4, +6 and +7, of which the most stable are +2 +4 and +7. The ionic radii of Mn^{2+} and Mn^{4+} are 0.80 and 0.54 A.U. respectively. Twenty three isotopes and isomers are known. A number of minerals of manganese exists in nature (~ 300 minerals) giving rise to an overall abundance of 0.106%. Twelve of the important among them are economically exploited and the most important of these are pyrolusite (MnO_2), manganite ($Mn_2O_3.H_2O$), hausmannite (Mn_3O_4) rhodochrosite ($MnCO_3$) and manganese(ocean) nodules. Much of the (85-90%) manganese is consumed in the manufacture of ferromanganese alloys. The other uses are: manganese coins, dry cell and alkaline batteries and glass. It is an essential trace element for all forms of life.

Octahedral complexes of Mn(III) are prone to Jahn-Teller distortion. It is of interest, therefore, to compare the structures of Cr(acac)₃ with Mn(acac)₃ since the former is a regular octahedron while the latter is prone to dynamic Jahn-Teller distortion.

17.1. EPR spectra of manganese compounds

1. Manganese(II): Manganese(II), being a d^5 ion, is very sensitive to distortions in the presence of magnetic field. Mn(II) has a total spin, S = 5/2. The six spin states labeled as ±5/2>, ±3/2> and ±1/2> are known as the three Kramers' doublets; in the absence of external magnetic field, they are separated by 4D and 2D respectively, where D is the zero-field splitting parameter. These three doublets split into six energy levels by the application of an external magnetic field. Transitions between these six energy levels give rise to five resonance lines. Each of these resonance lines, in turn, splits into a sextet due to the interaction of the electron spin with the nuclear spin of ^{55}Mn, which is 5/2. Thus one expects a 30- line pattern. However, depending on the relative magnitudes of D and A (hyperfine coupling constant of manganese), these 30 lines appear as a separate bunch of 30 lines or 6 lines (if D = 0). The separation between the extreme set of resonance lines is approximately equal to 8D (first order). If D is very small compared to hyperfine coupling constant (A), the 30 lines are so closely packed that one could see only six lines corresponding -1/2 to +1/2 transition. If D = 0. the system is perfectly octahedral. Deviation from axial symmetry leads to a term known as E in the spin- Hamiltonian. The value of E can be easily calculated from single crystal measurements. A non-zero value of E results in making the spectrum unsymmetrical about the central sextet.

Further, the following parameters have been calculated from the powder spectrum using the Spin- Hamiltonian of the form:

$$H = \beta 1 B_{3g} S + D\left(S_z^2 - \frac{1}{3}S(S+1)\right) + SA1 \tag{23}$$

Here the first term represents the electron-Zeeman interaction, the second term represents the zero field contribution and the third term represents the nuclear-Zeeman interaction. The extra set of resonances within the main sextet is due to the forbidden transitions. From the forbidden doublet lines, the Zero field splitting parameter, D is calculated using the formula,

$$\Delta H = \left(\frac{2D^2}{H_m}\right)\left[\frac{1+16\left(H_m - 8Am\right)^2}{9H_i H_m - 64Am}\right] \tag{24}$$

$$H_m = H_o - Am - \frac{\left[I(I+1)-m^2\right]A^2}{2H_o} \text{ or } H_m = H_0 - Am - \left(35-4m^2\right)\left(\frac{A^2}{8H_0}\right) \tag{25}$$

where H_m is the magnetic field corresponding to m $\leftrightarrow$ m in HF line; H_0 is the resonance magnetic field and m is the nuclear spin magnetic quantum number.

Percentage of covalency of Mn-ligand bond can be calculated in two ways using (i) Matumura's plot and (ii) electro negativities, X_p and X_q using the equation,

$$C = \frac{1}{n}\left[1-0.16\left(X_p - X_q\right)-0.035\left(X_p - X_q\right)^2\right] \tag{26}$$

Here n is the number of ligands around Mn(II) ion; $X_p = X_{Mn} = 1.6$ for Mn(II) and $X_q = X_{ligand}$.

Also hyperfine constant is related to the covalency by,

$$A_{iso} = \left(2.04C - 104.5\right)\times 10^{-4} cm^{-1} \tag{27}$$

Further, the g value for the hyperfine splitting is indicative of the nature of bonding. If the g value shows a negative shift with respect to the free electron g value (2.0023), the bonding is ionic and conversely, if the shift is positive, then the bonding is said to be more covalent in nature.

17.2. Typical examples

1. Manganese(II): The EPR spectrum of clinohumite contains a strong sextet at the centre corresponding to the electron spin transition +1/2> to -1/2>. In general, the powder spectrum is characterized by a sextet, corresponding to this transition. The other four transitions corresponding to ±5/2> $\leftrightarrow$ ±3/2> and ±3/2>$\leftrightarrow$ ±1/2> are not seen due to their

high anisotropy in D. However, in a few cases only, all the transitions are seen. Moreover, the low field transitions are more intense than the high field transitions. In addition, if E ≠ 0, the EPR spectrum will not be symmetrical about the central sextet. In clinohumite, the spectrum indicates the presence of at least three types of Mn(II) impurities in the mineral.

The extra set of resonances within the main sextet is due to the forbidden transitions. From the powder spectrum of the mineral, the following parameters are calculated:

Site I: g = 2.000(1), A = 9.15(2) mT; and D = 43.8(1) mT.

Site II: g = 2.003(2), A = 9.23(2) mT; and D = 44.1(1) mT.

Site III: g = 2.007(1), A = 9.40(2) mT; and D = 44.1(1) mT.

This large value of D indicates a considerable amount of distortion around the central metal ion. Since EPR is highly sensitive to Mn(II) impurity, three such sites are noticed. These two sites have close spin- Hamiltonian parameters. A close look at the EPR spectrum indicates a non-zero value of E, which is very difficult to estimate from the powder spectrum.

2.   Pelecypod shell EPR spectrum of powdered sample obtained at room temperature indicates the presence of Mn(II) and Fe(III) impurities. The spectrum contains a strong sextet at the centre of the spectrum corresponding to the electron spin transition +1/2> to -1/2>. Also, the powder spectrum indicates the presence of, at least, three types of Mn(II) impurities in the *pelecypod shell* which is noticed at the sixth hyperfine resonance line. The third Mn(II) site is of very low intensity. The extra set of resonances within the main sextet is due to the forbidden transitions. The variations of intensity are also due to the zero field splitting parameter. From the powder spectrum of the compound, the following parameters are calculated using the spin- Hamiltonian of the form:

$$H = \beta B g S + D\left[S_z^2 - S\frac{(S+1)}{3} + SAI \right] \tag{28}$$

where the symbols have their usual meaning.

Site I: g = 2.002(1), A = 9.33(2) mT; and D = 43.8(1) mT

Site II: g = 1.990(2), A = 9.41(2) mT; and D = 44.1(1) mT

Site III: g = 1.987(1), A = 9.49(2) mT; and D = 44.1(1) mT

This large value of D indicates a considerable amount of distortion around the central metal ion. A close look at the EPR spectrum indicates a non-zero value for E.

The hyperfine constant 'A' value provides a qualitative measure of the ionic nature of bonding of Mn(II) ion. The percentage of covalency of Mn-ligand bond is calculated using 'A' (9.33 mT) value obtained from the EPR spectrum and with Matumura's plot. It corresponds to an ionicity of 94%. Also, the approximate value of hyperfine constant (A) is calculated by using the equation (27).

The value obtained is 92×10^{-4} cm^{-1}. This calculated value agrees well with the observed hyperfine constant (93.3×10^{-4} cm^{-1}) indicating ionic character of Mn-O bond in the shell under study.

Using the covalency, the number of ligands around Mn(II) ion is estimated using the equation (26)

$$C = \frac{1}{n}\left[1 - 0.16\left(X_p - X_q\right) - 0.035\left(X_p - X_q\right)^2\right]$$

Where X_P and X_q are the electronagativities of metal and ligand. Assuming $X_P = X_{Mn} = 1.4$ and $X_q = X_O = 3.5$, the number of ligands (n) obtained are 18. This suggests that Mn(II) may be surrounded by eighteen oxygens of six CO_3^{2-} ions.

i.   Manganese (IV): This ion in biological samples gives rise to EPR signal around 3.30.
ii. (Mn^{7+} ion also gives EPR resonance signal at about 2.45 in ceramic materials and in biological samples.

17.3. Optical absorption studies

1.   Manganese(II): The free ion levels of Mn^{2+} are ^{6}S, ^{4}G, ^{4}P. ^{4}D and ^{4}F in the order of increasing energy. The energy levels for Mn^{2+} ion in an octahedral environment are ^{6}A$_{1g}$(S), ^{4}T$_{1g}$(G), ^{4}T$_{2g}$(G), ^{4}E$_g$(G), ^{4}T$_{1g}$(G) ^{4}A$_{1g}$, ^{4}T$_{2g}$(G), ^{4}E$_g$(D), ^{4}T$_{1g}$(P) respectively with increasing order of energy. The ^{4}E$_g$(G), ^{4}A$_{1g}$ and ^{4}E$_g$(D) levels are less affected when compared to other levels by crystal field. Hence, sharp levels are expected relatively in the absorption spectrum which is the criterion for assignment of levels of Mn(II) ion. Since all the excited states of Mn(II) ion will be either be quartets or doublets, the optical absorption spectra of Mn(II) ions will have only spin forbidden transitions. Therefore, the intensity of transitions is weak.

Energy level diagram of Mn(II) is extremely complex. Exact solutions for the excited state energy levels in terms of Dq, B and C may be obtained from T-S matrices. These matrices are very large (up to 10 x10) and ordinary calculations are not feasible. For this reason, the T-S diagrams given in many places in the literature are not sufficiently complete to allow the assignment of all the observed bands. Therefore a set of computer programmes is written to solve the T-S secular equations for any selected values of B, C and Dq. With the computer program, it is only necessary to obtain values of B and C and the complete scheme for any Dq can be quickly calculated. Fortunately B and C can be obtained analytically, if a sufficiently complete spectrum is obtained using the transitions given below:

$$^4A_{1g}, {}^4E_g\left(G\right) \rightarrow {}^6A_{1g} = 10B + 5C = v_1$$

$$^4E_g\left(D\right) \rightarrow {}^6A_{1g} = 17B + 5C = v_2$$

If ν_1 and ν_2 are correctly observed and identified in the spectrum, B and C can be calculated. Identification is particularly easy in these cases because of the sharpness of the bands of these levels and are independent of Dq.

2. Manganese(III): This ion has four 3d electrons. The ground state electronic configuration is $t_{2g}^3 e_g^1$. It gives a single spin-allowed transition $^5Eg \rightarrow {}^5T_{2g}$ corresponding to one electron transition. This should appear around 20000 cm⁻¹. Mn^{3+} cation is subject to Jahn-Teller distortion. The distortion decreases the symmetry of the coordination site from octahedral to tetragonal (D_{4h}) or by further lowering the symmetry to rhombic (C_{2v}). Under the tetragonal distortion, the t_{2g} orbital splits into e_g and b_{2g} orbitals whereas the e_g orbital splits into a_{1g} and b_{1g} orbitals. Hence in a tetragonal site, three absorption bands are observed instead of one. Further distortion splits the e_g orbital into singly degenerate a_{1g} and b_{1g} orbitals. Thus four bands are observed for rhombic symmetry (C_{2v}).

The transitions in the tetragonal field are described by the following equations:

$$^2B_{1g} \rightarrow {}^2A_{1g} : \left[6Dq - 2Ds - 6Dt - \left(6Dq + 2Ds - Dt \right) \right] = 4Ds + 5Dt \tag{29}$$

$$^2B_{1g} \rightarrow {}^2B_{2g} : \left[-4Dq - 2Ds - Dt - \left(6Dq + 2Ds - Dt \right) \right] = 10Dq \tag{30}$$

$$^2B_{1g} \rightarrow {}^2E_g : \left[-4Dq - Ds + 4Dt - \left(6Dq + 2Ds - Dt \right) \right] = 10Dq + 3Ds \tag{31}$$

In the above equations, Dq is octahedral crystal field and Ds and Dt are tetragonal field parameters. The same sign of Dq and Dt indicates an axial elongation and opposite sign indicates an axial compression.

The optical absorption bands observed for Mn(III) in octahedral coordination with rhombic distortion (C_{2h}) in montmorillonite are given in Table -14.

Assignment		Localities		
D₄h	C₂v	(Mexico)	(Gumwood Mine)	(California)
$^5B_{1g} \rightarrow {}^5A_{1g}$	$^5B_{1g} \rightarrow {}^5A_{1g}$	10480	10276	10542
$^5B_{1g} \rightarrow {}^5B_{2g}$	$^5B_{1g} \rightarrow {}^5A_{2g}$	19041	18751	18560
	$^5B_{2g}$	20660	20496	20605
$^5B_{1g} \rightarrow {}^5E_g$	$^5B_{1g} \rightarrow {}^5A_{3g}$	21837	22127	22143

Table 14. Assignment of bands for Mn(III) in montmorillonite

18. Iron

The atomic number of iron is 26 and its electronic configuration is [Ar]4s² 3d⁶. Iron has 14 isotopes. Among them, the mass of iron varies from 52 to 60 Pure iron is chemically reactive and corrodes rapidly, especially in moist air or at elevated temperatures. Iron is vital to plant and animal life. The ionic radius of Fe^{2+} is 0.76 A.U. and that of Fe^{3+} is 0.64 A.U. The most common oxidation states of iron are +2 and +3. Iron(III) complexes are generally in octahedral in shape, and a very few are in tetrahedral also.

18.1. EPR spectra of iron compounds

The EPR spectra of powdered Fe^{3+} compounds may be described by the spin- Hamiltonian,

$$H = gBS + D\left(S_z^2 - \frac{1}{3}S(S+1) + E\left(S_x^2 - S_y^2\right)\right) \tag{32}$$

The second and third terms in the equation (33) represent the effects of axial and rhombic components of the crystal field respectively. When D=E=0, it corresponds to a free ion in the magnetic field, H and if E= 0, it implies a field of axial symmetry. If λ (E/D) increases, it results in the variation of rhombic character. Maximum rhombic character is seen at a value of λ=1/3 and further increase in λ from 1/3 to 1 results in the decrease of rhombic character. When λ =1, the axial field situation is reached. When λ=1/3, the g value is around 4.27 and when λ is less than 1/3, g value is 4. Hence, the resonance is no longer isotropic and the powder spectrum in that region is a triplet corresponding to H along each of the three principle axes. For Fe^{3+}, in fields of high anisotropy, the maximum g value is 9. If g values are limited to 0.80 to 4.30, the Fe^{3+} ion is under the influence of a strong tetragonal distortion.

1. Iron (III): The iron (III) samples exhibit a series of g values ranging from 0 to 9. This is due to the fact that the three Kramers' doublets of |S=5/2> are split into |S±5/2>, |S±3/2> and |S±1/2> separated by 4D and 2D respectively where D is the zero field splitting parameter. Depending on the relative populations of these doublets, one observes g value ranging from 0 to 9.0. The line widths are larger in low magnetic field when compared to high magnetic field. If the lowest doublet, |S±1/2> is populated, it gives a g value of 2 to 6 whereas if the middle Kramers' doublet |S±3/2> is populated, a g value 4.30 is expected. If the third doublet |S±5/2> is populated, it gives a g value of 2/7 to 30/7. A few systems are known which exhibit resonances from all the three Kramers' doublets.

The iron(III) in the natural sample enters the lattice in various locations which may not correspond to the lowest energy configuration. After heating the sample, the impurity settles in the lowest energy configuration and the EPR spectrum is simplified. Thus, it is observed that heating the sample results in a simplification of the EPR spectrum and gives a g value of around 2.

18.2. Typical examples

1. The EPR spectrum of powdered red sandal wood obtained at room temperature contains a series of lines of various intensity and width. The g values obtained for these are 6.52, 2.63 and 1.92. These three peaks are attributed to Fe(III) impurity in the compound.
2. The EPR spectrum of prehnite at room temperature consists of two parts. The first part consists of the two strong lines (absorption and dispersion) and the second part

comprises a weak doublet within the strong doublet. The weak doublet also consists of two lines, absorption and dispersion line shapes. The g values of the strong doublet are 4.48 and 3.78 whereas the g values of the weak doublet are 4.22 and 3.96. The data reveal that there are two different centres of Fe(III) which are magnetically distinct.

3. The EPR spectrum of nano iron oxalate recorded at room temperature reveals three sets of four lines in low, medium and high fields corresponding to g_1, g_2 and g_3 respectively. From the positions of the peaks in the EPR spectrum, the following spectroscopic splitting factors are evaluated: g_1 = 2.130, g_2 = 2.026 and g_3 = 1.947. The hyperfine structure constants are A_1 = 78 mT, A_2 = 46 mT and A_3 =26 mT. The EPR spectrum is characteristic of Fe(III) ion or HCO_2^- or in rhombic symmetry. For the rhombic symmetry, g values follow in the sequence as $g_1 > g_2 > g_3$. Using the relation, spin-orbit coupling constant, λ is calculated. Resonant value of the magnetic field is given by the relation,

$$H_R(mT) = \frac{21419.49}{g\lambda(cm)} = \frac{0.07144775}{g}v(MHz) \qquad (33)$$

λ calculated for each g tensor is 32.18.

For axial symmetry, λ is zero. If rhombic character in the crystal field is increased, it results in the increase of λ upto a maximum of $\frac{1}{3}$. In the present case, the observed λ is $\frac{1}{3}$ (32.13%). Thus the EPR studies indicate that the iron oxalate nano-crystal is in orthorhombic structure.

18.3. Optical absorption spectra of iron compounds

18.3.1. Trivalent iron

Trivalent iron has the electronic configuration of $3d^5$ which corresponds to a half-filled d - sub-shell and is particularly most stable. In crystalline fields, the usual high spin configuration is $t_{2g}^3 e_g^2$ with one unpaired electron in each of the orbitals and the low spin state has the t_{2g}^5 configuration with two pairs of paired electrons and one unpaired electron. The energy level in the crystal field is characterized by the following features. i) The ground state of d^5 ion, ^6S transforms into $^6A_{1g}$ - a singlet state. It is not split by the effect of crystal field and hence all the transitions are spin forbidden and are of less intensity. ii) In excited state, d^5 ion gives rise to quartets (^4G, ^4F, ^4D, ^4P) and doublets (^2I, ^2H, ^2G, ^2F, ^2D, ^2P, ^2S) . The transitions from the ground to doublet state are forbidden because the spin multiplicity changes by two and hence they are too weak. Thus sextet-quartet forbidden transitions observed are: $^6A_{1g} \rightarrow ^4T_{1g}$ and $^6A_{1g} \rightarrow ^4T_{2g}$. The transitions which are independent of Dq and which result in sharp bands are $^6A_{1g} \rightarrow ^4E(^4D)$ $^6A_{1g} \rightarrow ^4E_g + ^4E_{1g}$ etc., iii) The unsplit ground state term behaves alike in both octahedral and tetrahedral symmetries and gives rise to same energy level for octahedral, tetrahedral and cubic coordination with usual difference,

$$\left[Dq_{Octa} : Dq_{Tetra} : Dq_{Cubic} = 1 : \frac{4}{9} : \frac{8}{9} \right]$$

18.3.2. Divalent iron

In divalent iron (d^6), the free ion ground term is ^5D and the excited terms are triplet states (^3H, ^3P, ^3F, ^3G, ^3D) and singlet states (^1I, ^1D). In an octahedral field, the ^5D term splits into an upper ^5E_g level and a lower $^5T_{2g}$ level of which the latter forms the ground state. The only allowed transition is $^5T_{2g} \longrightarrow {}^5E_g$ which gives an intense broad absorption band. This band splits into two bands due to Jahn-Teller effect. The average of these two bands is to be taken as 10Dq band. The transitions arising from the excited triplet states are spin forbidden and hence are weaker than the 10Dq band.

18.3.3. Typical examples

1. For Fe^{3+}, there are three transitions: $^6A_{1g}(S) \rightarrow {}^4T_{1g}(G)$ (v_1), $^6A_{1g}(S) \rightarrow {}^4T_{2g}(G)(v_2)$. v_1 occurs between 10525 cm^{-1} and v_2 occurs between 15380 to 18180 cm^{-1} usually as a shoulder. The bands corresponding to $^6A_{1g}(S) \rightarrow {}^4A_{1g}(G)$, $^4E_g(G)$ (v_3) appear around 22000 cm^{-1} . The last transition is field independent. The ligand field spectrum of ferric iron appears as if the first (v_1) and the third (v_3) bands of octahedral symmetry are only present. The analysis of general features of the spectrum of Fe^{3+} containing plumbojarosite is discussed here. The first feature observed in the range 12000 to 15500 cm^{-1} is attributed to $^6A_{1g}(S) \rightarrow {}^4T_{1g}(G)$, the third band at 22730 cm^{-1} is sharp and is assigned to $^6A_{1g}(S) \rightarrow {}^4A_{1g}(G)$, $^4E_g(G)$ transitions respectively. A broad and diffused band at 19045 cm^{-1} is assigned to the $^6A_{1g}(S) \rightarrow {}^4T_{2g}(G)$ band. The other bands are also assigned to the transitions with the help of Tanabe-Sugano diagram. The assignments are given in the Table -15.

2. Optical absorption spectrum of prehnite recorded in the mull form at room temperature (RT) shows bands at 9660, 10715, 12100, 12610, 15270, 16445,17095, 23380 and 24390 cm^{-1} in the UV –Vis region. For easy analysis of the spectrum, the bands are divided into two sets as 12100, 15270, 23380 cm^{-1} and 12610, 16445, 17095, 24390 cm^{-1}. Accordingly the two bands observed at 12100 cm^{-1} in the first set and 12610 cm^{-1} in the second set are assigned to the same transition $^6A_{1g}(S) \rightarrow {}^4T_{1g}(G)$ whereas 15270 cm^{-1} in the first set and 16445,17095 cm^{-1} in the second set are assigned to $^4T_{2g}(G)$ transition. The third at 23380 cm^{-1} and 24390cm^{-1} is assigned to $^4A_{1g}(G)$, $^4E(G)$ (v_3) transitions respectively. These two sets of bands are characteristic of Fe(III) ion occupying two different sites in octahedral symmetry. The broad and intense band observed at 10715 cm^{-1} with a split component at 9660 cm^{-1} is assigned to the transition $^5T_{2g} \rightarrow {}^5E_g$ for divalent iron in the sample. Using the Tree's polarization term, α = 90 cm^{-1}, the energy matrices of the d^5 configuration are solved for various B, C and Dq values. The evaluated parameters which give good fit are given in Table 15. A comparison is also made between the calculated and observed energies of the bands and these are presented in Table -15.

Prehnite						Plumbojarosite			Transition from $^6A_{1g}$
Site I			Site II						
Dq= 930, B= 600 and C =2475 cm⁻¹, α= 90 cm⁻¹			Dq= 900, B= 600 and C =2500 cm⁻¹ α= 90 cm⁻¹			Dq= 900, B= 700 and C =2800 cm⁻¹ α= 90 cm⁻¹			
Wave length (nm)	Wave number (cm⁻¹)		Wave length (nm)	Wave number (cm⁻¹)		Wave length (nm)	Wave number (cm⁻¹)		
	Observed	Calculated		Observed	Calculated		Observed	Calculated	
827	12100	793	793	12610	12528	800 650	12500	-- 15194	$^4T_{1g}(G)$
						525	15385	19379	
655	15270	608	608	16445	16331	440	19045	22766	$^4T_{2g}(G)$
430		585	585	17095		410	22730		
	23380	427	427	24390	23276	385		24815	$^4A_{1g}(G)$,
						330	24390	26474	$^4E(G)$
						265	25975	30656	
						240	30300	37710	$^4T_{2g}(D)$
							37735	41125	
							41665		$^4E_g(D)$

Table 15. Band headed data with assignments for Fe(III) in various compounds

19. Nickel

Nickel is the 7th most abundant transition metal in the earth's crust. The electronic configuration of nickel is [Ar]$4S^2 3d^8$. Nickel occurs in nature as oxide, silicate and sulphide. The typical examples are garnierite and pentlandite. Nickel exhibits +1 to +4 oxidation states. Among them divalent state is most stable. Nickel compounds are generally blue and green in color and are often hydrated. Further, most nickel halides are yellow in color. The primary use of nickel is in the preparation of stainless steel. Nickel is also used in the coloring of glass to which it gives a green hue.

19.1. Electronic spectra of nickel compounds

The electronic distribution of Ni(II) ion (d^8) is $t_{2g}^6 e_g^2$ which gives rise to ^3F, ^3P, ^1D and ^1S terms of which ^3F is the ground state. In a cubic crystal, these terms transform as follows:

$$^3F \rightarrow {}^3T_{1g}\left(F\right) + {}^3T_{2g}\left(F\right) + {}^3A_{2g}\left(F\right)$$

$$^3F \rightarrow {}^3T_{1g}\left(P\right)$$

$$^1D \rightarrow {}^3T_{2g}\left(D\right) + {}^1E_g\left(D\right)$$

$$^1G \rightarrow {}^1T_{1g}\left(G\right) + {}^1T_{2g}\left(G\right) + {}^1E_g\left(G\right) + {}^1A_{1g}\left(G\right)$$

$$^1S \rightarrow {}^1A_{1g}\left(S\right)$$

Of these crystal field terms, $^3A_{2g}(F)$ is the ground state. Hence three spin allowed transitions are possible and the others are spin forbidden The three spin allowed transitions are: $^3A_{2g}(F) \rightarrow {}^3T_{1g}(P)$, $^3A_{2g}(F) \rightarrow {}^3T_{1g}(F)$ and $^3A_{2g}(F) \rightarrow {}^3T_{2g}(F)$. These transitions are governed by linear equations as given below:

$$^3A_{2g}(F) \rightarrow {}^3T_{1g}(P) = 15Dq + 7.5B + 6B(1+\mu)^{1/2} = v_1 \tag{34}$$

$$^3A_{2g}(F) \rightarrow {}^3T_{1g}(F) = 15Dq + 7.5B - 6B(1+\mu)^{1/2} = v_2 \tag{35}$$

$$^3A_{2g}(F) \rightarrow {}^3T_{2g}(F) = 10Dq = v_3 \tag{36}$$

Here μ is of the order of 0.01. Dq and B are of similar magnitude. The spin allowed bands are calculated using the above equations whereas the spin forbidden bands are assigned using Tanabe-Sugano diagrams.

19.2. Typical examples

The data chosen from the literature are typical and representative for each sample. For more complete information on any specific case, original references are to be consulted. X-band spectra and optical absorption spectra of the powdered sample are recorded at room temperature (RT) only.

Divalent Nickel [d⁸]: The optical absorption spectrum of falcondoite mineral recorded in the mull form at room temperature shows three intense bands at 9255, 15380 and 27390 cm^{-1} and a weak band at 24385 cm^{-1}. Using the equations 34 to 36, the calculated values of Dq and B are 925 and 1000 cm^{-1} respectively. Using these Dq and B values and T-S diagrams for d⁸ configuration, the cubic field energy matrices and Racah parameter, C are evaluated (4.1B) .

Ni^{2+} also gives absorption bands in the NIR region. These bands suggest that Ni^{2+} is in tetrahedral site. In some of the samples, Ni^{2+} exbits both octahedral and tetrahedral coordination. Several examples are available in the literature. Some of them are given in the Table-16.

Compound	$^3T_{1g}(P)$ (v_1)	$^3T_{1g}(F)$ (v_2)	$^3T_{2g}(F)$ (v_3)	$^1T_{1g}(G)$	$^1T_{1g}(D)$	$^1Eg(D)$	$^1T_{2g}(D)$	$^1T_{2g}(G)$	Dq	B	C
Falcondoite	27390	15380	9255				24385		925	1000	4100
Ullmannite	24993	14966	8618	25967	21546	12252	21546		860	840	3350
Takovite	26665	15380	8200 10000				14930	24095	910	940	4.25B
(Zn,Ni)KPO 46H₂O	25967	15500	8770				14080	22216	900	890	3800
Ni(II) HZDT											
Garnierite	26300	15200	9100				13000				
Gaspeite	22730	13160 14705	7714 8685				20410	30300	810	800	3200
Annabergite		13885	8330								
Zartite	23805	14285	8195				21735		820	899	4.1B

Table 16. Assignment of bands for Ni(II) with $^3A_{2g}(F)$ as the ground state. All values are given in cm^{-1}.

19.3. EPR spectra

$Ni^{2+}(d^8)$ has no unpaired electron (square planer) in its orbit. Therefore it does not exhibit EPR signal at room temperature.

But in certain conditions, it shows EPR signal. The EPR data could be related with the optical data by the following equation $g = 2.0023 - \dfrac{8\lambda}{\Delta}$ where Δ is the energy of the transition of the perfect octahedral site. λ is 324 cm^{-1} for free Ni^{2+} ion.

20. Copper

Copper is one of the earliest known elements to man. The average percentage of copper in the earth's crust is 0.005%. Pure copper is soft and malleable. An important physical property of copper is its color. Most people refer copper colour as reddish-brown tint. Copper-63 and copper-65 are two naturally occurring isotopes of copper. Nine radioactive isotopes of copper are also known. Among them two radioactive isotopes, copper-64 and copper-67 are used in medicine. Copper easily reacts with oxygen and in moist air, it combines with water and carbon dioxide forming hydroxy copper carbonate ($Cu_2(OH)_2CO_3$).

Animals like crustaceans (shellfish like lobsters, shrimps, and crabs) do not have hemoglobin to carry oxygen through the blood but possess a compound called hemocyanin. This is similar to hemoglobin but contains copper instead of iron. Copper is an essential micronutrient for both plants and animals. A healthy human requires not more than about 2 mg of copper for every kg weight of the body. The main body parts where copper is found in animals are the tissues, liver, muscle and bone.

20.1. Copper compounds

Copper exists in two ionic states, Cu(I) and Cu(II). The ionic radius of Cu(II) is 0.73 A.U. The electronic configuration of Cu(I) is [Ar] $3d^{10}$ and hence has no unpaired electron in its outermost orbit. Hence it exhibits diamagnetism. The electronic configuration of Cu(II) is [Ar]$3d^9$ and has one unpaired electron which is responsible for its para magnetism. The main resources of copper are its minerals. Structural properties could be explored using electronic and EPR spectra which provides information on bonding between ligands and metal ion.

20.2. Electronic spectra of copper compounds

In optical spectroscopy, transitions proceed between the split orbital levels whereas in EPR spectroscopy they occur between spin sub- levels that arise due to the external magnetic field. Thus EPR spectroscopy is a natural sequel to optical spectroscopy.

20.3. Optical spectra

In octahedral crystal field, the ground state electronic distribution of Cu^{2+} is $t_{2g}^6 e_g^3$ which yields ^2E_g term. The excited electronic state is $t_{2g}^5 e_g^4$ which corresponds to $^2T_{2g}$ term. Thus only one single electron transition, i.e., $^2E_g \rightarrow {}^2T_{2g}$, is expected in an octahedral crystal field. The difference is 10Dq. Octahedral coordination is distorted either by elongation or compression of octahedron leading to tetragonal symmetry.

Normally, the ground ^2E_g state is split due to Jahn-Teller effect and hence lowering of symmetry is expected for Cu(II) ion. This state splits into $^2B_{1g}(d_{x^2-y^2})$ and $^2A_{1g}(d_{z^2})$ states in tetragonal symmetry and the excited term $^2T_{2g}$ also splits into $^2B_{2g}(d_{xy})$ and $^2E_g(d_{xz}, d_{yz})$ levels. In rhombic field, ^2E_g ground state is split into $^2A_{1g}(d_{x^2-y^2})$ and $^2A_{2g}(d_{z^2})$ whereas $^2T_{2g}$ splits into $^2B_{1g}(d_{xy})$, $^2B_{2g}(d_{xz})$ and $^2B_{3g}(d_{yz})$ states. Thus, three bands are expected for tetragonal (C₄ᵥ) symmetry and four bands are expected for rhombic (D₂ₕ) symmetry. Energy level diagram of d-orbitals in tetragonal elongated environment is shown in Fig. 5.

The transitions in the tetragonal field are described by the following equations:

$$^2B_{1g} \rightarrow {}^2A_{1g} : \left[6Dq - 2Ds - 6Dt - \left(6Dq + 2Ds - Dt \right) \right] = 4Ds + 5Dt \tag{37}$$

$$^2B_{1g} \rightarrow {}^2B_{2g} : \left[-4Dq + 2Ds - Dt - \left(6Dq + 2Ds - Dt \right) \right] = 10Dq \tag{38}$$

$$^2B_{1g} \rightarrow {}^2E_g : \left[-4Dq - Ds + 4Dt - \left(6Dq + 2Ds - Dt \right) \right] = 10Dq + 3Ds - 5Dt \tag{39}$$

In the above equations, Dq is octahedral, Ds and Dt are tetragonal crystalfield parameters. The same sign of Dq and Dt indicates an axial elongation [Fig. 5] and opposite sign indicates an axial compression .

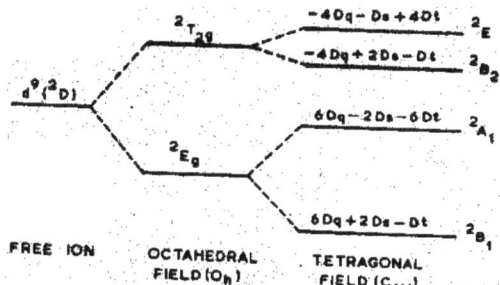

Figure 5. (a) Energy level diagram of Jahn-Teller distortion in d-orbital in octahedral and tetragonal elongation

The Jahn-Teller distortion is either tetragonal elongation along the Z axis or contraction in the equatorial xy plane which may ultimately result in a square planar environment in extreme cases as in D₄ₕ.

The optical absorption bands observed for Cu(II) in octahedral coordination with rhombic (D₂ₕ) symmetry are: $^2A_{1g}(d_{x^2-y^2}) \rightarrow {}^2A_{2g}(d_{z^2})$, $^2A_{1g}(d_{x^2-y^2}) \rightarrow {}^2B_{1g}(d_{xy})$, $^2A_{1g}(d_{x^2-y^2}) \rightarrow {}^2B_{2g}(d_{xz})$, $^2A_{1g}(d_{x^2-y^2}) \rightarrow {}^2B_{3g}(d_{yz})$ states respectively. This is shown in Fig.6. In rhombic (D₂ₕ) field, i.e., C₂ᵥ symmetry, the strong band $^2A_{1g}(d_{x^2-y^2}) \rightarrow {}^2B_{1g}(d_{xy})$ gives 10Dq value which depends on the nature of the compound.

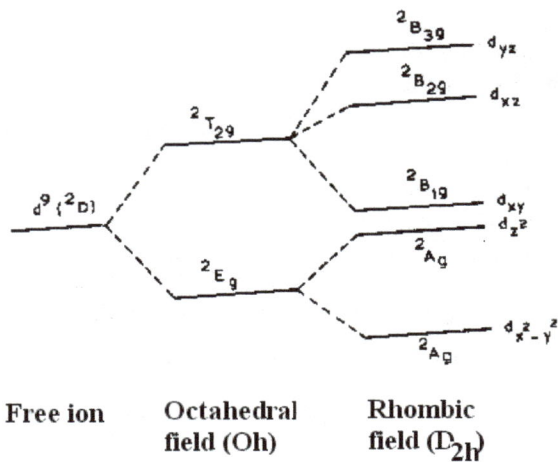

Figure 6. Energy level diagram of d-orbitals in rhombic distortion.

20.4. EPR spectra of copper compounds

When any Cu(II) compound in the form of powder is placed in a magnetic field, it gives a resonance signal. The signal is of three types. They are shown in Fig.-7:

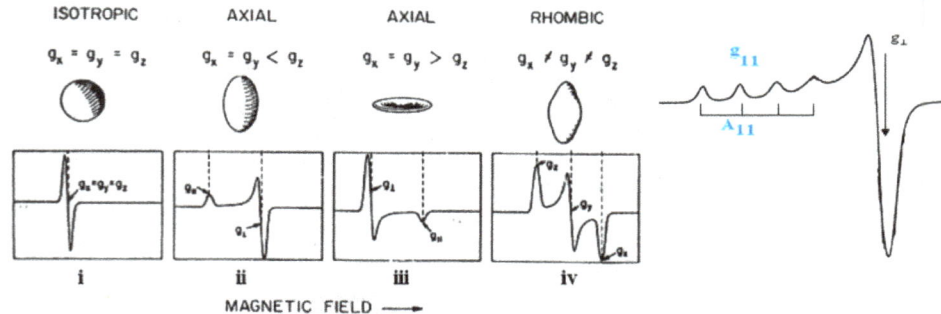

Figure 7. Different forms EPR spectra of Cu(II)

Fig.7(i) is due to high concentration of copper; if the copper content in the compound is high, it gives a broad resonance line. Therefore the hyperfine line from either ^{63}Cu or ^{65}Cu cannot be resolved. The g value for this resonance is around 2.2. (ii) Compression in the equatorial plane results in the elongation of Z axis .Elongation in the equatorial plane results in the compression of Z-axis. Thus there are two types of possibilities in the EPR spectrum. Hence an EPR spectrum similar to Fig. 7(ii) & (iii) is obtained. If $g_{11} > g_\perp$, the ground state is $^2B_{1g}$, [Fig. 7(a)] whereas if $g_\perp > g_{11}$ or $g_{11} = 2.00$, the ground state is $^2A_{1g}$ [fig.7(ii).]. The highest-energy of the half occupied orbital is $d_{x^2y^2}$ as it has the largest repulsive interaction with the ligands in the equatorial plane. Here g_{11}(corresponding to the magnetic field oriented along the z axis of the complex) $> g_\perp > 2.00$. This is a characteristic feature of $d_{x^2-y^2}$ ground state. Additionally, copper has a nuclear spin of (I)) 3/2 which couples with the electron spin to produce a four line hyperfine splitting of the EPR spectrum. This is shown in Fig-7(ii) and 7(v). Tetragonal cupric complexes generally have large A_{11} value than those of complexes with D_{4h} symmetry. If $g_{11} > g_\perp$, the ground state is $^2B_{1g}$ whereas if $g_\perp > g_{11}$ or $g_{11} = 2.00$, the ground state is $^2A_{1g}$. EPR results give rise to a new parameter, G which is defined as

$$G = \frac{(g_{11} - g_e)}{(g_\perp - g_e)} \tag{40}$$

If G value falls in between 3 and 5, the unit cell contains magnetically equivalent ions. If G value is less than 3, the exchange coupling among the magnetically non- equivalent Cu(II) ions in the unit cell is not very strong. If G is greater than 5, a strong exchange coupling takes place among the magnetically non -equivalent Cu(II) ions in the unit cell. Truly compressed structures are relatively rare when compared to elongated structures. In other words, $g_\perp > g_{11}$, is an unusual observation and this implies two possibilities:

i.   The concentration of copper in the complex is very high which results in the interaction between Cu(II) $\leftrightarrow$ Cu(II) ions.
ii. The Cu(II) ion is a compressed octahedron. If the complex contains low copper content, it is assumed that Cu(II) ion is a compressed octahedron. Hence the ground state is $^2A_{1g} (d_{z^2})$.
iii. Further lowering of symmetry gives rise to EPR spectrum which is similar to the one shown in Fig. 9(iv). This spectrum consists of three sets of resolved four lines in low, medium and high fields corresponding to g_1, g_2 and g_3 respectively. The hyperfine structure constants (A values) are designated as A_1, A_2 and A_3 respectively. Line width is estimated for simple cubic lattice using dipole-dipole equation;

$$H_p = 2.3 g_0 \beta \rho \sqrt{s(s+1)} \tag{41}$$

where β is the Bohr magneton, s = spin, g_0 = average value of g factor, ϱ = density (2.22 x 10^{21} spins/cc).

The calculated g values provide valuable information on the electronic ground state of the ion. If $g_1 > g_2 > g_3$, the quantity R value is given by $(g_2 - g_3) / (g_1 - g_2)$ which is greater than unity and the ground state is $^2A_{2g}(d_{z^2})$; if it is less than unity, the ground state is $^2A_{1g}(d_{x^2-y^2})$. A large value of g_1 is indicative of more ionic bonding between metal and ligand. Further the structure of the compound is an elongated rhombus. From the spin –Hamiltonian parameters, the dipolar term (P) and the Fermi contact term (k) are calculated using the following expressions:

$$P = 2\gamma_{Cu} \beta_O \beta_N \left(r^{-3} \right) \tag{42}$$

$$k = \left(\frac{A_O}{P} \right) + \Delta g_e \tag{43}$$

Here γ_{Cu} is the magnetic moment of copper, β_0 is the Bohr magneton, β_N is the nuclear magneton and r is the distance from the central nucleus to the electron, A_0 is the average A value and $\Delta g_0 = g_0 - g_e$ where g_0 is the average g value and g_e is the free electron g-value (2.0023). The Fermi contact term, k, is a measure of the polarization produced by the uneven distribution of d-electron density on the inner core s-electron and P is the dipolar term. By assuming either the value of P or k, the other is calculated. Using these values, the hyperfine constant is calculated. This is the average value of g_1, g_2 and g_3.

Using the data of EPR and dipolar term P, the covalency parameter (α^2) is calculated .

$$\alpha^2 = \frac{7}{6}\left[\left(\frac{A_3 - A_1}{P}\right) - (g_e - g_1) + \frac{11}{14}(g_e - g_3) - \frac{6}{14}(g_e - g_2)\right] \tag{44}$$

Thus the important bonding information is obtained. The bonding parameter, α^2, would be closer to unity for ionic bonding and it decreases with increasing covalency. Further the term, k, is calculated using the EPR data,

$$A_{11} = k\alpha^2 + P\left[-\frac{4}{7}\alpha^2 + \Delta g_{11} + \frac{3}{7}\Delta g_\perp\right] \tag{45}$$

$$A_\perp = k\alpha^2 + P\left[\frac{2}{7}\alpha^2 + \frac{11}{14}\Delta g_\perp\right] \tag{46}$$

20.5. Relation between EPR and optical absorption spectra

The optical absorption and EPR data are related as follows. In tetragonal symmetry, EPR studies are correlated with optical data to obtain the orbital reduction parameter in rhombic compression.

$$g_1 = g_e + \frac{8a^2 k_1^2 \lambda}{\Delta E_{xy}} \tag{47}$$

$$g_1 = g_e + \frac{2k_2^2 \lambda \left(a + \sqrt{3}b\right)^2}{\Delta E_{xz}} \tag{48}$$

$$g_1 = g_e + \frac{2k_3^2 \lambda \left(a + \sqrt{3}b\right)^2}{\Delta E_{yz}} \tag{49}$$

Similarly for rhombic elongation,

$$g_1 = g_e - \frac{8a^2 k_1^2 \lambda}{\Delta E_{xy}} \tag{50}$$

$$g_1 = g_e - \frac{2k_2^2\lambda\left(a+\sqrt{3}b\right)^2}{\Delta E_{xz}} \tag{51}$$

$$g_1 = g_e - \frac{2k_3^2\lambda\left(a+3b\right)^2}{\Delta E_{yz}} \tag{52}$$

where $a = \cos\theta$ and $b = \sin\theta$ which are coefficients for the mixing of the z^2 and x^2-y^2 orbitals. $a^2 + b^2 = 1$ and k_1, k_2, k_3 are the orbital reduction parameters. λ is the spin- orbit coupling constant for free Cu(II) ion = -830 cm^{-1}.

In equations (48) to (50), when a = 0, tetragonal compression is obtained [ground state is $^2A_{1g}(d_{z^2})$].

$$g_{11} = 2.0023 = g_e \tag{53}$$

$$g_\perp = g_e - \frac{6\lambda}{\Delta E_{\perp(xy,yz)}\left(^2B_1 \to {}^2E\right) = \Delta_\perp} \tag{54}$$

Also in equations (51) to (53), when b is equal to zero, tetragonal elongation is obtained [ground state is $^2B_{1g}(d_{x^2-y^2})$].

$$g_{11} = g_e - \frac{8\lambda}{\Delta E_{11(xy)}\left(^2B_1 \to {}^2B_2\right) = \Delta_{11}} \tag{55}$$

$$g_\perp = g_e - \frac{2\lambda}{\Delta E_{\perp(xy,yz)}\left(^2B_1 \to {}^2E\right) = \Delta_\perp} \tag{56}$$

Further, if A_{11}, g_{11} and $g_\perp$ values are known, α^2 can be estimated using the equation [53]

$$\alpha^2 = -\left[\left(\frac{A_{11}}{0.036}\right) - (g_{11} - g_{e1}) + \frac{3}{7}(g_\perp - g_e) + 0.04\right] \tag{57}$$

20.6. Typical examples

EPR and optical absorption spectral data of selected samples are discussed. The data are chosen from the literature for each typical sample. However, it is to be noticed that the crystal field parameters, EPR parameters often depend on chemical composition, nature of ligands and temperature of the compound. The data should be considered as representative only. For more complete information on specific example, the original references are to be consulted. The X-band spectra and optical absorption spectra of powdered samples are mostly recorded at room temperature (RT).

1. The EPR spectrum of covellite is shown in Fig-9. It is similar to the Fig 8(i). It consists of a broad line with a small sextet. The g value for the broad line is 2.24 which is due to the presence of Cu(II) in the sample. The hyperfine line from either ^{63}Cu or ^{65}Cu could not be resolved since the copper content (Cu = 66 wt%) in the mineral is very high. Several copper compounds exhibit this type of EPR spectra.

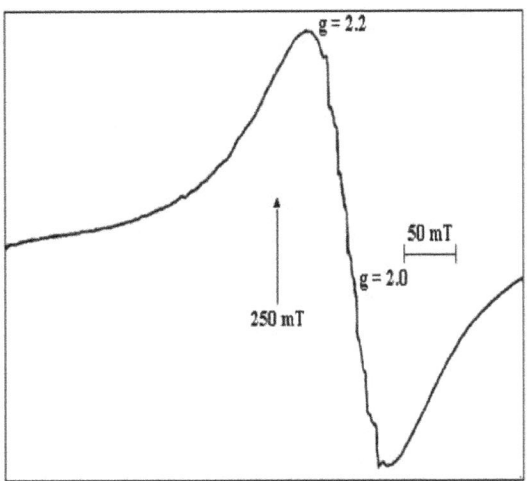

Figure 8. EPR spectrum of covellite at RT

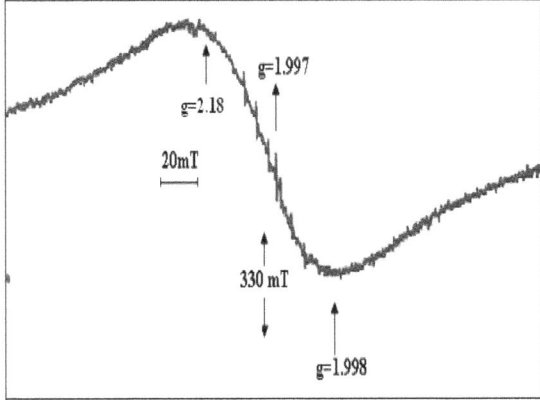

Figure 9. EPR spectrum of beaverite at RT

2. *Beaverite* [Pb(Fe^{3+},Cu,Al)$_3$(SO$_4$)$_2$(OH)$_6$]: X-band EPR spectrum of powdered sample recorded at RT is shown in Fig-9. This is similar to Fig-7(ii). The g values are: g_{11} = 2.42 and $g \perp$= 2.097. In addition to the above, a *g* value of 2.017 is observed which is due to Fe(III) impurity. Fig.9 indicates expanded form of EPR spectrum of Cu(II) and is not resolved because of high copper percentage. Tetragonal cupric complexes with D$_{4h}$ symmetry, possessing axial elongation have ground state ^{2}B$_{1g}$ (d$_{x^2-y^2}$).The EPR results are in the order of g_{11} > $g \perp$> g$_e$ and hence the ground state is ^{2}B$_{1g}$. Though the optical absorption spectrum shows two sites for Cu(II) with same ground state, the same is not noticed in the EPR spectrum because the percentage of copper is high in the sample.

A typical EPR spectrum of *enargite* is shown in Fig.10. The spectrum is symmetric with g_{11} = 2.289 and $g \perp$= 2.048 which are due to Cu(II). Since g_{11} > $g \perp$> g$_e$, the ground state for Cu(II) is ^{2}B$_{1g}$ (d$_{x^2-y^2}$). Using EPR and optical absorption results, the orbital reduction parameters are evaluated, i.e., K$_{11}$ = 1.03 cm^{-1} and $K_\perp$ = 1.93 cm^{-1}. Also G seems to be 5.0 which indicates that the unit cell of the compound contains magnetically equivalent ions.

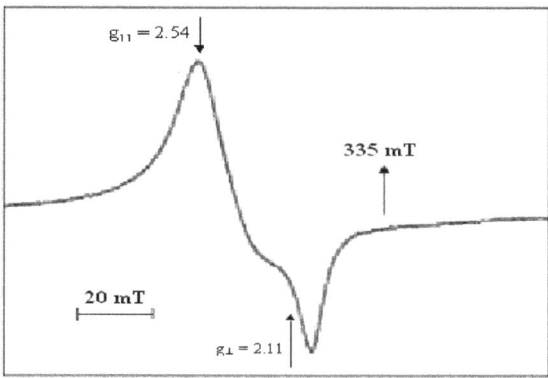

Figure 10. EPR spectrum of enargite at RT

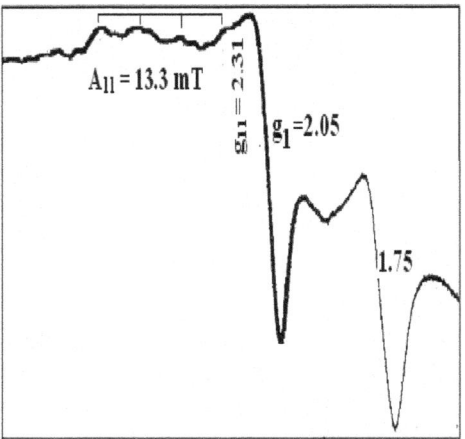

Figure 11. EPR spectrum of CuO-ZnO nano composite.

CuO-ZnO nano composite: EPR spectrum of CuO-ZnO nano composite recorded at room temperature is shown in Fig-11. The calculated g values are 1.76, 2.31 and 2.05. The g value of 1.76 is assigned to free radical of O^{2-}. Further g_{II} value of 2.31, $g_\perp$ value of 2.05 are due to Cu(II) in tetragonal distortion. Also it has A_{11} =13.3 mT. These results show that the ground state of Cu(II) as $d_{x^2-y^2}$. Further, the covalency parameter, α^2 (0.74) suggests that the composite has some covalent character.

3. *Atacamite* [$Cu_2(OH)_3Cl$]: The EPR spectrum is shown in Fig.12. The g values corresponding to three sets of the resolved four lines in low, mid and high fields are g_1 = 2.191, g_2 =2.010 and g_3 = 1.92. The corresponding hyperfine structure constants are A_1 = 11.0 mT, A_2 = 3,0 mT and A_3 = 5.0 mT respectively. Since $g_1 > g_2 > g_3$, the quantity $R = (g_2-g_3)/(g_1-g_2)$ = 0.50 which is less than unity. This indicates $^2A_{1g}(d_{x^2-y^2})$ is the ground state for Cu(II) which is in an elongated rhombic field. The optical absorption spectrum of the compound at RT shown in Fig-13 shows bands at 15380, 11083, 10296 and 8049 cm^{-1}. Using the EPR results, the energy states are ordered as $^2A_{1g}(d_{x^2-y^2}) < {}^2A_{2g}(d_{z^2}) < {}^2B_{1g}(d_{xy}) < {}^2B_{2g}(d_{xz}) < {}^2B_{2g}(d_{yz})$. Thus we have four bands with $^2A_{1g}(d_{x^2-y^2})$ as the ground state. Using the EPR results, the dipolar term (P) and the Fermi contact term (k) are calculated as 0.38 cm^{-1} and k = 0.3 respectively. The bonding parameter, α^2 is found to be 0.28 indicating reasonably high degree of covalent bonding between metal and ligands.

Synthetic copper doped *zinc potassium phosphate hexahydrate* (ZPPH), $ZnKPO_4$ $6H_2O$: It is similar to strubite, a bio-mineral. The g values are: g_1 = 2.372, g_2 =2.188 and g_3 = 2.052. The hyperfine structure constants are A_1 = 78 x 10^{-4} cm^{-1}, A_2 = 48 x 10^{-4} cm^{-1} and A_3 = 66 x 10^{-4} cm^{-1} respectively. It is seen that $g_1 > g_2 > g_3$ and the quantity $R = (g_2-g_3)/(g_1-g_2)$ = 0.85. This confirms that the ground state for Cu(II) is $^2A_{1g}(d_{x^2-y^2})$ (elongated rhombic field). Using the EPR data and substituting free ion dipolar term [P= 0.036 cm^{-1}] for Cu(II) and g_e value in equation (57), the bonding parameter, α^2 = 0.55, is obtained. It indicates a predominant covalency in compound.

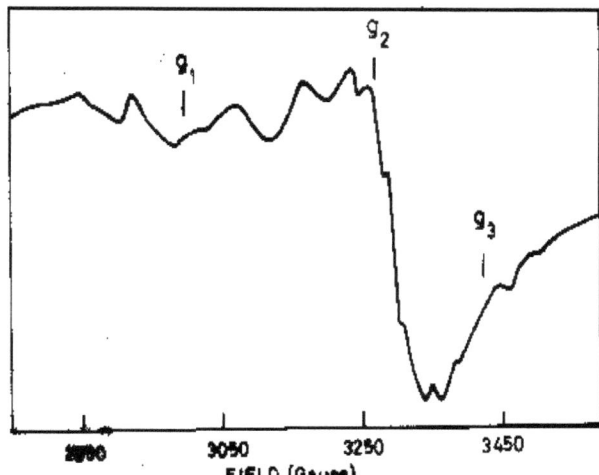

Figure 12. EPR spectrum of atacamite at RT

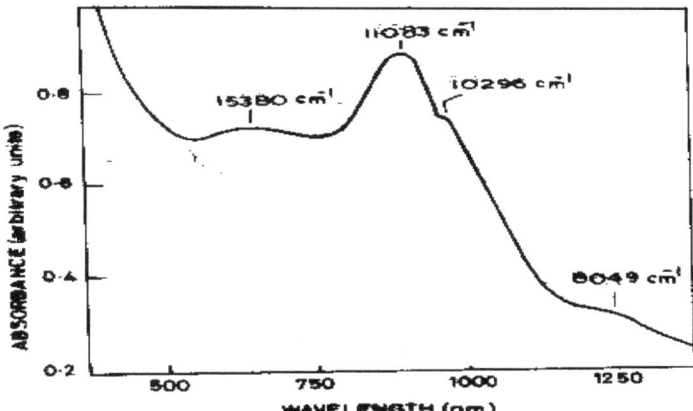

Figure 13. Optical absorption spectrum of atacamite

Author details

S.Lakshmi Reddy
Dept. cf Physics, S.V.D.College, Kadapa, India

Tamio Endo
Dept. cf Electrical and Electronics Engineering,
Graduate School of Engineering, Mie University, Mie, Japan

G. Siva Reddy
Dept. of Chemistry, Sri Venkateswara University, Tirupati, India

21. References

[1] B.N.Figgs,M.A.Hitchman, "Ligand Field Theory and Its Applications",Wiley-VCH, New York,(2000).

[2] A.Lund, M.Shiotani, S.Shimada,"Principles and Applications of ESR Spectroscopy", Springer New York (2011).

[3] C.J.Ballahausen, "Introduction to Ligand Field Theory", Mc Graw-Hill Book Co., New York (1962).

[4] P.B. Ayscough,"Electron Spin Resonance in Chemistry", Mathuen & Co., Ltd., London (1967).

[5] R.L.Carlin, "Transition Metal Chemistry", Marcel Dekker,New York (1969).

[6] Journal of "Coordination Chemistry Reviews".

[7] Journal of Spectrochimica Acta A Elsvier.

[8] J.S.Griffith, "Theory of Transition Metal Ions", Cambridge University Press,Oxford (1964).

[9] Journal of Solid State Communications.

Chapter 3

The Use of the Spectrometric Technique FTIR-ATR to Examine the Polymers Surface

Wieslawa Urbaniak-Domagala

Additional information is available at the end of the chapter

1. Introduction

The development of material engineering is accompanied by a growing demand for routine, nondestructive techniques for material and product testing. These techniques are to be used for the assessment of chemical and physical structure of new materials as well as for a systematic control of their manufacturing processes. Nowadays nanotechnologies fulfill a particular role in creating new materials of nanometric dimensions. The products of nanotechnology are made in various forms, mostly such as coatings and fibers. Coatings are of great practical importance while deposited on conventional substrates, such as metals, ceramics and polymers to impart new functions, e.g. anticorrosive, reflexive, sensory properties, etc., to them. Coatings are mostly made of polymers and hence their functional properties and durability mainly depend on polymer chemical and supermolecular structure. The current control tests of the chemical properties and supermolecular characteristic of materials are carried out with the use of IR absorption spectroscopy. Currently, these are dedicated to test the surface of materials. This paper concerns the spectroscopic technique FTIR used to test the surface of polymeric materials and coatings formed on polymeric substrates. The general characteristics, advantages and drawbacks of this technique in testing polymer surfaces have been presented.

2. Technique of infrared absorption spectroscopy

Material testing by the technique of IR spectroscopy consists in making a spectrum of radiation energy absorbed by material molecules and interpreting the spectrum obtained. IR radiation within the wavelength range from 2.5 mm to 15 mm (the wave number from 4000 cm^{-1} to 666 cm^{-1}) is selectively absorbed by material molecules and converted into their oscillatory energy. The oscillations of molecules are of various characters, connected with their chemical structure, and depend on the type of bonds (frequency increases with increasing bond energy), relative atomic weights (frequency decreases with increasing atomic weight), spatial position of atoms in a molecule, intra- and intermolecular interaction forces. During absorption, various vibration modes are generated that can be ranged with respect to energy in the following order: stretching vibration > bending vibration > oscillatory/torsional vibration. Vibration modes are active in IR only when the frequency of radiation coincides with the own frequency of molecule oscillation (resonance) and the dipole moments of molecules change in the same direction as the electric vector of IR radiation wave.

In the absorption spectroscopy techniques, IR radiation, after passing the material, where molecules selectively absorb radiation quanta, the absorption spectrum is recorded in the form of changes in the IR spectra radiation intensity as a function of radiation energy. The intensity of a beam after passing through sample (I), transmittance (T) or absorbance (A) is assumed as a measure of absorption. If the intensity of the primary incident beam on a sample is equal to I_0, the relation between intensity, transmittance and absorbance is as follows: $T = I/I_0$, $A = \log (I_0/I) = -\log T$. Energy is expressed in eV, but mostly practical parameters such as IR radiation wavelength (λ, nm), wave number ($v = 1/\lambda$, cm^{-1}) and radiation frequency (v, Hz) are used to express energy.

Originally, tests and recording the IR radiation absorption spectra of samples were performed by means of two-beam diffraction spectrometers. Modern technical solutions of IR spectrometers consist in replacing the reticular monochromators with interferometers, which considerably increases the sensitivity of spectrometers (a high value of the signal to noise ratio is obtained), making it possible to shorten the spectrum recording and to obtain its good definition. Moreover, there occur the transformation and ordering of the interferogram obtained to the frequency domain by the use of Fourier Transform (FTIR). The high resolving power of spectrometer makes it possible to record complicated spectra of materials, spectra mixing, the distinction of band derived from crystalline and non-crystalline areas and performing static and dynamic tests.

The IR spectrometry technique can be used in two variants: transmission and reflection. The transmission version is used to test the effects of IR radiation absorption in the volume of sample. It is possible to test samples in any form: solid, liquid and gaseous with the use of an appropriate procedure. Gases and liquids are placed in special cuvettes with windows, made of transparent materials for IR radiation (e.g. ionic crystals: KBr, NaCl). The spectra of solids can be measured using previously prepared specimens on quartz plates, in a suspension in liquid paraffin or in the form of tablets made of KBr. If the object tested is sufficiently thin and transparent, its spectra are measured directly on a sample. The transmission technique cannot be used for materials that strongly absorb IR radiation and to test local areas of sample such as surface. In the sixties of the last century, the reflection variant was developed, so-called Attenuated Total Reflection (ATR), which makes it possible to test specific version of samples. The ATR-IR uses the phenomenon of a complete reflection during the transition of IR radiation from an optically denser medium (prism) to thinner medium (sample). A sample is placed on the IR-transparent prism surface with a refractive index being always higher than that of the sample (Figure 1). The radiation beam is directed by one of the prism wall to the prism-sample interface at angle θ higher than the limiting. Under these conditions, a complete reflection occurs at the internal prism side and the beam reflected comes out through the second prism wall, where the beam intensity and absorption spectrum are recorded.

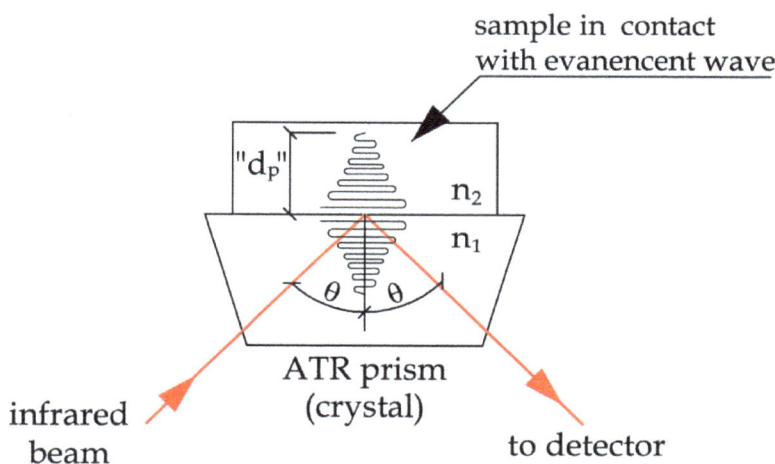

Figure 1. The schematic representation of infrared beam reflected on the crystal - sample interface in FTIR-ATR spectrometer. (on the base http://www.sprpages.nl/SprTheory/SprTheory.htm)

During the total internal reflection in the optically thinner medium (sample) is formed an electromagnetic wave, so-called evanescent wave that fulfills the condition of the continuity of electromagnetic field vectors at the interface of media with various wave refractive indices, n_1 and n_2 (Fornel, 2000). The IR evanescent wave has two wave vector components: parallel component to the interface of the contact between materials, under influence of which the wave propagates along surface resulting in the formation of so-called Goos-Hänchen's displacement (Goos&Hänchen, 1947), and perpendicular component, under the influence of which the wave propagates in the optically thinner medium in the direction perpendicular to the surface and exponentially disappears. The evanescent wave penetration depth, „dp", in sample depends on the IR radiation wavelength (λ), incident angle, (θ), prism refractive index, (n_1), and sample refractive index in relation to the prism ($n_{2,1}$) and is expressed by the following equation (Dechant, 1972):

$$d_p = \frac{\lambda / n_1}{2\pi\sqrt{(\sin^2 \theta - n_{21}^2)}} \tag{1}$$

Along the path of IR evanescent wave the sample selectively absorbs energy to decrease the intensity of radiation. The weakened wave returns to the prism and then to an IR detector.

There the system generates an FTIR-ATR absorption spectrum characteristic of the given sample. The FTIR-ATR absorption spectrum slightly differs from that obtained by the transmission method. The differences concern the intensity and frequency of absorption peaks characteristic of chemical groups in view of the phenomenon of reflection, e.g. Goos-Hänchen's displacement. Thus it is necessary to take corrective action that can be realized automatically. The penetration depth of IR beam can be controlled within some range by selecting an appropriate prism (selection of the refractive index) and the incident angle of beam. The commonly used prisms are made of diamond, germanium, silicon and ZnSe, whose refractive indices are equal to 2.4, 4.0, 3.4 and 2.4, respectively, and the beam penetration depths: 2.03 μm, 0.67 μm, 0.84 μm and 2.03 μm, respectively, at $v = 1000$ cm^{-1} (Material Thermo Scientific Smart ITR). During testing sub-micrometric coating, the beam penetrates a higher depth than the coating depth and also passes to the substrate, on which the coating is deposited. The absorption spectrum then constitutes a superposition of the spectrum of coating material and substrate. In such cases, qualitative analysis is carried out, which takes into account the absorption spectrum of substrate.

The basic requirement for ATR technique is to place a sample in direct contact with the prism as only such conditions allow the IR evanescent wave to penetrate the sample surface layer. Moreover, there should be a considerable difference between the refractive indices of prism and sample to get the phenomenon of internal reflection occurred.

The drawback of ATR technique is a relatively low sensitivity and susceptibility to the effect of environmental conditions, which makes it necessary to calibrate the IR spectrum. Modern spectrometers have an option of automatic computer-aided spectrum correction. ATR technique has numerous advantages. FTIR-ATR shows the features of a routine method for testing the chemical and physical surface structure of materials such as polymers, films and membranes provided that these well adhere to the crystal. Tests with a modern instrumentation are characterized by a high reproducibility (better than 0.1%) (Urbanczyk, 1988). FTIR-ATR makes it possible to record spectra within a wider frequency range of IR radiation than transmission spectroscopy owing to the lack of limitations caused by the absorption of cuvette windows. An important advantage of this technique is the possibility of recording spectra *in situ* and *in vivo*, e.g. in testing biological objects and using it as a diagnostic tool in medicine.

In this work, the FTIR-ATR technique was used to analyze the surfaces of modified polymers and to test the polymeric layers deposited on substrates.

3. Examples of testing polymers by FTIR-ATR

Tests were carried out by means of a single-beam FTIR-Nicolet 6700 spectrometer from Thermo Scientific, equipped with a diamond crystal (refractive index n = 2.4). IR spectra were recorded as changes in absorption as a function of wave number ranging from 600 cm^{-1} to 4000 cm^{-1}. A DTGS KBR detector was used. The following measurement technical conditions were used: measurement recording accuracy - 4 cm^{-1}, mirror travel rate - 0.31 cm^{-1}/s, aperture - 50, minimal scans number – 32.

3.1. Assessment of the modification effects on the surface of polypropylene (PP) films and nonwovens

Polypropylene products are commonly used in commodity production due to their special chemical properties (resistance to organic and inorganic solvent, hydrophobic properties) and physical characteristics (lightness, mechanical strength, electro- and thermal insulating capabilities). In the methods of making products such as composites with the use of PP films or fibers as reinforcing components, a serious drawback of these materials is their low free surface energy, which results in weak molecular interactions between the composite components. The free energy of PP material surface can be increased by creating new functional chemical groups in the material surface layer. This task has been fulfilled by exploring different approaches such as chemical, electrochemical, physical and plasma methods . The effectiveness of the methods used was assessed by means of the FTIR-ATR technique.

The moleculare structure of polypropylene is the same in the use of films and nonwovens products:

$$\left[\begin{array}{c} CH_3 \\ | \\ {}_nCH{-}CH_2 \end{array}\right]_n$$

In one unit of PP molecule chain are tree atoms of carbon, in the form of different groups: $-CH_2-$; $>CH-$; and $-CH_3$. Each of them is correlated in IR spectra with the suitable absorption peak by definite wavenumber values (Figure 2). The proper characteristic, concerning this correlation is presented in Table 1.

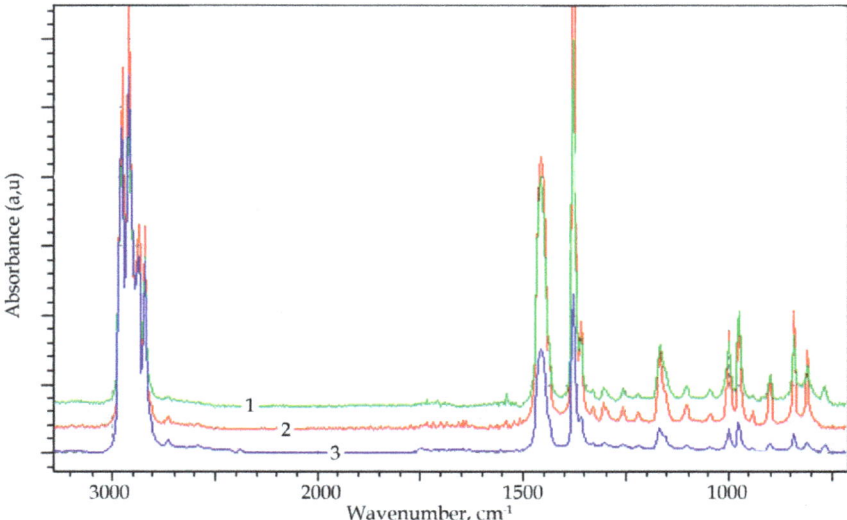

Figure 2. FTIR-ATR spectra of commercial PP films : non-oriented, non-crystalline PP Cast (Sample 1); bidirectionally oriented, crystalline PP AG (Sample 2) and PP nonwoven (Sample 3).

IR spectra of PP for film and nonwoven can differ between themselves only in defiles (the differences in shape and intensity of peaks), because in ATR technique, the contact of the samples with measure crystal, can be different for different structure of samples (film – continuous structure, nonwoven – porous structure). The explanation of this problems shown as an example at Figure 3.

Wave number, cm⁻¹	Absorbing group and type of vibration
2916	va (CH_2)
2959	va (CH_3)
2881	vs (CH_3)
2841	vs (CH_2)
1460	δa (CH_3)
1376	δs (CH_3)
1357,	γw (CH_2- CH)
1328	γw (CH_2 - CH)
1302, 1224, 941	Carbon lattice pulsation
1170, 1153	γw (CH_3), δ ($CH2$), δ (CH)
975, 899,	γr (CH_3), vr ($CH2$), vr (CH)
841, 810	γr ($CH2$), vr (CH), vr (CH_3)
765	γw (CH_2)

*) vs - stretching vibration symmetrical and va -asymmetrical, δs - deformation vibration symmetrical and δa – asymmetrical, γw - wagging vibration, γr – rocking vibration

Table 1. IR absorption bands of Polypropylene (Urbanczyk, 1988; Rau, 1963)

Sample A   Sample B

Figure 3. SEM images of type surface structure : PP nonwoven fabrics (Sample A), PP Cast film (Sample B)

3.1.1. Effects of PP film modification

Commercial PP, non-oriented, non-crystalline (PP Cast) and bidirectionally oriented, crystalline (PP-AG) films were modified in media of strong oxidants, such as: 3M nitric acid, 30% hydrogen peroxide, and a saturated solution of potassium dichromate in 70% sulfuric acid ($K_2Cr_2O_7+H_2SO_4$). The electrochemical oxidation was carried out with the use of anolyte ($AgNO_3$ solution in nitric acid) and catholyte (nitric acid solution).   The physical modification of PP was performed by means of a Xenotest apparatus, irradiating PP film with UV radiation according to EN ISO 105-B02:2006 (Urbaniak-Domagala, 2011). Plasma modification processes were carried out with the use of RF glow discharge of special gases under decreased pressure (Urbaniak-Domagala, 2011). Figures 4, 5 show the FTIR-ATR spectrograms of the PP film surfaces after oxidation compared with unmodified PP films. In the FTIR-ATR spectrograms of the PP film surface layer, one can observe absorption bands that are consistent with those of isotactic PP obtained by the authors mentioned in Table 1.

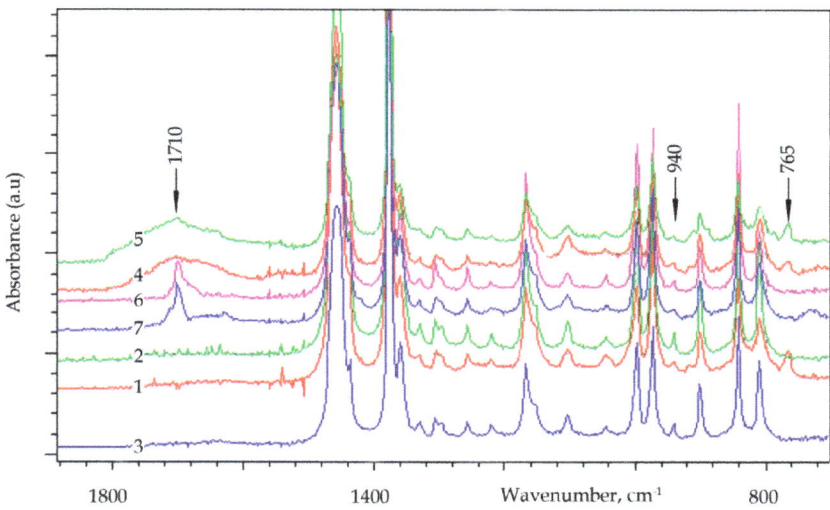

Figure 4. FTIR-ATR spectra of commercial PP films before and after oxidation. 1- PP Cast untreated, 2 - PP AG untreated. Samples 3÷7 PP Cast modified: by using electrochemical method, current intensity: 100 mA/cm², 30 min. (Sample 3), UV treatment (Xenotest) 170h (Sample 4), $K_2Cr_2O_7+H_2SO_4$ solution at 70°C, 3 min. (Sample 5), 3M nitric acid at 20°C, 24 h (Sample 6), 30% hydrogen peroxide at 20°C, 1 h (Sample 7).

Moreover, the spectrograms of PP surface layer oxidized by chemical methods show a new absorption band within the wave number range of (1730 – 1680) cm⁻¹ that corresponds to a carbonyl group formed in a oxidizing medium as a results of the nucleophilic substitution of PP, mainly at the tertiary carbon atom: - CH_2 – C < R H – CH_2 – (the substitution susceptibility of the tertiary, secondary and primary carbon is 7000: 1100: 1, respectively) (Wiberg & Eisenthal, 1964). The absorption maximum of carbonyl group is slightly shifted depending on the type of oxidizing medium.

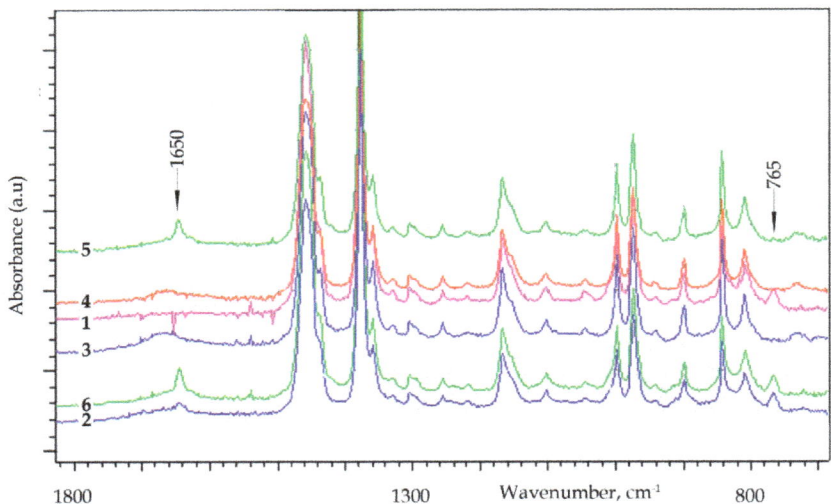

Figure 5. FTIR-ATR spectra of PP Cast films before and after plasma oxidation. Sample 1 - PP untreated, Sample 2. PP treated Ar plasma p=0.05Tr, power 300W, t=10 min., Sample 3 - PP treated Acetic Acid Vapour plasma: p=0.05Tr 300W, t=10 min., Sample4 - PP treated Water Vapour plasma: p=0.05Tr; 300W, 10min, Sample 5 - PP treated Air plasma: p=0.05Tr; 300W. t= 5min., Sample 6 - PP treated Air plasma: p=0.05Tr; 300W, t= 10min.

The absorption band of carbonyl group in the PP spectrum is broad, which can indicate the presence of carbonyl group in various products of oxidation, such as aldehydes and ketones (Carlsson & Wiles, 1969): 1700 cm^{-1}absorption ($>$C = CH-CO-OH), 1710 cm^{-1}absorption (-CO-OH), 1715 cm^{-1}absorption ($>$C = O), 1718 cm^{-1} absorption (-CCH$_3$ –CH$_2$ –CO- CH$_2$- CH$_3$), 1726 cm^{-1}absorption (-CCH$_3$ –CH$_2$ –CO- CH$_3$).

The spectrogram of oriented and crystalline PP AG shows no absorption band of carbonyl group despite the fact that the determination of the contact angle of PP surface wetted with polar liquids indicated an increase in free energy (Urbaniak-Domagala, 2011). One may assume that the active center concentration is too low for the FTIR-ATR method. The spectrogram of PP oxidized by chemical methods also indicates changes in two bands at 940 cm^{-1} and 765 cm^{-1} (Figure 4).

The first one indicates the skeleton vibration of mer links with a relative phase shift of 2/3, being mainly characteristic of the crystalline phase (Rau, 1963). In the case of PP AG film, this band is intensive, while in PP Cast, it decreases and after oxidation is absent, which can indicate that the PP surface layer becomes amorphous due to the oxidation process. The absorption band at 765cm^{-1} is characteristic of non-crystalline PP, caused by the deformation vibration of methylene group (–CH$_2$ –) (Kazicina.&Kupletska, 1976). This band is absent in the spectrogram of PP after oxidation, which can be due to the decrease in the number of methylene groups caused by the degradation of the polymer in its surface layer. This band is also absent in crystalline PP due to spherical limitations caused by a long-range order.

FTIR-ATR absorption spectra (Figure 5) present the chemical effects of plasma on PP film. The gases used in this process included: argon and air and vapors of acetic acid and water under optimal conditions of plasma treatment (time and power applied to the system).

The spectra of the plasma-treated PP film show a new absorption band within the range of wave numbers of (1640 ÷ 1660) cm^{-1}, which can indicate the formation of carbonyl group, >C=O, (valence vibration) as well as –C=C- groups (valence vibration) (Kazicina&Kupletska, 1976). The prolongation of plasma treatment and increase in power leads to the increase in the IR radiation intensity of the band of new functional groups. The position of IR absorption maximum slightly shifts depending on the plasma composition. These new active centers can be regarded as a result of PP surface oxidation with plasma particles. In the case of Ar plasma, the effect of surface functionalization can result from the so-called post-treatment process (Guruvenket et al, 2004). The results obtained indicate a particular activity of air plasma as oxidizing medium for polypropylene.

3.1.2. Effects of PP nonwovens modification

FTIR-ATR was also used to assess the effects of plasma-treated PP nonwovens. PP melt-blown nonwovens (surface weight: 80 g/m^2, average thickness: 1.5mm) made of PP fibers with an average thickness of 2.12 μm were modified by means of synthetic air plasma to form chemically active centers on the PP fiber surface.

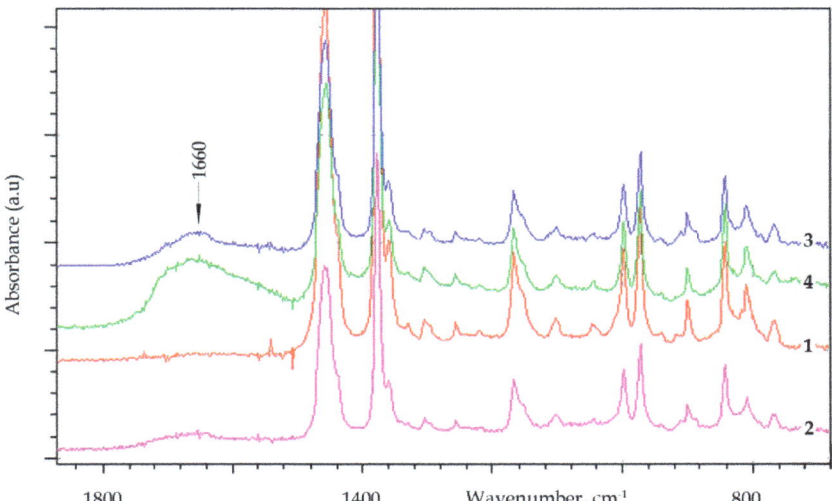

Figure 6. ATR IR spectra of PP nonwoves untreated (Sample 1) and air plasma treated, pressure 0.1Tr : Sample 2 - power 50W, time 5min. Sample 3- power 100W, time 5min., Sample 4 - power 100W, time 10min

The FTIR-ATR spectrogram of the air plasma-treated PP nonwoven shows two broad bands at 1660 cm^{-1} and 3320 cm^{-1} that indicate the formation of carbonyl group >C=O and hydroxyl group – OH (Kazicina.&Kupletska, 1976). One may assume that the air plasma oxidizes the fiber surface with the aid of reactive oxygen, peroxide and nitrogen groups that together with electrons react with the PP fiber surface causing not only the etching of surface layer but also its functionalization.

3.2. Polymeric coatings deposited on PP nonwovens

Nonwovens constitute a specific substrate for depositing thin polymeric layers. The nonwoven surface is developed to an extent dependent on the diameter of elementary fibers, density of their distribution and the formation technique used. The melt-blown PP nonwovens (see section 3.1.2.) were coated with thin layers of plasma polymers in a methane plasma and in hexamethyldisiloxane (HMDSO: O-(Si-(CH₃)₂) vapors [Urbaniak-Domagala et al, 2010). As a result of this process, the nonwoven surface was covered with a plasma polymer layer with a thickness of about 100 nm. SEM photographs (Figure 7) indicate that the coating obtained shows a character of a continuous film fitted to the uneven nonwoven surface covering only the elementary fibers in the near-surface nonwoven layer.

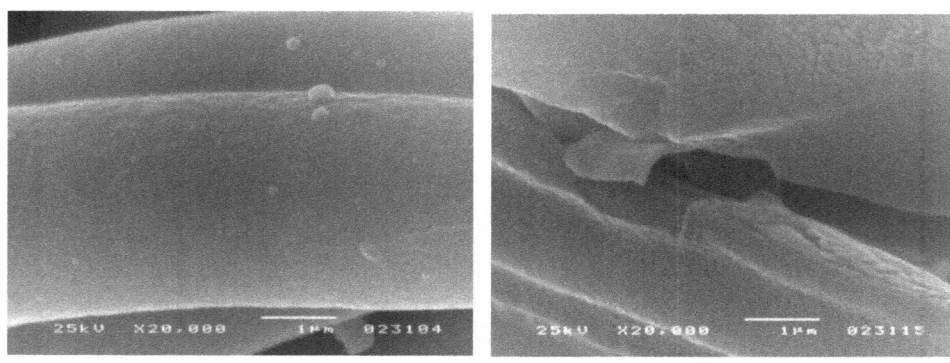

Sample A   Sample B

Figure 7. SEM images of PP fibers at nonwoven fabrics, treated methane plasma (Sample A), treated HMDSO plasma (Sample B). Plasma process time 10 min., pressure 0.05Tr, power 100W.

The FTIR-ATR spectrogram of the methane plasma-modified nonwoven surface (Figure 8) indicates that the layer chemical structure has a character of a hydrocarbon polymer as the PP substrate. The broad band with a maximal absorption at 1650 cm⁻¹ can be assumed as a post-treatment effect (Guruvenket et al, 2004).

The IR-ATR spectrogram of the HMDSO plasma-modified nonwoven indicates that the layer deposited has a chemical structure of a SiOC:H polymer (Creatore et al., 2002) and contains intensive absorption bands at 800 cm⁻¹, 841 cm⁻¹, 1040 cm⁻¹and 1256cm⁻¹ (Table 2) being characteristic of chemical groups containing silicon (Borvon et al., 2002).

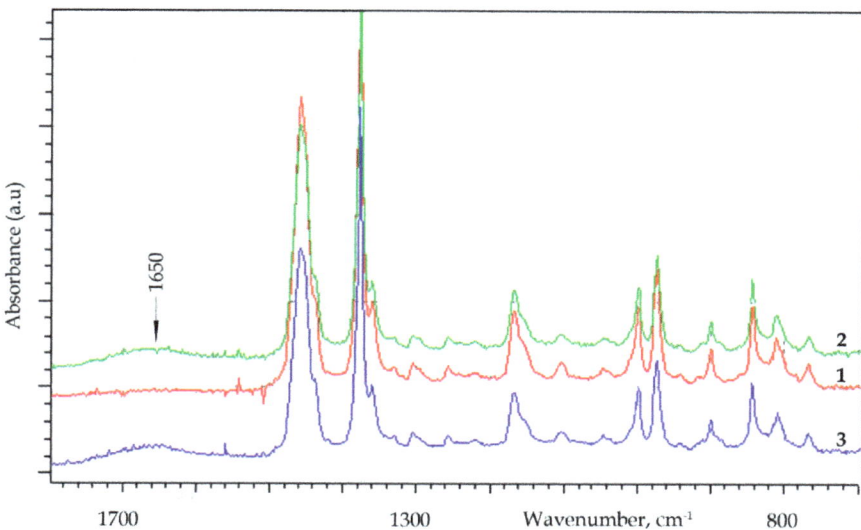

Figure 8. FTIR-ATR spectra of PP nonwoves. Sample 1- untreated, Sample 2 - methane plasma treated: power 25W, pressure 0.05Tr, Sample 3 - methane plasma treated: power 100W, pressure 0.05Tr.

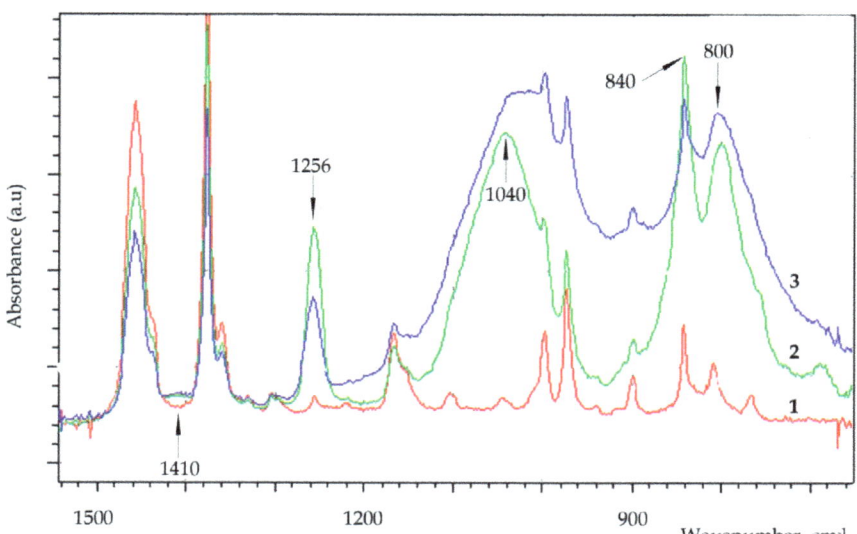

Figure 9. FTIR-ATR spectra of PP nonwoves. Sample 1 - untreated, Sample 2 - HMDSO plasma treated: power 25W, pressure 0.1Tr, Sample 3 - HMDSO plasma treated: power 100W, pressure 0.1Tr.

The assessment of the surface of samples was carried out in the diamond-sample system, in which the IR radiation beam penetrates the layer 2 μm in depth within the spectrum range discussed. The penetration depth of the IR radiation beam considerably exceeds the thickness of the p-HMDSO layer (~ 0.1 μm), hence characteristic bands of PP substrate also occur in the absorption spectrum.

Wave number, cm⁻¹	Absorbing group and type of vibration
1410	vs (CH₃), va in Si(CH₃)x
1256	ν CHx, δ (CH₃)in Si(CH₃)x
1040	va (Si-O-Si)
840	ν Si(CH₃)x , γr (CH₃)in Si(CH₃)₃
800	ν (Si-O-Si), γr (CH₃)in Si(CH₃)₂

*) vs - stretching vibration symmetrical and va -asymmetrical, δ - deformation vibration, γr – rocking vibration

Table 2. IR absorption bands of p-HMDSO plasma layers (Aumaille et al., 2002; Agres et al., 1996)

3.3. Testing polymeric coatings containing polypyrroles

Polypyrrole is a polymer widely used in commodity production owing to its high thermal stability, resistance to atmospheric conditions and biocompatibility. Its important advantages include electric properties. Using appropriate synthesis conditions, one can obtain electro-conductive, semi-conductive or electro-insulating polypyrroles. In view of processing difficulties, polypyrroles are produced directly on material surfaces in the form of coatings by *"in situ"* chemical, electrochemical or plasma methods. Moreover, polypyrroles are used to make composites as reinforcing and functional materials. In this work, the FTIR-ATR technique was used to monitor the results of polypyrrole synthesis by chemical and plasma methods and the preparation of pyrrole - containing composites.

3.3.1. Formation of latex-pyrrole composites

Polypyrrole (PPy) was used to make an electro-conductive composite as a backing of textile floor coverings (TFC). The TFC piles are fixed in a standard procedure with the use of dressing containing a synthetic rubber and vinyl-acrylic thickeners. The standard latex coating shows electro-insulating properties and impedes the leakage of static charges generated on the TFC pile during exploitation. A functional dressing was prepared to facilitate the leakage of static charges from TFC. PPy microspheres in the form of an aqueous dispersion, prepared by polymerization in an aqueous solution of ferric chloride, were added to an aqueous dispersion of butadiene-styrene-carboxyl copolymer (LBSK 4148 (Urbaniak-Domagala, 2005).

The dispersion components were intermixed by means of an ultrasonic stirrer and the resultant dressing was applied on the bottom of a raw TFC followed by the cross-linking process. The volume resistance tests of the latex-PPy coat confirmed its antistatic properties already with a 3% (by wt.) content of PPy in relation to the dry copolymer mass in the dressing.

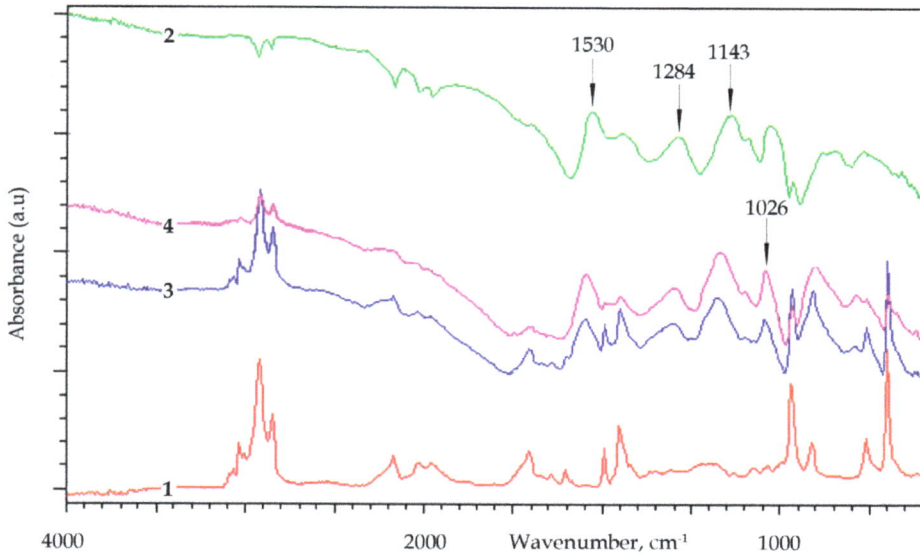

Figure 10. FTIR-ATR spectra pure latex LBSK 4148 (Sample 1), chemically synthesized polypyrrole (Sample 2), latex/PPy composite containing 2 wt% PPy (Sample 3), 3 wt% PPy (Sample 4)

The percolation of the coating electric conduction has a continuous character and the percolation threshold is relatively low. The coating formation on the TFC bottom was controlled by means of FTIR-ATR spectrometry

Changes in the IR radiation absorbance of the coatings were observed within the range from 500 cm^{-1} to 1700 cm^{-1}. Figure 10 shows the absorption spectra of pure LBSK 4148 latex (Sample 1), pure PPy (Sample 2) and two samples of LBSK-PPy composite containing 2% by wt. of PPy and 3% by wt. of PPy, respectively. The locations of absorption bands of LBSK, PPy and LBSK-PPy composite containing 3% by wt. of PPy are listed in Table 3.

The spectrum of LBSK indicates the presence of three types of butadiene isomeric units (1,4-cis, 1,4-trans and 1,2-vinyl), styrene PS and carboxyl (Molenda et al.,1998; Munteanu & Vasile, 2005) have carried out fundamental research of FT-IR spectra of butadiene-styrene copolymers with various structural arrangements (block and linear copolymers, block copolymers of the star type and statistic copolymers). The type of spectrum found for LBSK 4148 latex indicates the architecture of statistic copolymer. The spectrogram of the microspheres of chemically synthesized PPy (powder) indicates PPy rings in the polymer structure and groups connected with the ring being consistent with the results of authors Eisazadeh, 2007; Cruz 1999; Ji-Ye Jin et al.,1991).

LBSK 4148		Chemically synthesized polypyrrole (PPy)		LBSK 4148/PPy
λ, cm^{-1}	Absorbing group and type of vibration	λ, cm^{-1}	Absorbing group and type of vibration	λ, cm^{-1}
-		659	δ(C-N) out of plane ring	659
699, 758	δ (CH) out of plane in the aromatic ring, PS units	-		699, 758
-		770	δ (C-H) out of plane pyrrole ring, γ(NH$_2$), v(C-N-C), δ(C-N-C)	790
911	δ (CH) out of plane near the double bond of the vinyl-PB units	-		911
966	δ (CH) out of plane near the double bond in trans-PB units	-		966
-		1026	δ (N-H), δ(C-H), pyrrole rings pulsation v(C-N) secondary amines	1040
-		1143	δ (C-H)	1170
-		1284	v(C-N) secondary amines, v (C-N) in pyrrole ring	1300
1451	δ (CH) in cis-PB, trans-PB, vinyl-PB units	-		1451
1492	v (C=C) in aromatic ring PS units	-		-
-		1530	v(C=C) in pyrrole ring v(C=N) in pyrrole ring , pyrrole ring pulsation	1548
1600	v (C=C) in aromatic ring PS units	-		-
1638	v (C=C) PS units v (C=C) in vinyl-PB	-		-
1700	v (C=O)	-		1700

Table 3. The FTIR absorption bands for latex LBSK 4148, chemically synthesized polypyrrole and LBSK 4148/PPy composite containing 3 wag.% PPy (Kazicina.&Kupletska, 1976, Molenda et al.,1998; Munteanu & Vasile, 2005; Bieliński et al., 2009; Eisazadeh, 2007; Cruz 1999; Ji-Ye Jin et al.,1991)

The spectrograms of LBSK-PPy composite samples indicate the superposition of characteristic bands of the composite components: PPy and latex. As the PPy content in the composite increases, one can observe an increase in the intensity of characteristic peaks of PPy, but the quantitative analysis of the composite is difficult to perform due to great differences in the absorbance of the composite components (latex is white, PPy is black). The band maxima shown by the PPy powder are delocalized in the spectrum of latex-PPy composite. The band indicating the pyrrole ring vibration at 1530cm^{-1} is shifted towards a higher frequency to 1548 cm^{-1}. The bands of groups linked up to the pyrrole ring are also shifted: for CH deformation vibration (out of plane quinol PPy) from 1026 cm^{-1} to 1040 cm^{-1}, for C-N deformation vibration from 1143 cm^{-1} to 1170 cm^{-1}, and for the valence vibration of CN in pyrrole ring from 1284 cm^{-1} to 1300 cm^{-1}. The shifts of bands can be due to the scattering of IR radiation in the structure of PPy powder, but they can also indicate the occurrence of PPy - latex intermolecular interactions, with which the oscillatory excitation of chemical groups in PPy requires a higher energy.

3.3.2. Synthesis of polypyrroles

Below are presented examples of using the FTIR-ATR technique to assess the progress in the synthesis of PPy. The polymerization of pyrrole was carried out by chemical and plasma methods. Thin layers of PPy were formed on the surface of a PP film by the *in situ* technique.

3.3.2.1. Chemical polymerization method

Polymer layers were formed by the polymerization of pyrrole according to the redox mechanism. Two media of pyrrole oxidation were used: an aqueous solution of ferric chloride and aqueous solution of ammonium sulfate with p-toluenesulfonic acid as dopant. Based on the FTIR-ATR spectrum of the polymer, its synthesis progress and chemical structure were characterized. Figure 11 shows examples of the spectra of PPy synthesized in both media for 2h and 5 h.

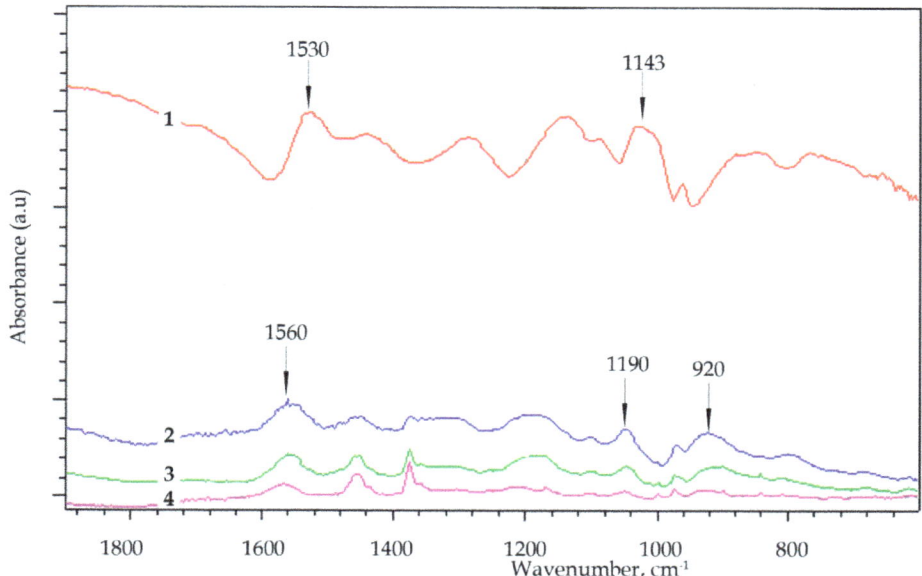

Figure 11. FTIR-ATR spectra chemically synthesized polypyrrole. Sample 1 - PPy powder (molar ratio of FeCl3:Py=2,3:1) - polymerization time 5h; PPy layers on the polypropylene foil: Sample 2 – molar ratio of FeCl3:Py=2,3:1, polymerization time 5h, Sample 3 – molar ratio of FeCl3:Py=2,3:1, polymerization time 2h; Sample 4 – molar ratio of $(NH_4)_2S_2O_8$: $CH_3C_6H_4SO_3H$: Py=0,2:0,25:1, polymerization time 2h.

The spectra of PPy are recorded on the PP substrate. Owing to the low thickness of layers (0.1 - 1µm), the spectrum additionally contains bands derived from the substrate. For comparison, the spectral characteristics of PPy synthesized in the form of powder were also presented. The absorption spectra of all the polymer samples within the wave number range of 600 cm^{-1} – 1800 cm^{-1} confirm the presence of pyrrole group (Table 3). The intensity of absorption bands increases with increasing polymerization time, which is due to the increased layer thickness. In the process of chemical synthesis, the aromatic character of pyrrole ring is maintained, which results in the formation of conjugated double bonds in the linear macromolecule chain. In the presence of admixtures intercalated to the system, the polymer is electro-conductive (incorporated dopants: Cl$^-$ and CH$_3$C$_6$H$_4$SO$_3^-$). The oxidized form of conductive PPy obtained shows a considerable absorption of IR radiation (black color of the polymer). The spectrogram of PPy synthesized with the use of two different oxidants shows no differences between the polymer chemical structures. Differences concern the progress rate of the synthesis: the higher intensity of pyrrole group bands in the polymer synthesized in the aqueous solution of ferric chloride indicates a higher polymerization rate, which is confirmed by the higher rate of layer building up.

3.3.2.2. Plasma polymerization method

Polymer synthesis performed in glow discharge of monomer vapors is a dry, ecological, energy- and material-saving process. The polymerization process is initiated by means of electrons and radicals formed in the gas discharge. The polymerization of pyrrole was carried out in a flow reactor, in glow discharge of the induction type by means of RF field 13.56 MHz (Urbaniak-Domagala, 2008). PP film substrate was centrally and axially placed on a glass carrier in the reactor. The film surface was preliminary purified by means of argon plasma followed by the deposition of the plasma polymer. The FTIR-ATR technique was used to examine the effect of process parameters, such as deposition time, pressure in the reactor and power input to the reactor, on the chemical structure of plasma polymer. In order to impart semi-conductive properties, the plasma PPy was doped after the deposition process by two methods: *in situ* in the reactor in glow discharge of the vapors of organic iodine compounds, and *ex situ* after removal from the reactor in crystalline iodine vapors.

Figure 12 shows the FTIR spectra of the plasma polymer within the range of (600-1850) cm^{-1}. The spectrogram shows the superposition of the absorption bands of plasma polymer (thickness 0.3 µm) and PP substrate. The broad band at (1500 – 1800) cm^{-1} indicates different structure of plasma PPy compared to that of PPy synthesized by the chemical method. This band points to a possible occurrence of primary and secondary amines , secondary amides (Kazicina.&Kupletska, 1976), and carbonyl groups (Ji-Ye Jin et al.,1991) in the polymer. The broad absorption band of the polymer indicates a complex absorption caused by the products of broken pyrrole rings that initiate the branching and cross-linking of the polymer followed by various substitutions. Thus the plasma spectrograms can show secondary and tertiary amines that complicate the absorption in this range. Moreover, one cannot exclude the occurrence of the band at 1710cm^{-1} that, according to authors (Ji-Ye Jin et al.,1991) indicates the presence of carbonyl groups. This band is often observed in neutral or weakly doped forms of PPy, mainly due to their susceptibility to oxidation in air. The spectrum of

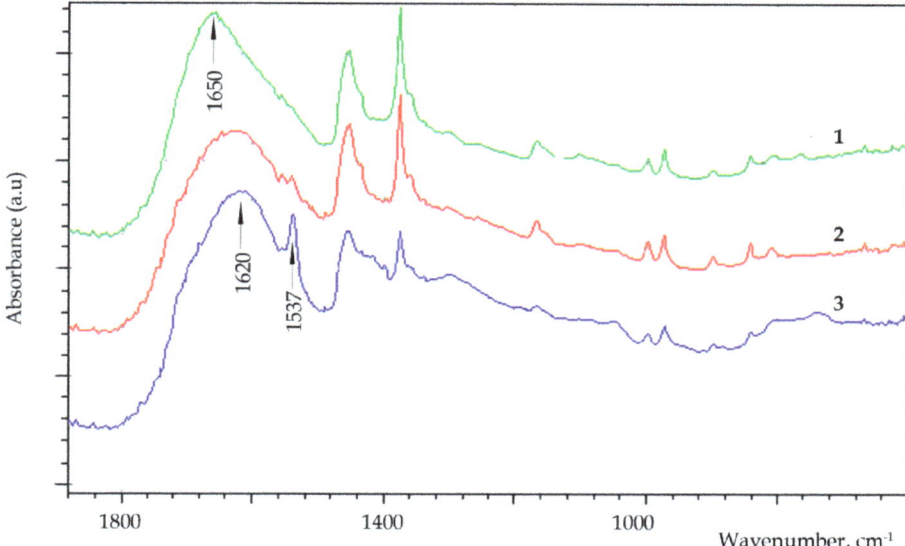

Figure 12. FTIR-ATR spectra plasma synthesized polypyrrole, pyrrole plasma: 50W, 10min, p=0,05Tr. Sample 1-polymer without dopand, Sample 2 - polymer dopanded at plasma CH₂J₂ : 25W, 30min., p= 0,05Tr, Sample 3- polymer dopanded at J₂ vapours, 30min

the polymer doped with iodine vapors contains an additional band at 1537cm⁻¹, induced by the vibration of pyrrole ring, especially intensive in the polymer doped with crystalline iodine vapors. (Groenewoud et al. 2002) observed an increase in the intensity of peak 1520 cm⁻¹ under the influence of iodine vapors, which is connected with the formation of a new CH₂=J group in the reaction of iodine with radicals present in the surface layer of the plasma polymer.

The spectrum of the plasma polymer synthesized in the presence of nitrogen as a carrier of pyrrole proves how significant is the influence exerted by the process gas on the chemical polymer structure. This is particularly evident in the polymer synthesized for a longer time (the spectrum of polymer after a 1 h process – Figure 13). The FTIR-ATR spectrum of the plasma PPy at (500-1000) cm⁻¹ contains numerous bands with a high absorption intensity that indicate the presence of primary amines (Kazicina.&Kupletska, 1976), and products of substituting chemical groups that were additionally formed in the polymer under the influence of the nitrogen plasma.

The absorption spectra obtained by the FTIR-ATR technique identify the chemical structure of PPy coatings and the structural changes that appear during changing the process parameters, such as pressure, power, type of doping agents, method of incorporating doping agents and the presence of process gas.

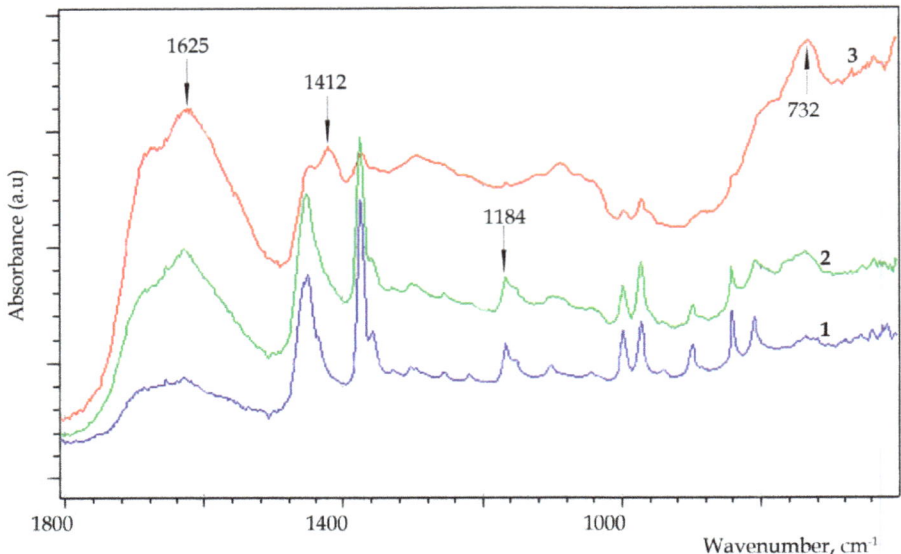

Figure 13. FTIR-ATR spectra plasma synthesized polypyrrole without dopands. Sample 1- plasma pyrrole 0,1Tr 10W, 15min, without processing gas, Sample 2- processing gas N_2 and pyrrole p= 0,15Tr. 10W, 15min. Sample 3 - processing gas N_2 and pyrrole p=0,15Tr 10W, 60min

4. Conclusions

The FTIR-ATR technique makes it possible to examine polymers in a simple, fast procedure avoiding sample destruction. It creates many opportunities for application to the chemical identification of the surface layer of polymers and thin polymeric layers. The analyses of test results of thin layers deposited on polymeric substrates can have rather qualitative character due to the penetration reach of the IR radiation beam being greater than the layer thickness. The examples of PP modification processes and deposition of coatings on polymeric substrates presented confirm that the FTIR-ATR method can be helpful in the examinations of the following:

-   the chemical structure of the polymer surface layer and changes in the layer structure caused by the effects of chemical agents and electrochemical factors, UV radiation and low temperature plasma on polymers;
-   the chemical structure of polymeric coatings deposited on substrates by chemical and plasma methods;
-   the dependence of the chemical structure of polymeric coatings on the parameters of technological process.

Author details

Wieslawa Urbaniak-Domagala
*Technical University of Lodz, Department of Material
and Commodity Sciences and Textile Metrology, Poland*

Acknowledgement

The author would like to thank Professor Barbara Lipp-Symonowicz for good advices and for helpful discussions.

5. References

F. de Fornel, (2000). *Evanescent waves—From Newtonian optics to atomic optics*. Berlin: Springer- Verlag, ISBN: 9783540658450

Goos F.; Hänchen H. (1947). Ein Neuer und fundamentaler Versuch zur Totalreflexion, Ann. Phys. (436) 7-8, 333-346

Dechant J. (1972) *Ultrarotspektroskopische Untersuchungen an Polymeren*, Berlin
Available from [http://www.sprpages.nl/SprTheory/SprTheory.htm]

Urbańczyk G.W. (1988) *Mikrostruktura Włókna - Badanie Struktury Krystalicznej i Budowy Morfologicznej*, WNT, ISBN 83-204-1014-2, Warszawa

Urbaniak-Domagala W. (2011) Pretreatment of polypropylene films for the creation of thin polymer layers, part 1: The use of chemical, electrochemical, and UV methods" *Journal of Applied Polymer Science* Vol. 122, No. 3, 2071–2080, 5 November 2011

Urbaniak-Domagala W. (2011) Pretreatment of polypropylene films for following technological processes, part 2: The use of low temperature plasma method. *Journal of Applied Polymer Science*, Vol.122, No. 4, 2529–2541, 15 November 2011

Urbaniak-Domagala W., Wrzosek H., Szymanowski H.,Majchrzycka K., Brochocka A., (2010) Plasma Modification of Filter Nonwovens Used for the Protection of Respiratory Tracts, *FIBRES & TEXTILES in Eastern Europe* , Vol. 18, No. 6 (83) pp. 94-99

Rau J.H. (1963). *Melliand Textilberichte*, 44, pp.1102, 1197, p. 1320

Wiberg K.B.; Eisenthal R. (1964) On the mechanism of the oxidation of hydrocarbons with chromic acid and chronyl chloride. *Tetrahedron* 20, 1151-1161

Carlsson D.J.; Wiles, D.M. (1969) The Photodegradation of Polypropylene Films. II. Photolysis of Ketonic Oxidation Products *Macromolecules* 2, 587.

Kazicina L., Kupletska N.:(1976) *Metody spektroskopowe wyznaczania struktury związków organicznych* 2nd edition; PWN: Warsaw

Guruvenket S.; Rao G.M.; Komath M.; Raichur A.M. (2004) Plasma surface modification of Polystyrene and Polyethylene. *Appl. Surf. Sc.*, 236, 278-284

Creatore M., Palumbo F., d`Agostino R., (2002) Deposition of SiOx Films from Hexamethyldisiloxane/Oxygen Radiofrequency Glow Discharges: Process Optimization by Plasma Diagnostics *Plasmas and Polymers*, Vol.7, No. 3, 291-310

Borvon G., Goullet A., Granier, A., Turban, G., (2002) Analysis of Low-*k* Organosilicon and Low-Density Silica Films Deposited in HMDSO Plasmas *Plasmas and Polymers*, Vol.7, No. 4, 341-352

Aumaille K., Vallee C., Granier, A., Goullet, A., Gaboriau, F., Turban, G., Turban (2000) A comparative study of oxygen/organosilicon plasmas and thin SiOxCyHz films deposited in a helicon reactor. *Thin Solid Films* Vol. 359,188-196

Agres L., Segui Y., Delsol R., Raynaud, P. (1996) Oxygen Barrier Efficiency of Hexamethyldisiloxane/Oxygen Plasma-Deposited Coating *Journal of Applied .Polymer Science*. Vol. 61, 2015-2022

Urbaniak-Domagala W. (2005) Modyfikacja właściwości elektrycznych klejonki lateksowej stosowanej do włókienniczych pokryć podłogowych" Konference *ENP`2005, Elektrotechnologie w nowoczesnym przemyśle*, Bialystok 2005

Molenda J.,Grądkowski M., Makowska M., Kajdas C. (1998) Tribochemical characteristic of some vinyl-type compounds in aspect of antiwear interactions. *Tribologia*, No. 3, 318-329

Munteanu S.B., Vasile C. (2005). Spectral and thermal characterization of styrene-butadiene copolymers with different architectures. *Journal of Optoelectronics and Advanced materials*, Vol.7, No.6, 3135-3148

Bieliński D., Głąb P., Ślusarski L. (2009) FT-IR internal reflection study of migration and surface segregation of carboxylic acid in butadiene-styrene rubber. *Polimery*, 2009, Vol. 54, No. 11-12, 706-711

Eisazadeh H. (2007). Studying the Characteristics of Polypyrrole and its Composites. *World Journal of Chemistry* 2 (2): 67-74

Cruz G.J., Morales J.,Olayo R. (1999) Films obtained by plasma polymerization of pyrrole. *Thin Solid Films*, 342, 119-126

Ji-Ye Jin, Kumi.T., Ando, Teramae N., Haraguchi H. (1991) FT-IR Spectroscopy of Electrochemically Synthesized Polypyrrole. *Analytical Sciences* Vol. 7, 1593-1594

Urbaniak-Domagala W. (2007) Morphology of polypyrrole films formed in low-temperature plasma. *Proceedings Conference IMTEX 2007*, October 8-9, Lodz, 96-99, ISBN 978-83-911012-6-1

Groenewoud L.M.N.,Engbers G.H.M., White R., Feijen J., (2002) On the iodine process of plasma polymerized thiophene layers" *Synth. Met.* Vol. 125 , 429-440

Advanced Spectroscopy

Chapter 4

Laser-Induced Breakdown Spectroscopy

Taesam Kim and Chhiu-Tsu Lin

Additional information is available at the end of the chapter

1. Introduction

Laser-induced breakdown spectroscopy (LIBS) is an atomic emission spectroscopy. Atoms are excited from the lower energy level to high energy level when they are in the high energy status. The conventional excitation energy source can be a hot flame, light or high temperature plasma. The excited energy that holds the atom at the higher energy level will be released and the atom returns to its ground state eventually. The released energy is well-defined for the specific excited atom, and this characteristic process utilizes emission spectroscopy for the analytical method. LIBS employs the laser pulse to atomize the sample and leads to atomic emission. Compared to the conventional flame emission spectroscopy, LIBS atomizes only the small portion of the sample by the focused laser pulse, which makes a tiny spark on the sample. Because of the short-life of the spark emission, capturing the instant light is a major skill to collect sufficient intensity of the emitting species. Three major parts of the LIBS system are a pulse laser, sample, and spectrometer. Control system is usually needed to manage timing and the spectrum capturing. Figure 1 illustrates those three major components and a computer in the conventional LIBS.

The LIBS has been used for the materials detection and analysis in various applications, such as steel and alloys[1-8], paints and coatings[9-15], wood pre-treatment[16], polymers [17], bacteria[18], molds, pollens, and proteins[19,20], and space exploration[21]. The great majority of LIBS results were consolidated in the reviews[22] and books[23, 24].

In spite of its advantage in analytical spectroscopy, LIBS application is still restricted within certain areas and propagation of the technology is not very wide. Many laboratory LIBS systems are built in schools, research labs and companies with discrete optical parts. Their pioneering approach in the new application seemed promising for a certain samples, but actual use in the application field is usually very limited. We can explain the situation with other analytical techniques, for example, Gas Chromatography (GC). The GC can separate the volatile species. However, one GC setup can work for a narrow range of species grouped in the sample. For different applications, the user must change the GC column, detector,

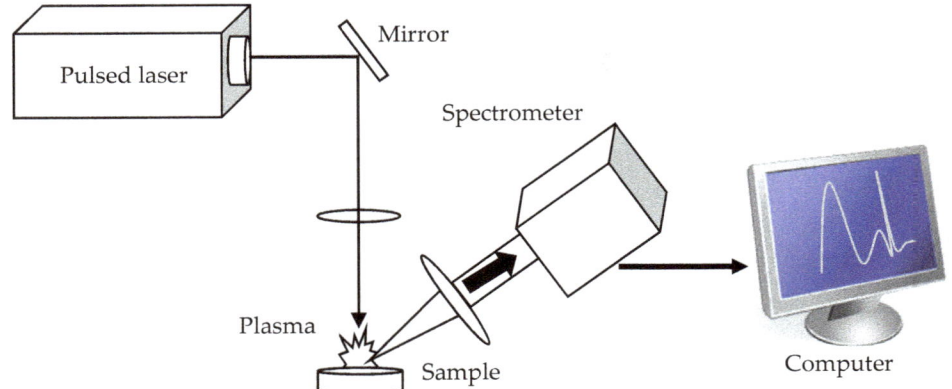

Figure 1. The conventional LIBS system configuration

carrier gas or at least use a new column temperature cycle. The application of LIBS also needs case-by-case adjustment. Many new applications start with looking at the advantages of LIBS and choosing a LIBS setup, and it still needs a detailed investigation for successful analysis.

This chapter describes how the LIBS system works and explains the major parts of LIBS to select specific functional requirements for its intended application. The three major parts: laser, sample and spectrometer are explained. The laser provides the breakdown energy and plasma generation. Analytical sample is the target of the laser shot and the source of emission species. The spectrometer comprises detection system with light detector and computer. Their disadvantages and limitations are discussed then suggesting how to select the equipment type and configuration to maximize the advantages of LIBS. This will provide a beginning inspiration of LIBS systems to install and apply the desired specific analytical purpose or application area.

2. LIBS system design with modern technology

a.   Laser as a breakdown energy source

LIBS uses pulsed-laser light and focuses it onto the sample surface to make a plasma plume that contains the highly excited species of the sample composition. For generating plasma, there is a threshold value of the energy density. The threshold level will depend on the absorption coefficient of the sample surface of the laser wavelength, which is highly different by the sample phase. Gas and liquid need more energy to make breakdown. Solids with a dark color surface easily make a strong breakdown compared to clear or highly reflective solids. Figure 2 shows the effect of laser energy to make breakdown by the relation of laser power and focusing. Starting with a laser beam as 1 cm diameter, this light beam can be condensed by a convex lens. The focused beam density becomes 160 J/cm² as in Figure 2(a). Also, the laser is operating in the pulsed mode, assuming a10 nsec duration, total power per unit of time will be 16 GWatt/ cm² as

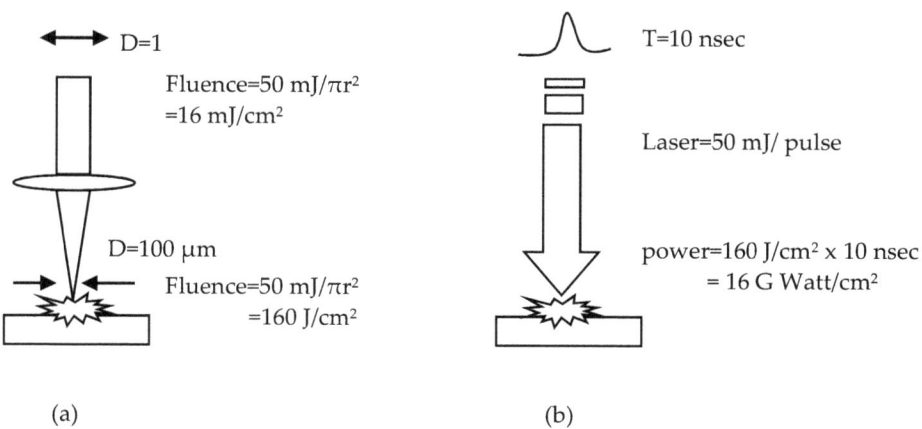

(a)   (b)

Figure 2. Laser energy delivery for breakdown condition. (a) focusing effect, (b) pulsing effect

in Figure. 2(b). Most of breakdown needs a few GW (10^6 Watt) of energy density, indicating that 50 mJ of laser energy is sufficient to make breakdown and evaporate most of material.

At the early stage of LIBS development, several types of pulse laser were used to make laser-induced breakdown plasma. An eximer laser was an important pulse laser especially for the UV light pulse. XeCl-eximer with 308 nm was used in the LIBS to measure elemental distribution on the paper coating[25]. The laser energy of 0.2 mJ was focused and made a crater of 30 μm diameter. This energy is corresponding to 10^8 W/cm². More than 90 % of ingredients in the paper coating are pigment, binder and other agents. The pigment's main component is usually aluminum oxide, silicon dioxide and calcium carbonate. The mass of coating material ablated by single laser pulse was estimated to be about 2 ng by a laser shot. A typical nitrogen laser has a wavelength 337.1 nm and a pulse duration of 10 nsec. Just like the eximer laser, the nitrogen laser LIBS configuration in Figure. 3 also includes discharge from the wide shape of the electrode. The laser beam is usually a few cm wide, so a tight focusing is needed. The surface of solar cell was measured by nitrogen laser breakdown and only a 40-nm-thick TiO_2 layer was detected[26]. The very popular pulse laser is Nd:YAG laser because it has a solid laser oscillator in a small size and light weight. The fundamental wavelength is 1064 nm with a pulse duration of 10 nsec typically. The Nd:YAG laser does not require any gas supply. The laser model for LIBS size usually has a closed loop water cooling that excludes external connection. A typical LIBS setup was shown in an earlier paper[27] as in Figure 4. A 50 mm focal length convex lens makes a simple optics configuration to make plasma on the sample.

3. Optical arrangement for laser–induced breakdown spectroscopy

When a laser shoots on the sample surface, a plasma plume arises from the inner to the outer surface. The actual size of plasma plume made by a 100 mJ laser pulse will be few millimeters. During the plasma propagation from the sample surface, the time profile features can be observed. The very initial emission is generated at the bottom of the plasma

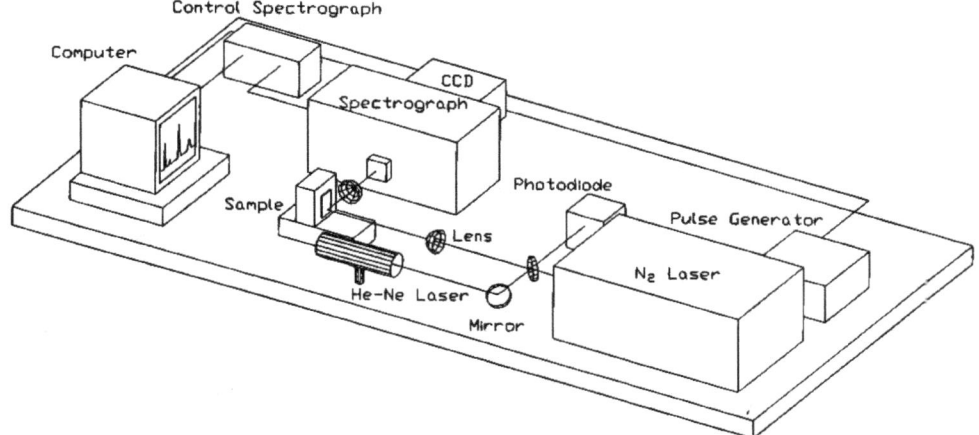

Figure 3. A LIBS setup with nitrogen laser.

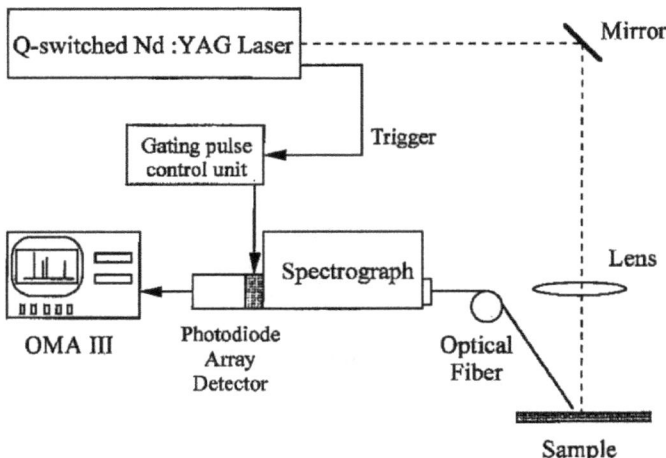

Figure 4. Schematic diagram of the LIBS setup with ND:YAG laser

plume, and then expanded to the outer plume. Depending on the light collecting optics, plasma propagation is captured at the different time. At the initial LIBS design uses a side-view emission collection as in the Figure. 5 (a). The angle between the laser light path and the collecting optic can be any angle, but is typically 30-45 degrees. Some experiments use 90 degrees, which is a complete side view of the plasma and will lose some portion of emission by the shadow of the sample itself. This configuration is occasionally used for plasma physics study. The collateral view design Figure. 5 (b) is a useful optical configuration for non-fixed sample distances. Laser light path shares emission collection optics. A selective wavelength reflector or prism can be used to separate laser light and emission through the light path. This design has several optical parts and needs complicated adjustments for optimum light measurement. The collateral configuration has many advantages. Collecting optics looks at the plasma in front of the plasma (or top of the plasma)at every point in the light axis and in the focus cone, which means they capture every light emitting species during plasma propagation to the space. Because some elements have different propagation profile than others, propagation height changes the signal significantly at the angled collection. The next advantage is that the optical part can be integrated in the compact enclosure, and it allows the operator to move the optics (detector head of LIBS) more freely. Remote monitoring LIBS, hand held design LIBS, should be compact and have a mostly collateral optics configuration.

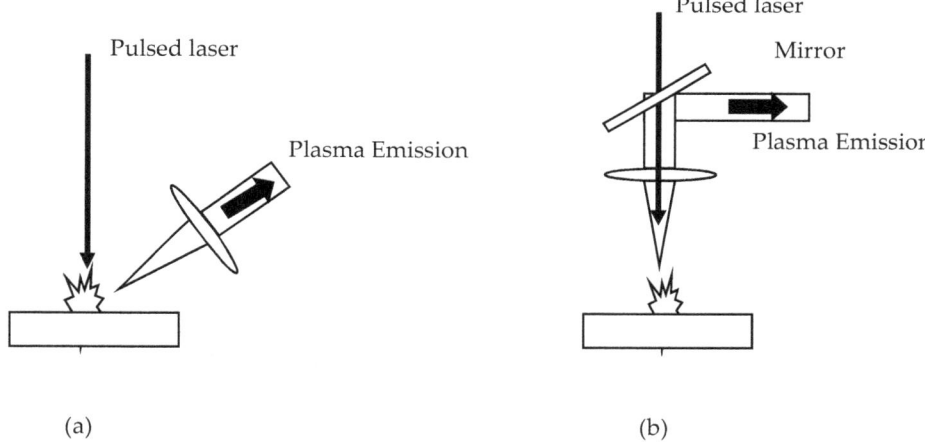

(a)   (b)

Figure 5. Side collection and (b) collateral collection configuration of plasma emission

e.   Sampling technique

The first mentioned advantage of LIBS has been no-sampling step. In the very beginning review in the *Encyclopedia of analytical Chemisstry*, Yueh, Singh and Zhang described it as "LIBS uses a very small amount of samples, and no sample preparation is necessary. It has the ability to perform real-time analysis because it prepares and excites the sample in one step". They then consecutively mentioned, "The disadvantage of LIBS is that the plasma conditions vary with the environmental conditions as well as the laser energy fluctuation."We can infer from the description of LIBS that no-sampling is both an advantage and a disadvantage. Most of analytical the techniques need a certain sampling procedure to bring the sample to the technique (or machine). During the sampling procedure, like the acid digestion in the flame analysis, the sample is homogenized and their matrices become concordant. However, if LIBS analyzed the sample without any pre-treatment process, then the irregular homogeneity is inevitable. As a result LIBS will include severe matrix effects at the real sample. It will mitigate the biggest advantage, i.e., no-sample process. In other words, if the sample is measured as it is, the species in same concentration do not make a consistent signal, the analytical result will be severely diverted.

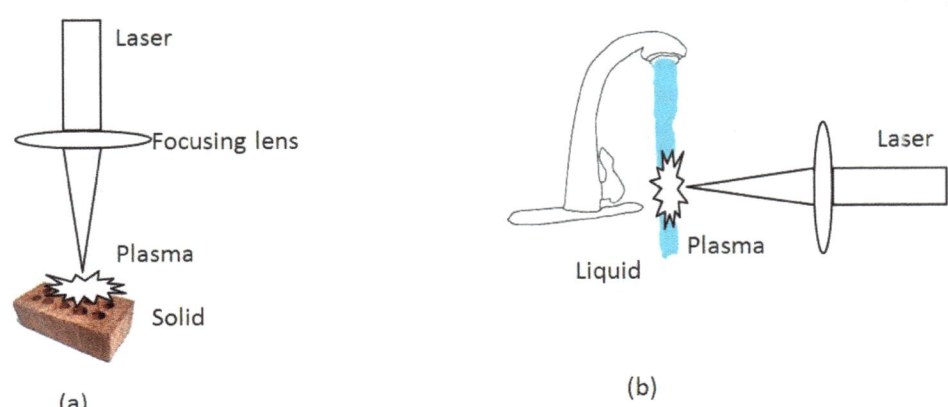

Figure 6. Solid sample and liquid sample under LIBS measurement

The fluctuation will be more serious because LIBS takes only a small amount of the sample, usually a micron sized spot. Two possible sample types are depicted in the Figure. 6. A solid sample is the most convenient sample type. Metal and ceramic samples include elements with strong atomic and ionic emission. Their emission spectra are measured at the range from UV to visible light, which is feasible by the most spectrometers. The spectra from many elements from the tool steel are shown by the Nd:YAG laser excitation[28]. In this research, the microscopic view of the ablated holes made on tool steel is about 10 microns in diameter. This resolution indicates that any inhomogeneity more than 10 microns will be clearly observed from each laser pulse measurement. The intensity of element-specific spectra provides a simple qualitative analysis. Their method was sufficient to characterize the nature of the defect by a simple estimation of the elemental composition between the basic material and the defect.

d.   Capturing emission light

The LIBS signal is instantaneous and decays quickly. Temporal control of the detecting device is very important. In spite of the fact, overall emission can be captured by opening entire time of the spectrometer, most of LIBS measurement is controlled by time gate operation. Time control improves the signal-to-noise ratio by eliminating the continuum emission. A typical emission profile shown in the Figure 7, recorded at the different heights from the sand/ soil mixture sample[29]. As soon as the laser fires with the duration of a few nsec of pulse width, the plasma intensity is propagating outward from the sample surface. At about 0.5 μsec, plasma is observed at 0.3 mm away from the surface. At the propagating distance is 3 mm at 12 μsec, then plasma cool down with decreasing intensity until 20- 30 μsec. The plasma size will be much smaller and life is shorter when a weak laser power is used. The experiment uses aluminum[30] with a diode-pumped Nd:YAG laser, which can run at a faster repetition rate (kHz) with a laser energy of 80 μJ, was setup under the microscope excitation and detection optics. Like other flash lamp pumped lasers, the temporal profile of continuum emission is shown in Figure. 8 for aluminum atom (Al 396.1 nm) and aluminum ion (Al II at 358.6 nm) emission lines. The laser pulse was fired at the zero time of the x- axis. This profile indicates, the broad band continuum emission, which comes from high temperature heated plasma and regardless of the species in the plasma, has a lifetime of about 13 nsec. The ionic line from Al ion has shorter lifetime about 24 nsec. The neutral lines stay much longer, up to 80 nsec.

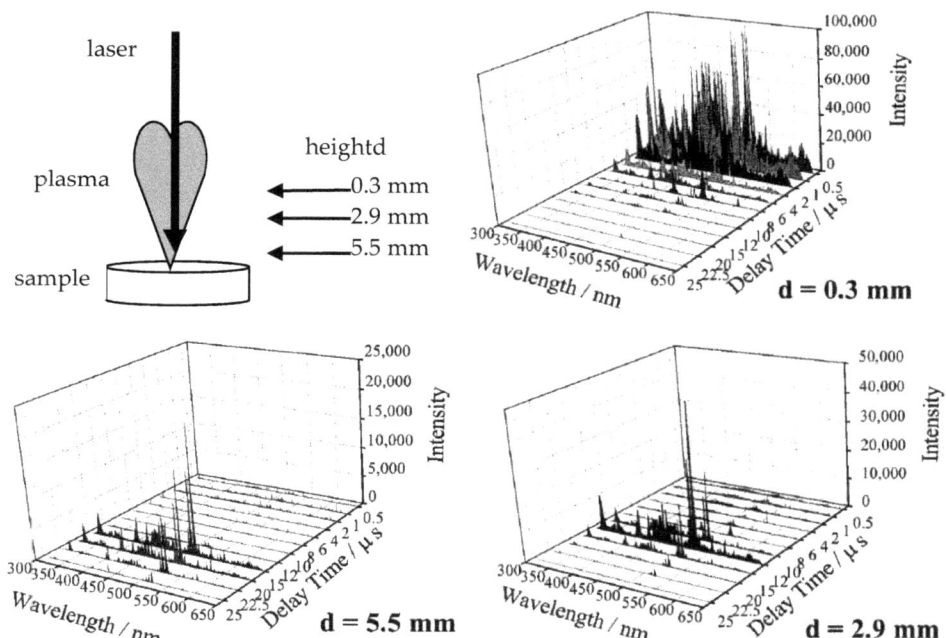

Figure 7. Spectra as a function of decay time measured at three observation distance from the sample surface. The original figure is rearranged to indicate observation height more clearly.

The lifetimes of laser-induced plasma are easily compared at various excitation energies from the silicon sample[25]. The time profile shows the plasma emission signal depends on the excitation pulse energy. Absolute intensity of the signal will increase by increasing laser pulse energy. The decay plot in the Figure 9 is normalized to a maximum intensity for comparison. This research explains the decay time dependence by the excitation energy that the probability of excitation to higher energy level is increased and more populated, leading to a longer decay time. Also, the upper state of the monitored transition receives population from this higher state at later times and lengthening of the rise time of the signal

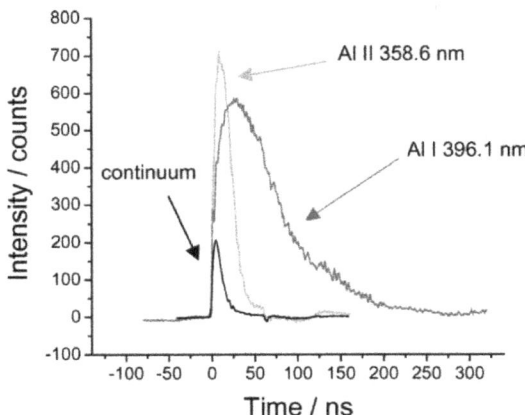

Figure 8. Temporal profile of continuum emission and aluminum (atom and ion) emission.

will result. As a result, the lifetime point of LIBS will be changed by the system setup, especially using laser power. Capturing time of emission signal should be determined empirically by looking at the profile, usually at peak intensity point.

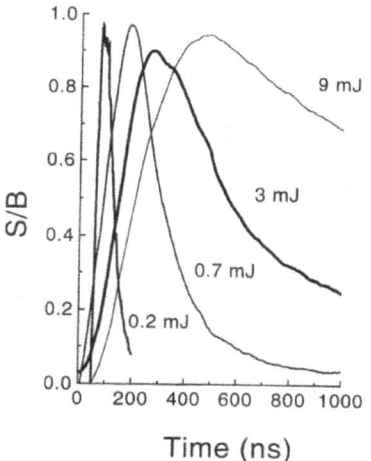

Figure 9. Time resolved signal-to-background ratio of the silicon line at 251 nm at various excitation energies.

e.   Spectrometer and detector

Spectrometer completes the detecting part with a photo sensor and a manipulating computer. The spectrometer must have proper resolution and sensitivity. Also, in many cases the plasma emission needs to be separated from the continuum background signal, the detector has to be operated by timing control or gating operation. Various types of spectrometer with CCD array detector are available in the market. The wavelength range needed for LIBS is UV to visible range to have detection of most elements. If the dispersion of the spectrometer is 0.3 nm to measure 1 nm peak with three pixel, the pixel to pixel dispersion should be 0.3 nm. Total of 1024 pixel CCD array can have coverage 1024 x 0.3 = 307 nm, which can assign the range as 250 nm to 557 nm span. In many cases, the sample will have mixed elements and the emission lines will be overlapped and difficult to distinguish with 0.3 nm resolution. A conventional CCD array detector may not provide sufficient resolution and coverage to measure LIBS.

Figure 10. Echelle spectrometer dispersion image (a) Hg lamp, (b) LIBS spectrum of Sn metal

A correction of the array detector resolution is accomplished using multiple stacked spectrometers. For example, 5 spectrometers with 1000 array CCD stacks will cover a 500 nm span, in which each spectrometer covers a 100 nm range with 0.1 nm resolution. Echelle spectrometer uses very high orders of dispersion. One or two prisms are used to separate each diffraction order. As a result, the spectra are dispersed in two dimensional surfaces as shown Figure 10. The CCD detector in the Echelle spectrometer should be a two dimensional, the same as in the image camera. The continuum emission from the spark also engages in the Echelle spectrometer, so the detector must have gated operation. To satisfy those requirements, such as two dimensional, sensitive and gated operation, the cost of CCD detectors for the Echelle spectrometer is still significantly high.

3. Sample type and their application

This section illustrates various application examples that have performed from the authors' research group. As we have mentioned in the previous section, the LIBS technique needs individual verification for an application, because it does not need sampling. Three typical applications are explained in this section.

a.   Paint and coating identification

Materials and techniques of paints and coatings require an appropriate verification process to achieve the desired property of the protective finishing. The organic coating involves a multi-step process in which the quality of the metal finish required for an industrial product would determine the number and the type of steps in a given process[31]. These multi-step coating processes include the selection and composition verification of substrates, surface cleaning, surface pre-treatment, primer, topcoat, and the application of paint curing methods. The paint formulation is a mixture of multi-ingredients, composing: resins, solvents, pigments, fillers, corrosion inhibitors, and other rheological additives. The organic coating in metal finishing practice is extremely complex. The complexities of paint compositions, paint types, and painting processes make their chemical analysis very difficult. In spite of some elemental analysis methods that have been well established for the general purpose in chemistry, the determination of metallic components in paint has been relied on the indirect analytical methods. For example, the metallic zinc dust in the Zn-rich epoxy primer was determined by differential scanning calorimetry (DSC)[32]. The DSC method measured the apparent heat of fusion of the paint sample, and compared this measured value to the standard value of pure zinc as an indirect measurement of zinc composition in paint. Infrared absorption spectroscopy is useful sometimes for the composition analysis if the paint ingredients contain any specific functional groups which are spectroscopically active[33], such as the isocyanate group in the urethane. The direct analysis of these functional groups may be possible only if the paint sample is uncured, and contained a relatively simple composition. In practice, there is no direct way for identifying a cured paint film. Once the paint, e.g. epoxy or urethane, is applied and fully cured, no more epoxy or isocyano functional group would remain in the paint film. Even though, the researchers have attempted to characterize the fully cured paint products by identifying the hydroxyl or amino groups, and use them for differentiating the epoxy paint or urethane paint. The results are generally inconclusive because the majority of other cured paints also have those functional groups as reaction products. The LIBS technique described in this section shows the capability of coating identification at the specimen surface. [34]

3.1. Materials of paints and coatings

The substrates selected for spectral fingerprinting by LIBS technique are: (i) aluminum alloys (2024-T3, 3003, 7075-T6 from Advanced Coating Technologies, Inc. (ACT), Hillsdale, MI) and pure aluminum foil (Aldrich Fine Chemicals), and (ii) cold-rolled steel (CRS from Q-PANEL, Cleveland, OH and Caterpillar's OEM facility) and pure iron (Aldrich). The surface pretreated substrates used for LIBS studies are: (i) Al 2024-T3/Clad, a ultra thin layer of pure aluminum is treated on 2024-T3 aluminum alloy, (ii) Al 2024-T3 Bare/Alodine 1200, the surface of 2024-T3 aluminum alloy is treated with Alodine 1200 solution which contains chromates (i.e., hexavalent chromium), (iii) phosphated (Bonderite 1000, or B-1000) and phosphated/chromated (B-1000/P-60) from ACT, and (iv) galvanized (electroplated and hot-dipped) and galvalume steel plates that have a treated surface layer of Zn and Zn/Al, respectively. Eleven heavy-machine OEM paint samples (four urethane, three epoxy, and four alkyd) were used for the spectral fingerprinting by LIBS technique. The paints were applied on 2 x 4 inch steel panels using a spray coating method and cured thermally or by air-dry as directed by the paint manufacturer.

3.2. LIBS: An *in situ* and quasi-nondestructive analytical technique

LIBS technique is capable of carrying out a depth profile analysis of successive surface layers by controlling and calibrating the working parameters of LIBS system. A Q-switched Nd-YAG laser (Continuum, Minilite II) operating at a wavelength of 1064 nm was employed as the excitation source. The pulse laser has a power of 50 mJ per pulse and a pulse width of 3 ns. The laser beam was focused onto the sample with a 5 cm focal length lens. A fiber optic cable collected the breakdown plasma emissions at the sample surface and directed them to a portable, miniature, CCD array fixed-grating spectrometer. Figure 11 shows the optical microscope images of some LIBS-measured sites on a painted steel panel: (a) paint surface before analysis, (b) one laser pulse applied, (c) two laser pulses applied, and (d) five laser pulses applied. The scale bars are 50 m in length. The first shot of the focused laser beam (70 mJ/pulse) made a burn pattern on the paint surface (Figure. 11b). The successive laser pulses penetrated into the coating layers and eventually reached the metal substrate (Figure. 11d). In principle, the LIBS spectrum recorded after each laser pulse, or for each layer of the multilayer paint samples, should generate the characteristic breakdown spectral peaks of the corresponding chemical compositions. The affected coating area by the laser pulse is limited to less than 100 μm in diameter (Figure. 11b, 11c and 11d). The layer thickness of materials that each laser pulse can penetrate is a function of laser fluent at the focal point, optical geometry and material type. It is important to mention that a well-established elemental analysis method, such as EDX can also perform a similar analysis. However, the sample used in EDX analysis must be cut into a few millimeter sizes for fitting inside the detection stage in a vacuum chamber. Also, the cut samples need to be covered with a conductive coating for EDX analysis because paints are the dielectric materials. This film deposition of conductive layer is again done under another vacuum facility. These complicated sampling processes are eliminated in the LIBS analysis.

3.3. LIBS characterization of substrates

A less trivial experiment was performed to determine whether the LIBS system could be used to distinguish between different alloys of the same main metal content or between the same metal alloys obtained from different manufacturing sources. Aluminum has many alloys in common use, and these alloys frequently need specific protective coatings for aerospace applications. The 2024-T3 Al alloy contains copper as the main dopant (i.e., 4.4% Cu, 0.6% Mn, and 1.5% Mg). The 7075-T6 Al alloy contains zinc as the main dopant (i.e., 5.6% Zn, 1.6% Cu, 2.5% Mg, and 0.23% Cr). The 3003 Al alloy contains no specific main dopant (0.0-0.6% Si, 0.0-0.7% Fe, 0.05-0.20% Cu, and 0.0-0.10% Zn). The Al alloys, 2024-T3 and 7075-T6 have high surface protection strength, whereas Al alloy 3003 displays a good pitting corrosion resistance. All three alloys should show aluminum peaks in LIBS spectra, and 7075-T6 Al should display zinc and magnesium peaks and 2024-T Al should display

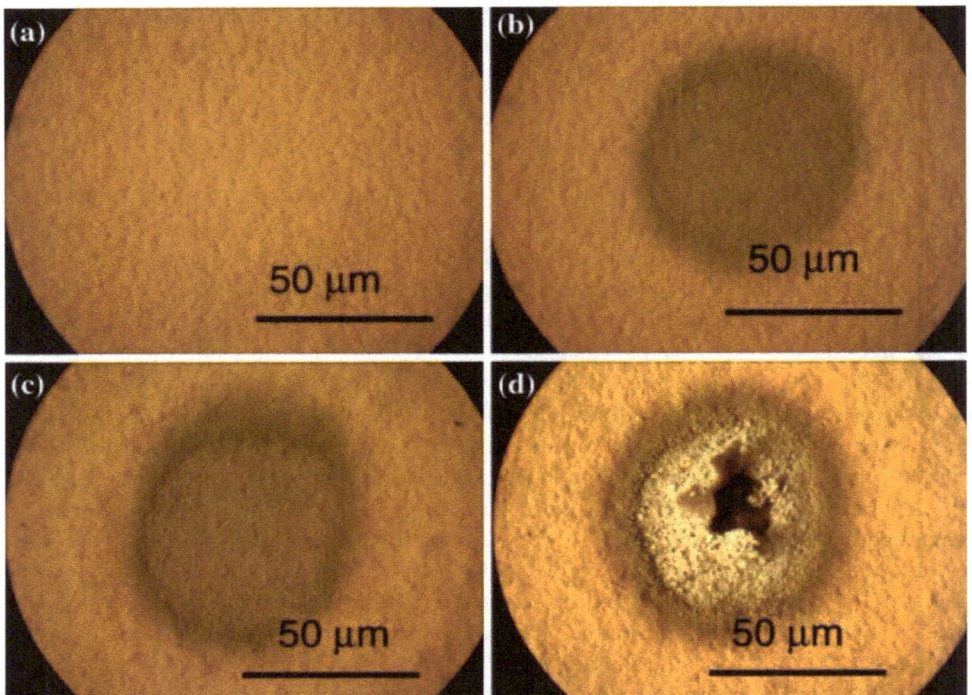

Figure 11. Microscopic images of laser burn patterns on paint film in LIBS experiment after (a) zero, (b) one, (c) two, and (d) five laser pulses

copper and manganese peaks in their breakdown spectra. Figure 12 compares the LIBS spectra recording from 250 nm to 450 nm for pure aluminum foil (spectrum a) and three Al alloys (spectra b, c, and d). As expected, spectrum 13a gives only aluminum peaks at 281.6 nm, 306.3 nm, 308.2/309.3 nm, 358.0 nm and 394.4/396.1 nm.

The spectrum of aluminum alloy shows, in addition to the aluminum peaks, three spectral peaks at 328.2 nm, 330.2 nm, and 334.8 nm are due to zinc (I) ionic states. The 7075-T6 Al alloy gives also the LIBS peaks at 278.6 nm, 285.2 nm and 383.5 nm for Mg and at 325.0 nm, 327.7 nm and 423.0 nm for Cu emission. The LIBS technique is not only able to identify the chemical compositions of alloys, but also capable of differentiating the possible contaminants in those alloys. For example, the contamination of Mn has been detected in Al 7075-T6 sample as illustrated in spectrum 13d. The contaminants of Mn and Mg are observed in spectrum 13b for Al 3003 sample.

The qualitative LIBS spectral assignment is also carried out for pure iron strip, cold-rolled steel, and industrial steel coupons used in the Caterpillar's OEM facility (referred to as CAT machine steel). The bare cold-rolled steel (CRS, SAE 1010) has a composition of 0.08-0.13 % C, 0.3-0.6% Mn, 0.04% P(max), and 0.05% S(max). The LIBS spectra recorded from 250 nm to

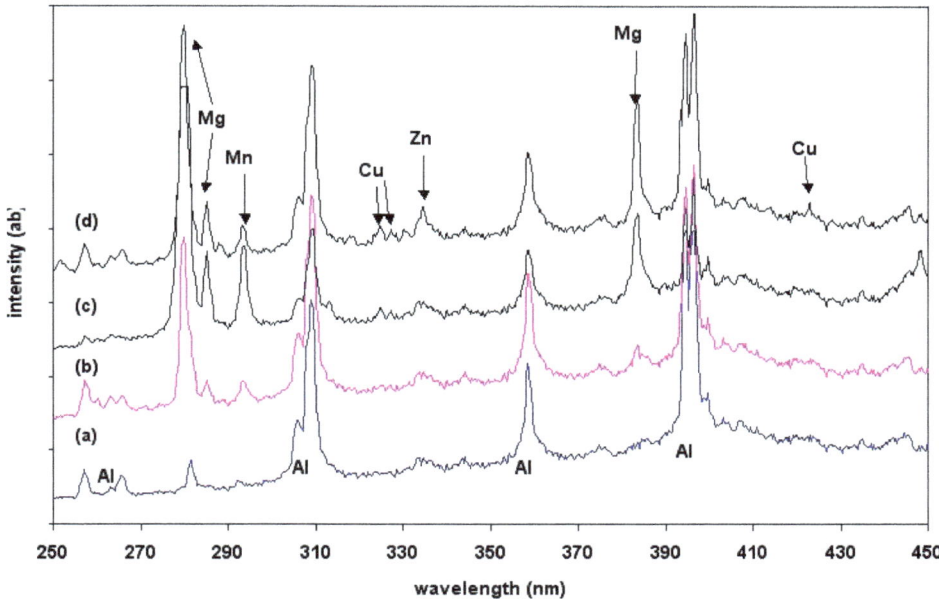

Figure 12. LIBS spectra of aluminum alloys, (a) pure Al foil, (b) 3003 alloy, (c) 2024-T3 alloy, and (d) 7075-T6 alloy

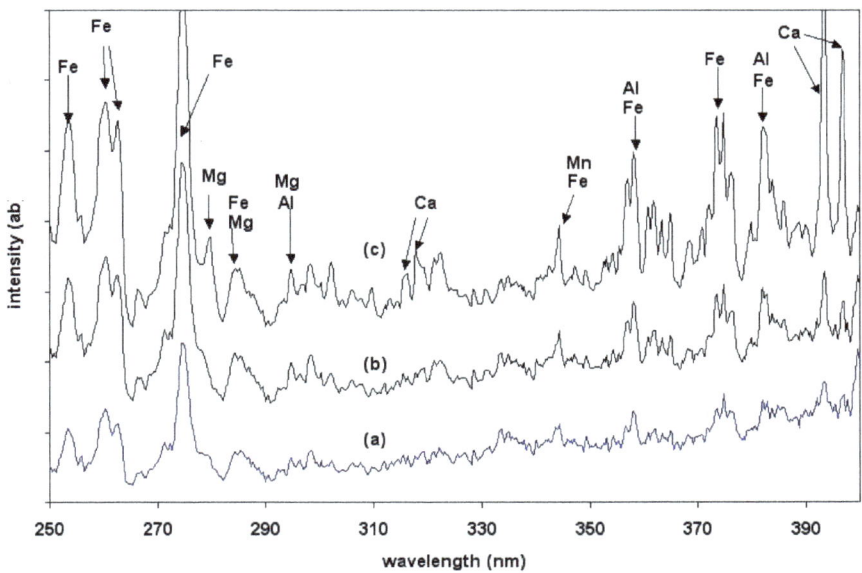

Figure 13. LIBS spectra of steel panels, (a) pure iron piece, (b) cold-rolled steel from Q-PANEL, and (c) CAT machine steel

400 nm are shown in Figure 13: (a) pure iron strip, (b) cold-rolled steel, and (c) CAT machine steel. Spectrum 13a shows LIBS peaks for the pure iron piece at 259.9 nm, 262.6 nm, 275.0 nm, 358.1 nm, 373.4 nm and 373.7 nm. The laser breakdown emission for CRS as shown in spectrum 13b is almost identical to that of the pure iron strip, except an additional peak at 344.3 nm which may be assigned to Mn as incorporated in the cold-rolled steel. The spectrum 13c indicates that CAT machine steel is not a pure iron piece or a standard CRS sample, but rather is a surface pretreated CRS. The surface layer of CAT machine steel contains Ca, Mg, Al, Mn, and P (at 589.1 nm), in addition to Fe. The results indicate that CAT machine steel is an iron phosphate treated CRS, containing a substantial quantity of Ca and Mg, and some small amount of Al, and Mn in the phosphating bath.

3.4. LIBS characterization of surface pretreatment layer on substrates

Another important part of this research is to establish the effectiveness of LIBS spectral fingerprinting technique for characterizing the composition of any metal surface pretreatment that may have been applied on the substrates. The common metal surface pretreatment used on aluminum alloys today is a chromium-based pretreatment (such as Alodine 1200 or Alodine 1000), which usually contains the chromates (i.e., the compounds contain hexavalent chromium). There are different processes used for surface pretreatment on aluminum alloys; some processes cause a color change of the metal surface to a yellowish color, and some cause no color change at all. In the latter case, it is almost impossible to tell, visually, whether the metal alloy has been pretreated. In this work, the different panels analyzed by LIBS are aluminum alloys of 2024-T3 bare, and 2024-T3 Clad (Clad: a thin layer of pure aluminum on 2024-T3 substrate). The surface pretreatment layer on 2024-T3 bare panel is Alodine 1200. The main active ingredient of Alodine solution is potassium dichromate or strontium chromate. Upon the deposition of a thin layer of pure aluminum on Al 2024-T3 bare, the LIBS spectrum of Al 2024-T3/Clad should give only the pure aluminum peaks which are the same as spectrum 12a for pure aluminum foil.

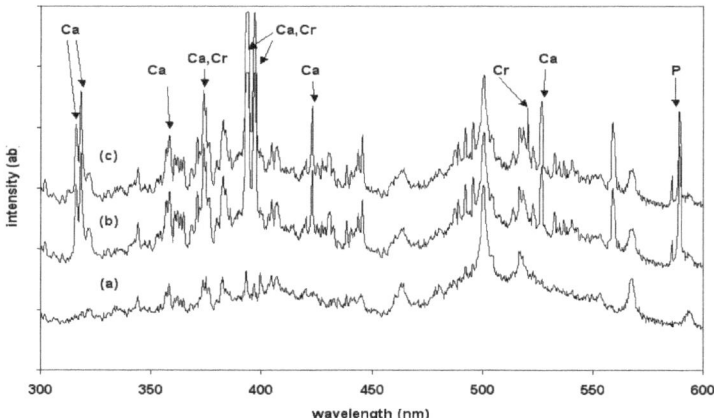

Figure 14. LIBS spectra of surface pretreated layers on CRS substrate, (a) untreated, (b) B-1000, and (c) B-1000/P60 panel

The surface pretreatment of metal prior to the application of a coating or adhesive is a conventional industrial practice to improve the coating adhesion and inhibit substrate corrosion. For cold-rolled steel, the phosphate conversion coating (e.g., Bonderite® B-1000) and phosphating/chromating (using parcolene 60) pretreatment (e.g., B-1000/P60) are commonly used. The LIBS technique is used to fingerprint the differences in chemical compositions of the surface pretreated layer. The LIBS spectra were taken at the first laser shot spot on (a) untreated CRS panel, (b) B-1000 CRS panel, and (c) B-1000/P60 CRS panel. The laser-induced breakdown spectra of untreated and different chemically treated CRS panels are clearly identifiable and their spectral assignments are marked in Figure 14. The LIBS peaks in spectrum 13a are assigned to Fe and Mn, and are similar to those in spectrum 14b. The phosphate treated B-1000 panel gives a few additional LIBS peaks in spectrum 14b, such as P at 589.1 nm and Ca at 315.8 nm, 318.2 nm, 358.3 nm, 394.0 nm, 423.0 nm, and 527.1 nm. In spectrum 16c, the additional P60 treatment on B-1000 CRS is evident by the appearance of chromium peaks at 373.9 nm, 396.8 nm, and 527.1 nm. When these LIBS spectra are compiled in the software system as a standard library file, they can be used to determine if the surface pretreatment processes (including composition, uniformity, and thickness) have been done in according to the products specification.

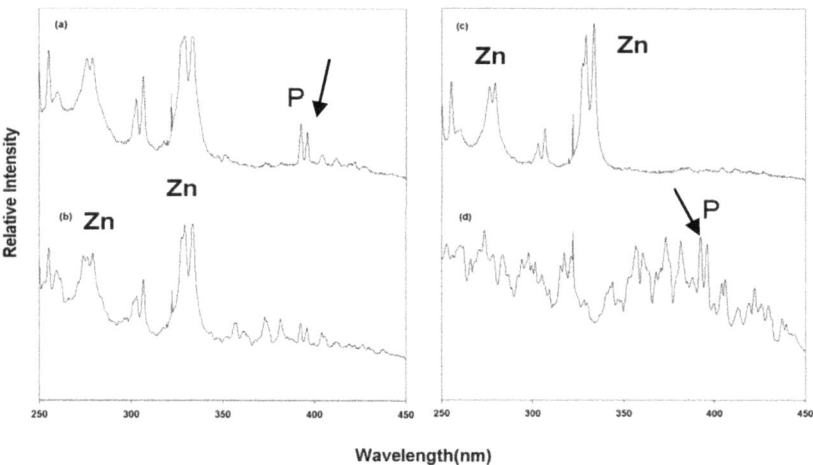

Figure 15. LIBS spectra of EZG panel, (a) from ACT Laboratories, Inc., (b) from China Steel Corp., Taiwan, (c) pure Zn metal piece, and (d) B-1000 CRS panel

Zinc-coated steel (such as Zn/B-1000) is known to inhibit iron corrosion, similar to the effect of zinc anodes. The addition of aluminum to zinc is highly beneficial in improving its corrosion resistance and has resulted in the development of coatings with aluminum contents between 5 and 55% (i.e., "galvalume" zinc-coated steel). Zinc coatings may be applied to steel panel by hot dipping (i.e., hot dipped galvanized steel, HDG) and electroplating (i.e., electrogalvanized steel, EZG). Due to the high degree of variations in the processing of EZG, HDG, and galvalume, it is critically important to have a versatile materials characterization technique, such as LIBS CoatID, to verify the manufacturing conditions of zinc-coated steel at the different factory sites. For a simple illustration, we use LIBS system to test two EZG panels (ACT Laboratories, Inc. vs. China Steel Corp., Taiwan), two HDG panels (ACT vs. Valspar Corp.), and two galvalume panels (Valspar Corp. vs. China Steel Corp.). Figure 15 compares the breakdown emission spectra (recorded from 250 nm to 450 nm) for (a) EZG panel from ACT, (b) EZG panel from China Steel Corp., (c) pure Zn metal piece, and (d) B-1000 CRS panel from ACT. The LIBS spectra of EZG panels (spectra 15a and 15b) should resemble those of the combined spectra of pure Zn (spectrum 15c) and B-1000 CRS (spectrum 15d), depending on the thickness of both phosphate layer on bare CRS and Zn-galvanized layer on B-1000 CRS panel. By comparing the EZG panels processed at ACT Laboratories, Inc. (spectrum 15a) and that processed at China Steel Corp. (spectrum 15b), it shows that both EZG panels have been subjected to the electrogalvanizing process as stated in their products data sheet. However, the LIBS was able to distinguish a thinner Zn-galvanized layer in the Taiwanese sample, because the steel plate was not covered fully by the Zn-layer and thus the B-1000 steel peaks are still quite visible as shown from 350 nm to 450 nm in spectrum 15b. On the other hand, both Zn-galvanized layer and B-1000 phosphate layer in ACT sample are thicker than those in the Taiwanese sample, as indicated by the appearance of a strong P emission doublet and also several intense Zn peaks. In ACT sample, the thicker Zn and phosphate layers give a higher coverage on the steel panel, and thus almost no steel peak is observed in Figure. 15a.

3.5. LIBS identification of paints and coating ingredients

Eleven paints from Caterpillar's OEM coating facility were selected for the identification test by LIBS technique, and listed in Table 1.It is noted that all paint samples have the same color (i.e., Caterpillar yellow) with only slightly different tint, the differences are hardly distinguishable with naked eyes. Samples 1 to 4 are two-pack urethane paints, 5 to 7 are two-pack epoxy paints, and 8 to 11 are one-pack alkyd paints. The paint samples 1 and 5-9 are primers, whereas those of 2-4 and 10-11 are topcoats. The processing methods used in coating applications, such as drying and thermal curing conditions are specified in the remark column of Table 1. The paint systems used in Caterpillar's OEM facility were specifically formulated by the paint manufacturers that have been successfully tested and verified for the required protection of heavy duty machines. Once the paint formulations were established, the manufacturer would strictly maintain the composition of paint ingredients in an effort to achieve a good quality control. This is the reason that LIBS technique may be effectively used for fingerprinting a specific brand of paint.

Figure. 16 displays the LIBS spectra for the eleven paint samples listed in Table 1. The topcoat paints (samples 2-4 and 10-11) display a relatively simpler LIBS spectrum than that of the primer paints (samples 1 and 5-9). In all spectra, the LIBS peaks grouped around 250 nm may be attributed to iron oxide as a dispersed pigment. The peaks originated from calcium at 393.3 nm and 396.8 nm are predominantly shown in the primer type paints. Calcium carbonate has been used at high levels for certain paints because of their low oil absorption. Calcium compound imparts some film structure to the wet paint by improving the stability to sedimentation of other heavier pigments in paint. It is not surprising that primer paint for CRS coating contains a rich calcium ingredient. The primer paints, samples 1, 6, 8, and 9, are shown to contain not only calcium carbonate but also magnesium silicate, as their corresponding LIBS peaks displayed at 279.8 nm and 333.5 nm.

In the previous section, the peak picking algorithm has been successfully used for characterizing substrates and surface pretreatment layers which contain only a few elements and have the well-characterized LIBS peaks. Since paint formulation contains a rather complex mixture of multi-ingredients, thus the decisions for paint identification could best be made by peak correlation algorithm. Any spectral pairs of identical samples must show a 100% correlation value. Due to the possible fluctuation in laser power density, the inhomogeneity of paint film compositions, and the variation in thickness of a paint film, the LIBS spectra for both testing and reference samples were measured at ten (10) different spots for each painted panel. A statistical average spectrum was made to achieve the reproducibility for the identification of a paint sample. The correlation values of identical samples show a 96-99% of reproducibility. On the other hand, the correlation values between two different types of paints, such as urethane and epoxy, show to be around 86.8 ± 0.7%. These correlation values give a clear discrimination between types of paints. Based on the correlation values, we can say that the test sample has a good match to the reference sample, if the correlation values are greater than 95%. We estimate from the use of peaks correlation algorithm, the LIBS system is capable of correlating the test paint samples to the standard paint films to give a 90-95% of perfect match. The remaining 5-10% near match or no match may due to the complex nature of paints and coatings, including the possible surface contaminations. In this case, a careful spectroscopic analysis is further required to achieve the proper paint sample identifications.

Sample No.	Resin	Type of paint	Remark
1	Urethane	Primer	2-part system, cured at 66 °C
2	Urethane	Top coat	2-part system, low temperature curing
3	Urethane	Top coat	2-part system, high temperature curing
4	Urethane	Top coat	2-part system, cured at 66 °C
5	Epoxy	Primer	Low temperature curing (54 °C)
6	Epoxy	Primer	Medium temperature curing (66 °C)
7	Epoxy	Primer	High temperature curing (82 °C)
8	Alkyd	Primer	Air-dry system
9	Alkyd	Primer	Baking system
10	Alkyd	Top coat	Air-dry system
11	Alkyd	Top coat	Baking system

Table 1. The sample paints obtained from Caterpillar Inc.

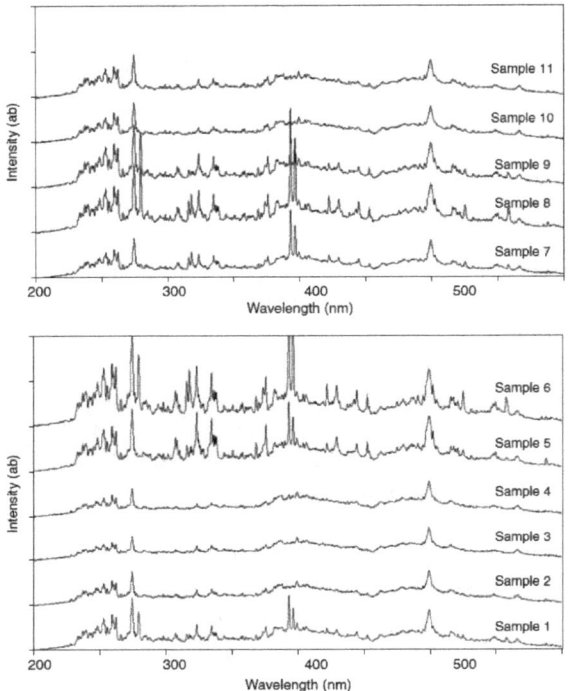

Figure 16. LIBS spectra for the eleven industrial paint samples obtained from Caterpillar Inc.

b.   Organic and bio material screening

Biomaterial application has two areas depending on the analytical goal. The first goal is the analysis of metallic component in the biomaterial. The conventional elements like Na, K, Ca, Mg are included in plant, wood, grain, tissue and bio-remains. Their analysis is similar to other solid samples except those samples include high level of carbon compound. The second application of biomaterial is characterization of biomaterial itself. Breakdown spectrum from LIBS can have information of specific sample group. One of the researches has been made for classification of bacterial strains by major components analysis with LIBS[35]. A pulsed Nd:YAG laser (Continum, Powerlite8000, 10-ns pulse width) was focused on the sample solution by using a 20-cm-focal-length lens. The frequency-doubled laser output at 532 nm was used for plasma generation. The laser power used on the bacteria analysis was 50 mJ/pulse. A light collecting optical fiber was placed near the sample surface to detect the plasma emission which was sent to a spectrometer (Acton research, 1200 grooves/mm grating). The spectra were captured with a photodiode-array detector (OMA IV, EG&G, 1024 array) with a spectral resolution of 0.061 nm. The available spectral range is limited to about 50 nm from the full OMA coverage of 76 nm because of the shadow of optical components in the monochromator. The OMA output was processed and stored by using a personal computer.

Several bacterial strains have been classified depending on their major components analyzed by laser-induced breakdown spectroscopy (LIBS). The bacteria studied were *Bacillus megaterium, B. Subtillis, B. Thuringiensis, and Escherichia Coli*. Each strain was streaked on the cultivating plate and grown to prepare the colonies of vegetative or spore forms. The major inorganic components of the bacteria samples, including Ca, Mn, K, Na, carbon, and phosphorus, were clearly identified from the LIBS data. The vegetative forms of bacteria, beginning step of bacteria life, represent the similar quantities of analytical components between bacteria. After the bacteria have used up the available food supply the bacillus enter into their non-vegetative spore form. The bacteria spores accumulate a lot of calcium on the spore shell which showed strong emission of 393.7 nm and 396.9 nm in the LIBS spectrum. The diverse emission from phosphate at 588.1 nm and 588.7 nm provides a fingerprint of the bacteria. The relative change of inclusions of bacteria was clearly distinguished on the 2-dimensional chart of the bacterial components. This work demonstrates the potential of this method for the rapid and precise classification of bacteria with minimum sample preparation. The quick process of LIBS expected to be used in the real-time analysis of intentionally cultured bulk bacteria in the industrial or weaponized microorganism.

3.6. Preparation of microorganism

Five types of bacterial samples were prepared from the biology lab. Te laboratory stock stains used were *Bacillus Megaterium* QM B1551 (seven indigenous plasmid), *Bacillus Megaterium* PV361 (QM B1551 with all plasmid removed), *Bacillus Subtillis* 168M, *Bacillus Thuringiensis* T34, and *Escherichia Coli* carrying pHT315. QM and PV of *B. Meg* are closely related on their genetic origin and *B. thu*. is a divergence of *B.sub*. Only *E. Coli* is a gram-negative genus among them with antibiotic and enzyme resistive cell wall, and *E. Coli* does not make dormant spore on the contrary to other *bacillus*. All bacterial genera used are biosafety level 1 (non pathogenic). Each stain was streaked on *Luria-Bertani* (LB) plates (10.0 g tryptone, 5.0 g yeast extract and 5.0 g of NaCl, 15% agar, in 1 L double distilled H_2O) and grown overnight for both tests. LB plates were then spread with 0.1 mL of culture and grown for 24 hours for confluent plate test. The same set of vegetative bacteria were kept more than 5 days at room temperature to be spore forms after consuming nutrient and drying .

The series of bacteria cultured on the plastic dish are measured on the LIBS system without any pretreatment. The bacteria colonies on the top of the culturing medium (LB) are grown to roughly a 0.5 mm thickness in the wet condition. The areas of the colonies are wide enough for manual mounting on the sample stage and focusing into less than 50 micron diameter of breakdown diameter. The LIBS spectra from bacteria and culturing medium in the UV and visible spectral range have higher background level compare to solid metallic samples such as aluminum, copper and steel. The lack of light absorption on the sample requires a more intense laser for the breakdown. The threshold intensity of the laser pulse for stable and sufficient breakdown was 40 mJ. This value is bigger than 10-20 mJ of the solid LIBS application because the solid samples are not transparent and thus absorb more of the light.

3.7. LIBS spectra of bacterial strains

Although the entire spectrum, from the UV to visible range was initially scanned it was experimentally determined that the three ranges mentioned contained most of the peaks of possible interest. The presence of certain elements was investigated in the micro-organisms such as chlorine, sulfur, phosphorous, calcium, sodium, and potassium. Also of interest were the possible trace elements such as zinc, magnesium, manganese, cadmium, nickel, cobalt, and strontium. Listed in the table are peaks that could be candidates for these possible elements. Because of the atomic elemental nature of the LIBS we expect to match the major or minor peaks to the above mentioned elements certain series of peaks. The peaks were matched to the possible elements in microorganism using the library available in the NIST data base In our preliminary studies we were unable to match several trace elements such as Mg, Cd, Ni, Co, and Sr. The complexity of the iron emission can make it difficult to distinguish it from other elemental peaks of interest. Despite the different outward appearance and life cycle of the bacterial species they all shared a similar elemental composition. Strong peaks were found at 252.8 nm, 279.7 nm, 393.7 nm, 396.9 nm, 398.3 nm, 578.9 nm, 588.1 nm, and 588.7 nm. Using a spectrometer with 1200 gr/mm grating at 50 nm blocks of the spectrum four of the major peaks of interest were identified. The spectra shown in Figure 17 are emissions from bacteria around 400 nm. The peaks at 393.7 nm and 396.9 nm attributed to the calcium atomic transition $4s^2S_{0,1/2} - 4p\ ^2P_{1,1/2}$, and $4s^2S_{0,1/2} - 4p\ ^2P_{0,1/2}$, the strongest emissions from calcium used in many other atomic spectroscopy. These calcium emission were verified by using $CaCl_2$, $CaPO_4$ as spectral references.

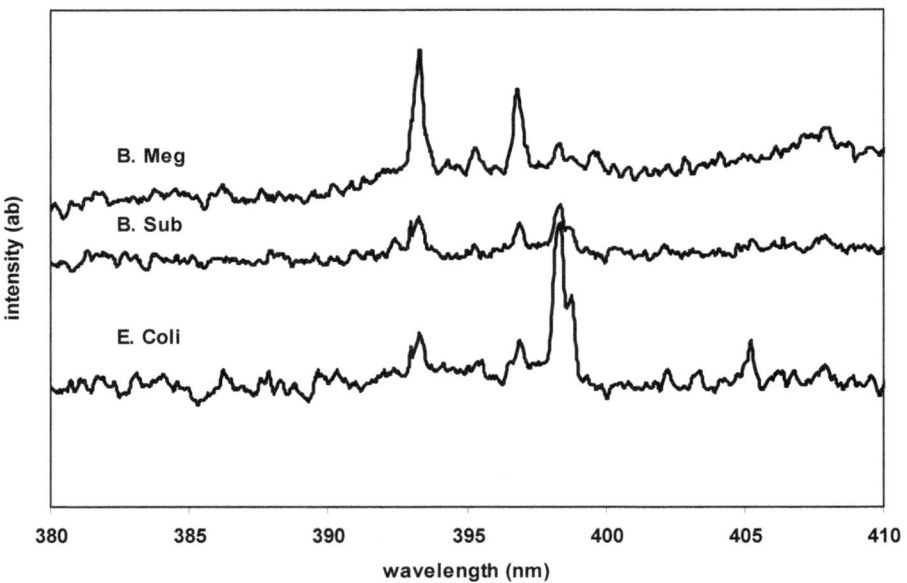

Figure 17. Spectrum after cell death and sporulation of the bacillus has occurred.

At the spore stage of the bacteria, after remaining for 5 days at room temperature to consume all of the nutrients, most of the surface water evaporated. It is known that the water content of spores is only about 10-30% of the water content of vegetative cells (active bacteria) to survive spores at levels of dehydration that would kill vegetative cells. The low water content also provides the spore with chemical resistance (to chemicals such as hydrogen peroxide) and it causes the remaining enzymes of the spore cell to become inactive. This inactivity makes the immunological detection hard to improve sufficient sensitivity. One chemical produced by spores that is thought to lend to their high resistance is dipicolinic acid. Dipicolinic acid interacts with calcium ions to form calcium dipicolinate, which is the main substance believed to lend spores their resistance and represents about 10% of the dry weight of a spore. The intensity of calcium is strong on the spore sample of B. Meg, B. Sub., and B. Thu. colonies. E. Coli colonies have low calcium content and the composition does not change after aging. This is a proper result that because E. coli does not make spores. The spore cell also contains special spore proteins. These protective wall structures are highly resistant to heat desiccation, chemical disinfection and radiation. These functions are to protect DNA from harsh environments, but also disturb measurement chemical property on the conventional spectroscopy. Another component of spores that contributes to their resistance to chemical agents is the strong spore coat, which is composed of highly cross-linked keratin. Laser breakdown is strong enough to break the protective shell of the spores and take out inner component of the cell. The peak at 398.3 nm is overlapped to Mn emission from library data. The strong intensity in this wavelength could not be assigned as Mn because of relative intensity of Mn in other wavelengths. On the other hand, the peak on this wavelength is observed on the samples of organic compounds such as LB of culture medium, cellulose, and many organic polymers. This result leads to assigning this peak as carbon compound fragment.

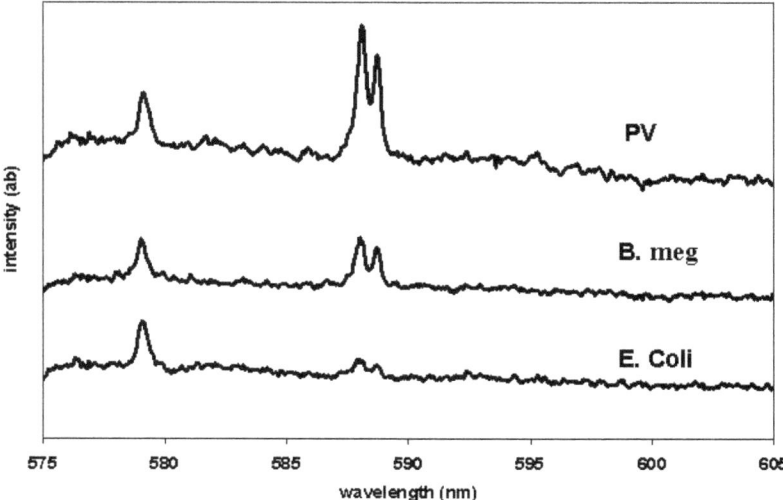

Figure 18. Spectrum to show the amount of phosphate compound in three different bacteria samples.

Figure 18 shows LIBS spectra of 3 bacteria colonies. The doublet peaks appeared at 588 nm is due to phosphate functional compound in the bacteria. In spite of phosphorus elemental emission library doesn't have significant emission on this wavelength, we assign this peak as phosphate because of strong peak observed from some other phosphate compound examination. *PV* shows strongest peak at the phosphate emission. *E. coli* shows weaker intensity. The other bacteria have intermediate peak height, like *B. meg.* in the middle. The amount of phosphate seems to be related to the strain of the bacteria not to the life progress of the microorganism. The phosphate intensities are always weak at the vegetative step and then increase at the spore. The increment of phosphate is biggest for *PV* bacteria. The peak at 578.9 nm is not assigned properly for a certain component of our culturing system. Every bacterial sample and culturing medium shows similar intense peak at this wavelength. With the result of LIBS experiments for several organic compounds, this component is identified as an organic functional group. The proper identification is still on the research.

Figure 19 shows distribution of bacteria on the spread chart of intensity ratio. The X axis is the intensity ratio to represent calcium amount on the bacteria samples. *B. meg* and its plasmid treated *PV* bacteria strains are at high amount of calcium. *B. sub* and *B. thur* are at a relatively lower than *B. meg* strains. *E. coli* does not store calcium whether vegetate or aged colonies. The outstanding feature of phosphate intensity is on the difference of *B. Meg* and *PV*. These two strains are genetically same organism but only *PV* have been removed their plasmid from *B. Meg* type bacteria. The modification of *PV* leads more storage of phosphate in the spore cells.

To identify the selective intake of certain element from culturing medium, we examine the components of the culture medium. The calcium in the medium was lower level than LIBS detectable. This result shows the sporation of bacteria absolutely needs calcium and they collect calcium very efficiently.

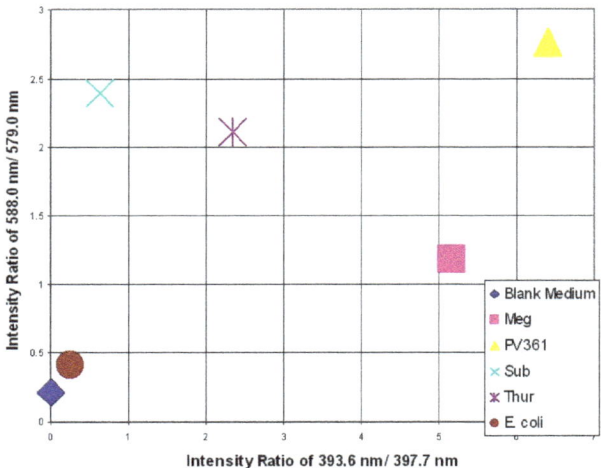

Figure 19. Intensity ratio of two selected wavelength range plotted against each other. The values for X axis are made the intensity of the samples at 393.6 nm divide by their intensity at 398.24 nm. The values for Y axis are made by intensity of the samples at 58

c.   Application of LIBS for metallic component in aqueous solution

As mentioned at the LIBS property, solid samples are most convenient and strong LIBS signal. Liquidor gas samples need more specific optical arrangement to generate breakdown and emitting light collection. There are several ideas to overcome the sampling difficulty of gas and liquid samples. One of the publications here is a typical sample type conversion from liquid sample to solid with concentrating effect. Ion-exchange resins are conventional substances used to capture metal ions and hold them in the solid resin matrix. Chemically activated microporous membranes functionalized with polycarboxylic acid are typically employed[36]. The matrix encapsulation technique has been applied to collect trace metals from water, and converting metal ions to a solid form. The pre-concentration of analyte from a liquid sample into the ion exchange membrane was extensively studied for LIBS measurement by Schmidt and Goode[37]. The captured Cu on the small area of the ion exchange membrane has shown that the pre-concentration can provide a large volume of liquid filtration. Many of the test elements gave results in the sub mg/L range of detection limits by liquid filtration method. Total chromium elements were captured in the membrane and measured by LIBS[38] in the range of ng/mL detection limit, where Cr(VI) was chemically converted to Cr(III). A fast analysis technique with automated LIBS analyzer is configured for monitoring of metal ions in water[39]. The ion exchange membrane is used to develop Copper solutions were used to establish pre-concentration parameters. The chelating resin based filter membrane was used to capture Cu ions in water. A series of standard solutions were filtered through the ion exchange membrane by using a vacuum suction system to reduce the filtration time. The LIBS signals of copper absorbed on the layers of the membrane were investigated to determine parameters for practical analysis.

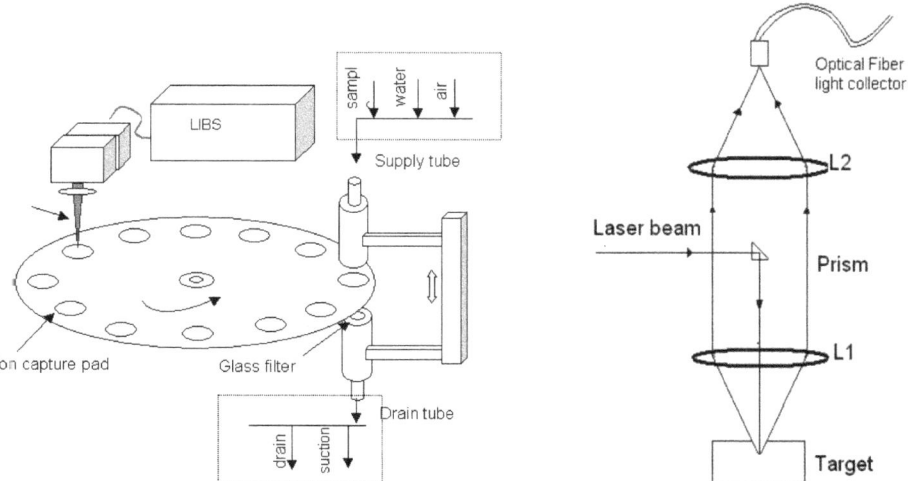

Figure 20. Experimental set up for ion-capture filtering and LIBS (top) and detail of collection optics (bottom).

3.8. Solution samples and instrumentation

The stock solutions were prepared using copper salts as the source of the copper ions from, cupric nitrate ($Cu(NO_3)_2 \cdot 3H_2O$, Fisher), cupric chloride and cupric sulfate, at a 500 mg/L concentration, which were then diluted to make a series of test solutions with well-defined concentrations. A copper sheet (99.9% Cu) was obtained. A commercially available extraction membrane was obtained from 3M filtration Products (St. Paul, MN). The original purpose of using the ion-exchange membrane was to extract multivalent metal ions for environmental analysis by chelating in the PTFE matrix. The ionic selectivity of the membrane is known to follow roughly along the EDTA complex formation constants. The membrane was affixed between the edges of two Teflon tubes with a 12 mm inner diameter. The upper tube was connected to the supply manifold valve for switching sample solutions, flushing water and drying air. The lower tube was depressurized for filtering and suction through a drain reservoir. The drainage tube held a round glass filter to support the membrane during solution filtering.

For this LIBS system, an ion-exchange membrane concentrator was assembled as shown in the Figure 20 for use in the experiment. A fiber optic cable collected the breakdown emissions through collection optics (L1 and L2) and directed them to a fixed-grating spectrometer. The spectrometer has a 3600-sensor array and covers a spectral range of 250 nm – 800 nm. The samples on the motorized stage were moved 0.5 mm to 1 mm stepwise to collect an averaged spectrum from a membrane sampler.

3.9. Operation conditions for membrane concentrator

A 10 mg/L copper solution prepared from $Cu(NO_3)_2$ was used to test the filtering conditions of the membrane concentrator. The LIBS spectra were obtained in the measurable intensity range. The inset in Figure 21 shows the major copper peak that was used for this test. The metal capturing function of the ion exchange membrane is known as that of the EDTA chelating process, therefore the chelating speed must be fast enough to capture all the copper ions during filtration. The estimated filtering time through a 0.2 mm thickness of membrane filter is 3 min when 20 mL of solution is passed through a 12 mm diameter filter at the vacuum suction pressure of 20 kPa. The amount of copper captured on the membrane filter should be proportional to the observed LIBS intensity. As shown in Figure 21, the average intensity for the LIBS peak at 324.75 nm is almost constant throughout the changes of suction pressure which control the filtering speed of test solution. The experimental results show that there is a large uncertainty in capturing Cu on membrane filter (a large error bar in Figure 21) when the sample solutions are filtered very slowly with low suction pressure. If the filtration process took more than 30 minutes at a pressure of less than 10 kPa, the error in measurement increased. It was also found that soaking for extended periods of time in test solution would lead to wrinkles on the membrane filter, due to membrane swelling. On a wrinkled membrane, the liquid filtration path is biased in certain areas of the filter paper resulting in uneven dispersion of captured ions on the membrane surface. On the other hand, when the test solution was filtered at a very high speed, i.e., the suction pressure is more than 80 kPa, the ions just pass through the membrane. A 20 mL solution takes only 10 sec to filter and ions begin to pass through without being captured which leads to a weaker intensity on the LIBS spectra. Based on this membrane operation test, a suction pressure of 30 kPa was maintained as the standard condition, so 20 mL of solution can be completely filtered within 2 minutes.

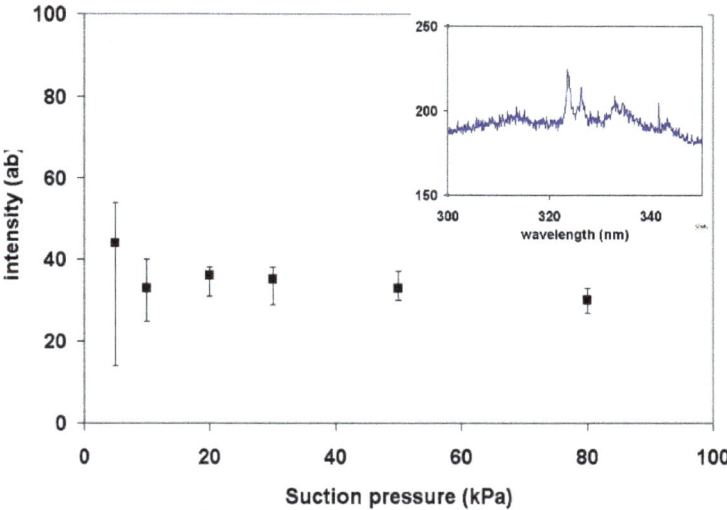

Figure 21. LIBS intensity of Cu at 324.7 nm sampled by filtering at different suction pressure. The inset shows the Cu emission that was concentrated from a 10 mg/L of Cu solution.

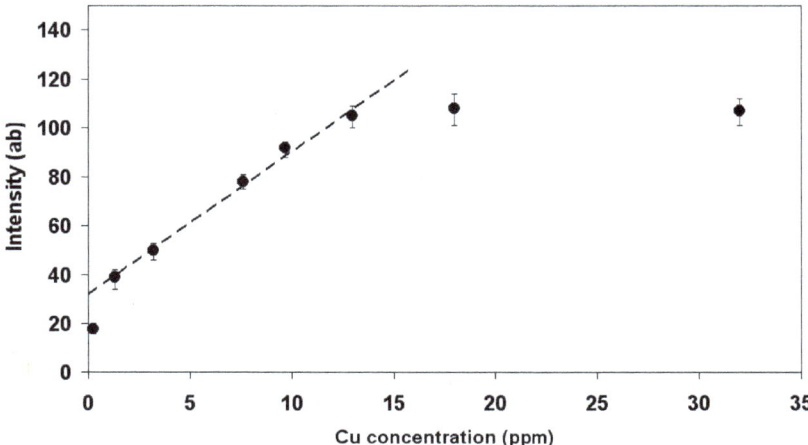

Figure 22. LIBS intensity measurements for low concentration samples. Each of the data points were obtained by averaging five individual measurements on a single filter surface at five different spots. The line was made using the data points from 0.5 mg/L to 15 mg

3.10. LIBS intensity from the top membrane surface

Figure 22 shows LIBS intensity of low concentration samples (a few mg/L Cu solutions) for the peaks at 324.75 nm. All data points were obtained by averaging five individual measurements at five different spots on a single membrane filter. Since the sample surface required for a single laser breakdown shot is less than 0.1mm in diameter, multiple measurements and their average values are easily obtained over the surface of the membrane filter. The LIBS intensity of copper displays a linear correlation over concentration ranges below 15 mg/L. The line in Figure 22 was made by using the data points from 0.5 mg/L to 15 mg/L of Cu solutions with the exclusion of higher concentration data. The results give a relatively too narrow dynamic range for general analytical use. However, the correlation of the line is R²=0.9926 and it can be an acceptable analytical calibration concentration. The reason for the extremely limited dynamic range is investigated further in the next section.

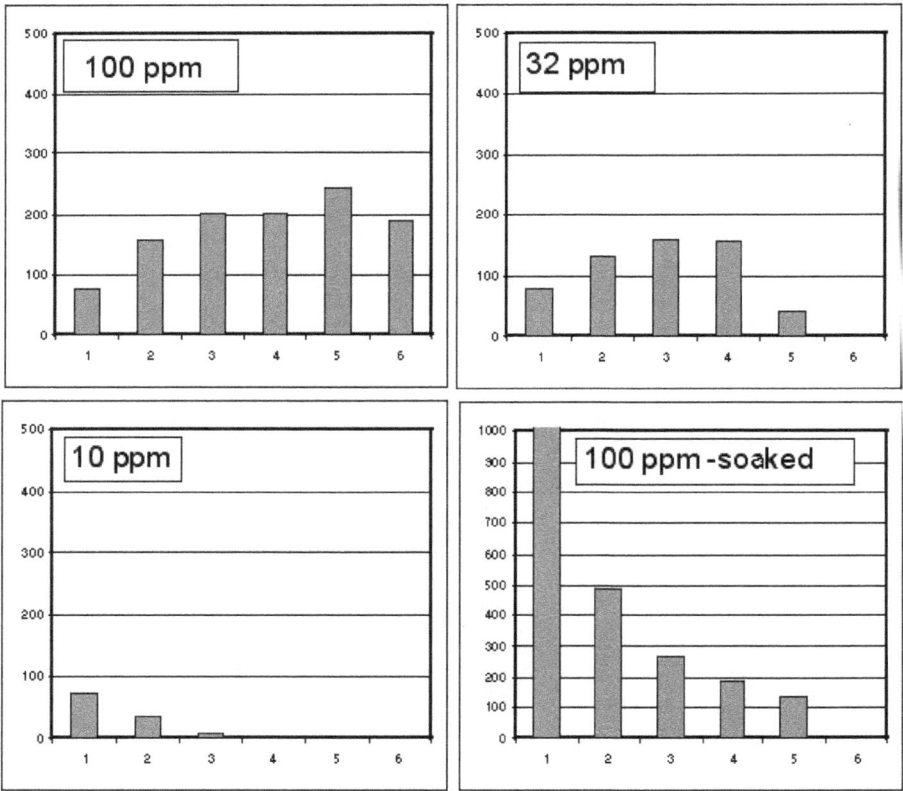

Figure 23. Intensity change by consecutive laser shots. The Cu concentration in the sample solutions for the membrane filtration is 100 mg/L, 32 mg/L and 10 mg/L. The membrane filter soaked for 24 hours in 100 mg/L solution (lower right) shows intensity greater

3.11. Depth effects on ion-capture membrane

In Figure 22, when the concentration of Cu solution is higher than 20 mg/L, the LIBS intensity of Cu at 324.75 nm seems to lose its proportionality relationship to the concentration and remain constantly extended to 35 mg/L. Initially, the retention capability of the membrane was suspected but this was shown not to be the case because the filtrate (drain) from high concentrations did not contain copper ions. To verify the effects of depth profiling on ion-capture membrane, further investigation was done using multiple laser shots at the same spot of membrane. Figure 23 shows the LIBS intensity change by consecutive laser shots at a single membrane point. The intensity of the first laser shots from the filtered samples of 100 mg/L, 32 mg/L and 10 mg/L solutions are unexpectedly similar, and then the next laser shots show intensity change relative to the solution concentration. The 100 mg/L sample makes strong LIBS intensities until the 10th laser shot (top-left graph in Figure 23, after the 6th shot is not shown). On the contrary, the LIBS intensity seems to disappear at the 3rd shot from 10 mg/L sample. A LIBS intensity comparison after the 10th pulse was not possible because the laser pulse had already penetrated through the membrane and a hole was generated. This observation can be explained by the thickness of membrane filter as resulted from the effects of depth profiling on the ion-capture membrane. The Cu ion in the sample solution is drawn inward on the ion-capture membrane during filtration and captured at a certain depth. It is clear that our laser power, 50 mJ/pulse, can ablate the Cu-membrane layer by layer. A well prepared calibration curve from other literature[37], using stronger laser power, also showed that the calibration began to taper off at around 10 mg/L (similar to Figure 22). The authors suggested a linearly regressed calibration curve, simply, for the entire concentration range. The effects of depth profiling on the ion-capture membrane is also proven by the sampling of passive extraction. For the passive extraction, the membrane filter was soaked for 24 hours in 100 mg/L solution and the LIBS intensity changes of Cu at 324.75 nm by the consecutive laser shots are shown in the graph of lower-right in Figure 23. The most intense spectrum was obtained from the first laser shot. None of the ions were drawn physically into membrane during the passive extraction, so they were mostly captured on the surface and gave the strongest LIBS intensity at the first laser shot. It is clear that the membrane captures ions at the deeper layer if the solution is drawn in by suction. As a result, the total ions through the entire thickness should be counted to get the proportional values to determine concentration. A modified calibration curved is made from the integration of 10 laser-shot intensities as in Figure 24 using a 2nd order equation.

3.12. Analysis of tap water using ion-capture membrane

Tap water was analyzed by using the ion-capture membrane concentrator and LIBS. The tap water to our lab at Northern Illinois University is supplied throughout the building by copper pipe. There were many studies which showed the copper contamination in the water supply from the pluming. The local government which supplies the tap water to this lab declared that the source of the city water is collected from the active public water supply wells. Inorganic contaminants, such as salt and metals, can be naturally occurring or

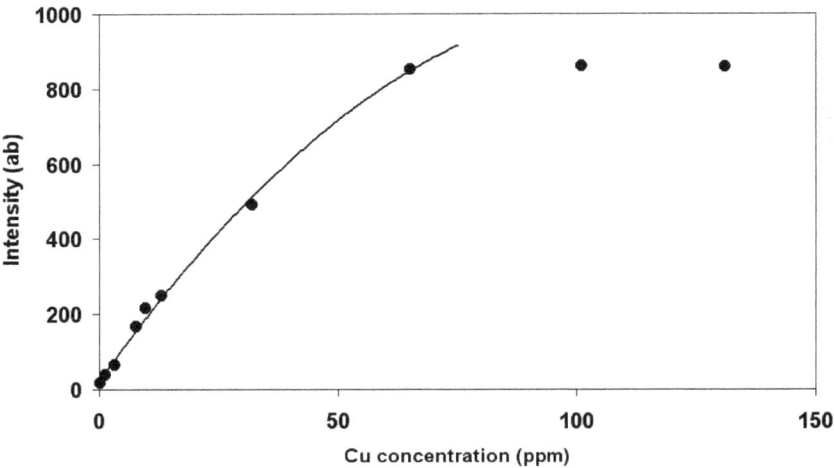

Figure 24. Modified calibration curve made from the integrated intensity of 10 laser shots and Cu concentration up to 70 mg/L. The obtained curve is y = -0.079x2 + 17.81x +23.46

resulted from urban storm water runoff, industrial, or domestic waste water discharges. The ions of Ca, Mg, Na, K, in the city water are known to be of reasonable concentrations within the EPA regulations. The annual water report from city shows that the concentration of Cu is 1.3 mg/L, and this amount is expected to maintain reasonably throughout the day and week. To make an illustration on the effects of pluming for copper contamination from the copper pipe, we took water samples during a specific time schedule, early Monday morning and Tuesday evening. Each sample contains 20 mL of water and filters through an ion-capture membrane. Figure 25 shows the LIBS spectrum of tab water sample captured by the membrane. Copper emissions were identified along with strong peaks of Ca. The concentrations are obtained from the integration of the same number of breakdown shots by using the calibration curve as in Figure 24. In case the integrated intensity is greater than 300 counts, which is around the saturation range, we simply dilute the original sample and measure again. The Tuesday sample shows Cu concentration in the range of 5- 10 mg/L, which is expected due to the large use of water during the active days in the building and the species in the water should be similar to the source water. However, the Monday morning samples show higher concentrations mostly around 100 mg/L or more. Some water samples show up to 270 mg/L of copper, especially for hot water pipe line. We can infer for the high concentration of copper on the Monday sample that water remained in the pipes during the weekend and did not move. This result shows that the contamination of copper from the pluming of building is significant and is shown to depend on the retention time of water in the building pipe.

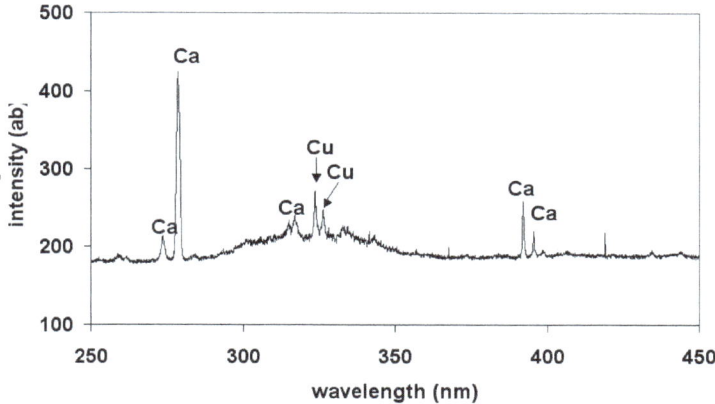

Figure 25. LIBS spectrum of ion captured membrane. The sample solutions are 20 mL of tap water filtered through membrane filter.

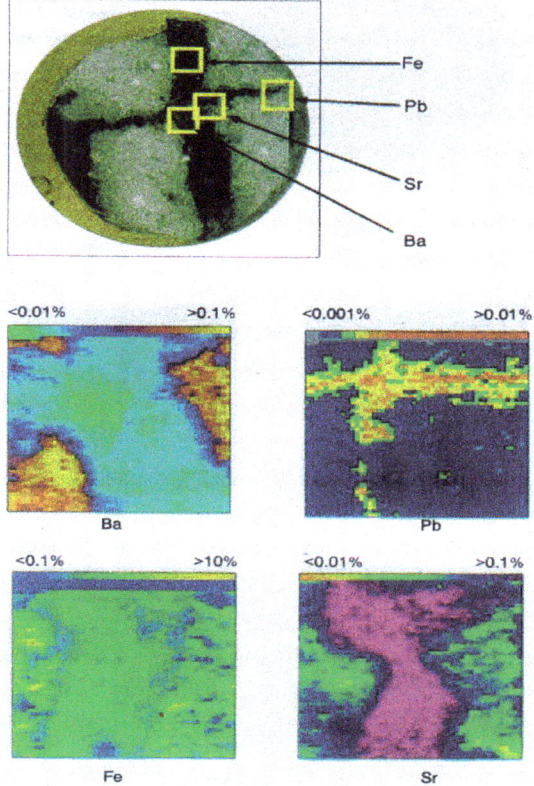

Figure 26. The elemental distribution of patterns for Ba, Pb, Sr, and Fe in a polished granite rock section.

d.   Ceramic and geology sample application

Laser induced plasma spectroscopy has been applied to the analysis of element distribution mapping of polished rock sections[40]. The plasma was generated by focusing a frequency-doubled second harmonic 532 nm Nd:YAG laser on the target under atmospheric conditions. The experimental parameters, such as laser energy, atomic emission line and time profile of the plasma spectrum, were characterized to obtain optimum experimental conditions and estimate the element composition of the target surface. For the element mapping of samples, an X-Y stage was used to move the sample and an element image of 50 x 50 mm could be made in 30 min. Using this technique, the element concentration distribution of Ba, Cu, Fe, Mn, Pb, Si, and Sr in polished rock sections were obtained. Quantitative analysis was achieved by analyzing standard rock samples. Calibrated concentration versus plasma intensity was used for the color grading for the mapping of element concentration distribution. The elemental mapping analysis for a granite sample is illustrated in Figure 26. The ore vein within the existing sample was selected to identify the different compositions of ore and surface element distribution. The element distribution differences were represented by color grading, where the upper first line represents the color scale. The region where a lode was crossed during the analysis is rich in Pb and Sr but the Ba content is low. Iron does not show differences and is nearly uniformly distributed across the sample.

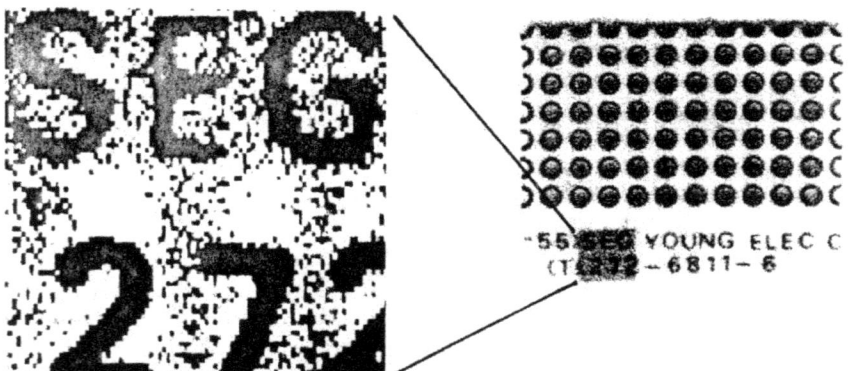

Figure 27. The mapping image of a commercial printed circuit board. Black circle and character on the right circuit board is copper layer for soldering electronic component. The measured values from LIBS constitute pixels on left map.

A sensitive optical technique for compositional mapping of solid surface using LIBS was described[41]. A pulsed Nd:YAG laser with second harmonic module was focused on the solid surface, giving a small ablation area, to produce plasma emission. Copper and magnesium emissions from a standard sample were carefully analyzed and assigned in the wavelength range 500-520 nm. The assigned spectral information was selected to construct an image of 100 x 100 pixels by mapping the measured emission intensity values from the analyzed points. The time required for image construction and image sharpness depends on he number of laser shots per point of analysis and the number of analyzed points per image. A clear image of a copper conductor pattern from a printed circuit board was generated. In addition, some copper contaminations around the conductor area are clearly visible in the scanning LIBS map. The contaminated copper salt probably resulted from the incomplete washing step during manufacturing that could cause a short circuit in an electronic device. A commercial printed circuit board is shown in the right portion of Figure 27. Circles on the top are patterned copper layer for soldering the electronic components. Characters on the bottom are the same copper layer for product identification. The laser-ablated area (gray shaded square) is 5 x 5 mm². The left portion of Figure 27 is a mapping image of copper corresponding to the gray shaded square. Each 100x100 pixels corresponds to a measured emission intensity value of LIBS on the ablated point.

4. Conclusion

The goal of LIBS development is to extend the analytical feasibility of LIBS for detecting organic, inorganic metals and ceramic material for the various applications. The sample types also not limited to the solid and expanded to liquid gas, aerosol, powder, bacteria, and industrial products. Several applications of LIBS were illustrated in this chapter. Detections for metallic components are usually accomplished easily from the measured spectral ranges from ultraviolet (230 nm) to visible (700 nm) of the plasma emission. Only a few seconds of measuring time is a great advantage and will be useful for screening and monitoring system for industry and security monitoring. Evaluation of analytical feasibility for detecting and identifying the sample should be decided by analyzing the LIBS spectra of specific components as well as matrix derived from the source of the samples.

Author details

Taesam Kim and Chhiu-Tsu Lin
Northern Illinois University, Illinois, USA

5. References

[1] Jurado-Lopez, A., Lque de Castro M.D. "Laser-Induced Breakdown Spectrometry in the Jewelry Industry. Part I. Determination of the Layer Thickness and Composition of Gold-Plated Pieces" *J. Anal.At.Spectrom.* 2002, 17, 544-547.

[2] Jurado-Lopez, A., Lque de Castro M.D. "Chemometric Approach to Laser-Induced Breakdown Analysis of Gold Alloys" *Appl. Spectrosc.* 2003, 57, 349-352.

[3] Sturm V., Peter. L., Noll, R. "Steel Analysis with Laser-Induced Breakdown Spectrometry in the Vacuum Ultraviolet" *Appl. Spectrosc.* 2000, 54, 1275-1278.

[4] Palanco S., Laserna, J. J. "Full automation of a laser-induced breakdown spectrometer for quality assessment in the steel industry with sample handling, surface preparation and quantitative analysis capabilities" *J. Anal. At.Spectrom.* 2000, 15, 1321-1327.

[5] Sattmann, R., Sturm, V., Noll, R. " Laser-induced breakdown spectroscopy of steel samples using multiple Q-switch Nd:YAG laser pulses" *J. Appl. Phys.*, 1995, 28, 2181-2187.

[6] Kraushaar, M., Noll, R., Schmitz, H. U." Slag Analysis with Laser-Induced Breakdown Spectrometry" *Appl. Spectrosc.*, 2003, 57, 1282-1287.

[7] Mateo, M. P., Cabalin, L. M., Laserna, J. J. "Automated Line-Focused Laser Ablation for Mapping of Inclusions in Stainless Steel" *Appl. Spectrosc.*, 2003, 57, 1461-1467.

[8] Thiem, T. L., Salter, R. H., Gardner, J. A., Lee, Y. I., Sneddon, J. "Quantitative Simultaneous Elemental Determinations in Alloys Using Laser-Induced Breakdown Spectroscopy (LIBS) in an Ultra-High Vacuum" *Appl. Spectrosc.*, 1994, 48, 58-64.

[9] Valdillo, J. M., Grcia, C. C., Palanco, S., Lasena, J. J. "Nanomertic range depth-resolved analysis of coated steels using laser-induced breakdown spectrometry with a 308 nm collimatied beam" *J. Anal. At.Spectrom.* 1998, 13, 793-797.

[10] Burgio, L., Clark, R. J., Stratoudaki, T, Doulgeridis, M., Anglos, D. "Pigment Identification in Painted Artworks: A Dual Analysis Approach Employing Laser-Induced Breakdown Spectroscopy and Raman Microscopy" *Appl. Spectrosc.* 2000, 54, 463-469.

[11] Anglos, D., Couris, S., Fotakis, C. "Laser Diagnostics of Painted Artworks: Laser-Induced Breakdown Spectroscopy in Pigment Identification" *Appl. Spectrosc.* 1997, 51, 1025-1030.

[12] Garcia C. C., Corral, M., Vadillo, J. M., Laserna, J. J. "Angle Resolved Laser-Induced Breakdown Spectrometry for Depth Profiling of Coated Materials" *Appl. Spectrosc.* 2000, 54, 1027-1031.

[13] Marquardt, B. J., Goode, S. R., Angel, S. M. "In Situ Determination of Lead in Paint by Laser-Induced Breakdown Spectroscopy Using a Fiber-Optic Probe" *Anal. Chem.*, 1996, 68, 977-981.

[14] Häkkänen, H. J., Korppi-Tommola, J. E. I. "UV-Laser Plasma Study of Elemental Distributions of Paper Coatings" *Appl. Spectrosc.*, 1995, 49, 1721-1728.

[15] Hidalgo, M., Martin, F., Lasema, J. J. " Laser-Induced Breakdown Spectrometry of Titanium Dioxide Antireflection Coatings in Photovoltaic Cells" *Anal. Chem.*, 1996, 68, 1095-1100.

[16] Moskal, T. M., Hahn, D. W. " On-Line Sorting of Wood Treated with Chromated Copper Arsenate Using Laser-Induced Breakdown Spectroscopy" *Appl. Spectrosc.*, 2002, 56, 1337-1344.

[17] Sattmann, R., Moüch, I., Krause, H., Noll, R., Souris, S., Hatziapostolou, A., Mavromanolakis, A., Fotakis, C., Larrauri, E., Miguel, R." Laser-Induced Breakdown Spectroscopy for Polymer Identification "*Appl. Spectrosc.*, 1998, 52, 456-461.

[18] Dixon, P. B., Hahn, D. W. " Feasibility of Detection and Identification of Individual Bioaerosols Using Laser-Induced Breakdown Spectroscopy" *Anal. Chem.*, 2005, 77, 631-638.

[19] Morel, S., Leone, N., Adam, P., Amouroux, J. " LIBS Applications - Detection of bacteria by Time-Resolved Laser-Induced Breakdown Spectroscopy" *Appl. Opt.*, 2003, 42, 6184-6191.

[20] Samuels, A. C., Delucia, F. C., McNesby, K. L., Miziolek, A. W. " LIBS Applications - Laser-induced breakdown spectroscopy of bacterial spores, molds, pollens, and protein: Initial studies of discrimination potential" *Appl. Opt.*, 2003, 42, 6205-6209.

[21] Knight, A. K., Scherbarth, N. L., Cremers, D. A., Ferris, M. J. " Characterization of Laser-Induced Breakdown Spectroscopy (LIBS) for Application to Space Exploration" *Appl. Spectrosc.*, 2000, 54, 331-340.

[22] Yueh, F., Singh, J. P., Zhang, H. "Laser induced Breakdown Spectroscopy, elemental analysis" in encyclopedia of Analytical Chemistry, John Wily & sons, 2000, pp2066 - 2087

[23] Sneddon, J., Thiem, T.L., Lee, Y. "Lasers in analytical Atomic Spectroscopy" 1996, VCH publish

[24] Miziolek, A. W., Palleschi, A., Schechter, I. "Laser induced Breakdown Spectroscopy"2006, Cambridge

[25] H.J. Hakkanen, J.E.I. Korppi-tommola, UV-Laser Plasma study of elemental distribution of paper coating. Appl. Spectrosc, v49,(12) 1995,p1721

[26] M. Hidalgo, F. Martin, J. J. Lasema, Laser induced Breakdown spectrometry of titanium dioxide antireflection coating sin photovoltaic cells. Anal. Chem. 1996, 68(7) pp1095-1100

[27] Demetriosanglos, StelionCouris, Costas Fotakis, Laser diagnostics of painted artworks: Laser-Induces breakdown spectoscopy in pigment identification, Appl. Spectrosco, 81(7),1997 1025-1030

[28] Klaus Loebe, Arnold Uhl, and HartmutLucht, Micro analysis of tool steel and glass with laser-induced breakdown spectroscopy, applie optics 2003, 42(30) 6166-6173

[29] Valery Bulatov, rivieKrasiker, and Israel Schechter, "Study of Matrix effect in laser plasma spectroscopy by combined multifiber spatial and temporal resolutions. Anal chem. 70(24) 1998, pp5302-11

[30] R Gabriele Cristoforetti, Stefano Legnaioli, Vincenzo Palleschi, Azenio Salvetti, Elisabetta Tognoni, Pier Alberto Benedetti, Franco Brioschi and Fabio Ferrario, Quantitative analysis of aluminium alloys by low-energy, high-repetitionrate laser-induced breakdown spectroscopy,J. Anal. At. Spectro. 2006,21 607-702

[31] Prefetti, B. M. *Metal Surface Characteristics Affecting Organic Coatings*, Federation Series on Coating Technology, FSCT, Blue Bell, PA, 1994.

[32] Weldon, D. G., Carl, B. M. "Determination of Metallic Zinc Content of Inorganic and Organic Zinc-Rich Primers by Differential Scanning Calorimetry" *J. coatings technology* 1997, 69, 45 – 49.

[33] Stamenkovic, J., Cakic, S., Konstantinovic, S., Stoilkovic, S. "Catalysis of the Isocyanate-Hydroxyl Reaction by Non-Tin Catalysts in Water borne Two Components Polyurethane Coatings", *Working and living environmental protection* 2004, 2, 243-250.

[34] Kim T, Nguyen B, Minassian V, Lin C.T Paints and coatings monitored by laser-induced breakdown spectroscopy. J. Coat.Technol. Res. 2007 4:242-255.

[35] Kim T, Specht Z, Vary P, Lin C. T. Spectral fingerprint of bacterial strains by laser-induced breakdown spectroscopy. J. Phys. Chem. 2004 108(17):5477-5482.

[36] Hestekin, J.; Sikdar, S.; Bhattachayya, D.; Bachas, L.; Cullen, L. *US Patent* 6139742, 2000.

[37] Schmidt, N.; Goode, S. *Appl. Spectrosco.*, 2002, 56, 370-374.

[38] Dockery, C.; Pender, J.; Goode, S. *Appl. Spectrosco.*, 2005, 59, pp. 252-257.

[39] Kim T, Ricchia M, Lin C. T. Analysis of copper in an aqueous solution by ion- exchange concentrator and laser-iduced breakdown spectroscopy. J. Chin. Chem. Soc. 2010 57(4B):829-835.

[40] Yoon,Y., Kim, T., Chung, K., Lee, K., and Lee, G., "Application of Laser induced Plasma Spectroscopy to the Analysis of Rock Samples", *Analyst* 1997, 22, pp. 1223-1227

[41] Kim, T., Lin, C., Yoon, Y., "Compositional Mapping by Laser-Induced Breakdown Spectroscopy, *J. Phys. Chem. B*, 1998, 102 ,pp. 4284-4287

X-Ray Photoelectron Spectroscopy for Characterization of Engineered Elastomer Surfaces

Lidia Martínez, Elisa Román and Roman Nevshupa

Additional information is available at the end of the chapter

1. Introduction

Among different types of polymers, elastomers, also called rubbers, are of special interest for many industrial applications. This interest resides in the high yield strength of these materials that makes possible deforming them manifold their original length without permanent residual strain. However, elastomers can suffer from surface deterioration when subject to rubbing, contacting with aggressive media, ultraviolet light and other. Oxidation of elastomers can produce degradation of its chemical, physico-mechanical, rheological and surface properties. In tribological applications, the quality of the elastomer surfaces is also of special concern since significant degradation of mechanical and tribological behaviour is usually associated with small changes in the surface composition and properties [1]. Therefore, studying the mechanisms of surface degradation of elastomers is very important for comprehension of the failure modes of elastomer components and improving their durability.

For improving the performance of material surfaces, different surface modifications have been developed so far. Properties of elastomer surfaces depend, to a large degree, on the chemical constitution of molecules in the surface layer [2]. Therefore, tailoring polymer surfaces has attracted much interest of researchers in polymer chemistry [3]. Polymer surface modification allows obtaining good performance of components at lower costs than using expensive advanced bulk materials [4]. Presently, halogenation, etching, grafting, oxidation, and other surface modification techniques are intensively used. Another alternative is the application of coatings onto the elastomer surface, although the application of coatings on deformable substrates without occurrence of interfacial delamination is not straightforward. Among various coatings, amorphous diamond-like carbon (DLC) is considered by various authors as a good candidate for application on elastomer surfaces [5]. Such coatings have excellent tribological behaviour, i.e., low friction coefficient and wear rate [6].

Ethylene-propylene-diene elastomer (EPDM) is one of the most widely used elastomers in various outdoor and industrial applications, such as waterproof coatings, electrical insulation, pipes, and mounts. In general, it is employed in applications which demand a material with good mechanical properties and with a retained elastic nature [7]. World production of EPDM is estimated to be 41% of all elastomers [8-11]. Also it has a good resistance to degradation at elevated temperature, sunlight, in oxygen and, in particular, ozone [12]. Acrylonitrile-butadiene rubbers (NBR) and hydrogenated acrylonitrile-butadiene rubbers (HNBR) form another widely used family of elastomers. In hydrogenated rubber the double bonds of butadiene ($CH_2=CH-C\equiv N$) are saturated yielding rubber with much higher chemical inertness. These elastomers are extensively employed in automotive industry, especially for lip seals, due to their moderate cost, excellent resistance to oils, fuels and greases, processability and very good resistance to swelling by aliphatic hydrocarbons [13].

Our study is focused on characterization of surface chemical composition of different elastomers subject to rubbing, surface modification and application of coatings. The main technique used for this study was X-ray photoelectron spectroscopy (XPS), which is a very powerful technique for characterizing the chemical composition of very thin (few nm) surface layers. XPS is particularly useful when analysing elastomers, as it provides information about the chemical environment of the elements, i.e. type of bonds, chemical state, etc. Thus, XPS is well suited for investigation the changes in binding energy of chemical elements situated within the first tens of nanometres of the material surface [4]. Elastomers are typically composed of carbon, hydrogen, oxygen and nitrogen. Their surface and bulk properties depend on the way these elements are combined rather than on the presence of other chemical elements. XPS allows detection of new functional groups [4] and evaluation the variation in the amount of existing functional groups, e.g. C-O, as function of surface tailoring, ageing [1], or rubbing [14]. However, often it can be difficult to distinguish between different functional groups having similar binding energies. Therefore, in many cases some complementary techniques should be used to elucidate chemical features of elastomer surfaces. One of these complementary techniques consists in measuring of contact angles (CA) of sessile drops of various liquids placed on the elastomer surface. This very simple method provides valuable information on the types of surface groups [15]. In particular, by using water, presence of polar groups, e.g. C-O, can be determined. So, the degree of surface activation due to surface modification can be determined from measurements of surface hydrophobicity [16]. Then, more information on the surface chemistry and Surface Free Energy (SFE) can be obtained from measuring CA of various liquids with different characteristics. In the following sections we present some fundamental aspects of these techniques and case studies of elastomer surfaces.

2. XPS for characterization of elastomer surfaces

2.1. Introduction to the XPS technique

XPS is an analytical technique that has its fundamental origin in the photoelectric effect, which was first explained by Einstein in 1905 [17]. This effect has become a powerful tool for studying the composition and the electronic structure of the matter [18]. A schematic drawing of typical XPS measurement device is shown in Figure 1a. The measurements are performed in ultrahigh vacuum (UHV) in order to control the surface cleanliness and to reduce the electron scattering on gas molecules. To provide a beam of photons with given characteristics the device is equipped with an X-ray source focused on the sample surface. The photoelectrons emitted from the sample material at characteristic energies are analysed by a suitable electron analyser. The kinetic energy, E_k, at which electrons are emitted follows the fundamental energy conservation equation in photoemission:

$$hv = E_B + E_K + \varphi_a \, , \tag{1}$$

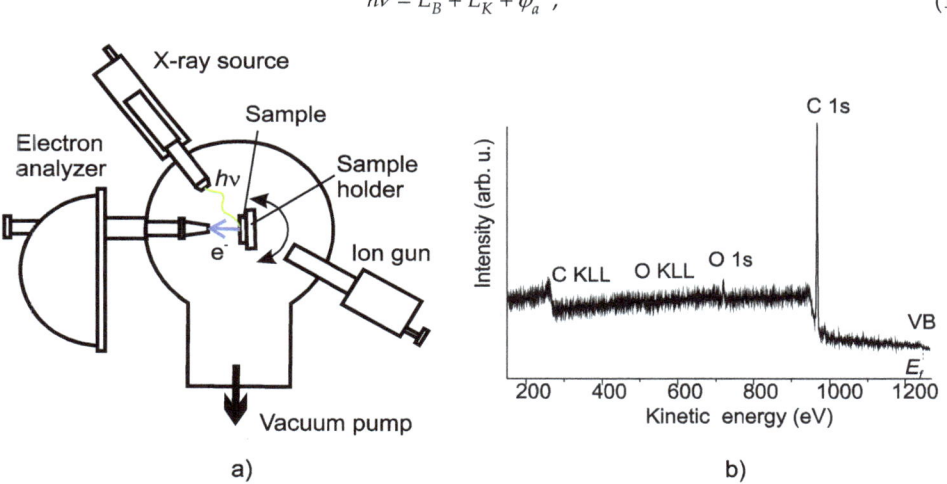

Figure 1. a) Schematic drawing of an experimental XPS system; b) typical photoemission spectra of an elastomer

in which h is Plank constant; v is the photon frequency and the product hv defines the energy of the incident photon; E_B is the binding energy of the electron in the atom. The origin of the binding energies is related with Fermi level, E_f, whereas the kinetic energies are referenced to the vacuum level. The difference between both levels corresponds to work function of the analyser, φ_a. By measuring the electron kinetic energies and knowing the spectrometer work function, it is possible to determine the binding energies of various inner levels (or core electrons), as well as those of the outer (or valence) electrons involved in chemical bonding. A typical photoemission (PE) spectrum, i.e. PE yield vs. kinetic energy of the emitted photoelectrons obtained from a photon-illuminated area, is shown in Figure 1b. The spectrum consists of a series of peaks on a background signal which generally increases at low kinetic energy due to secondary electrons, i.e. photoelectrons that are inelastically scattered in the way out of the sample. In summary, the XPS spectra consist of peaks at discrete kinetic energies corresponding to atomic core levels (CLs) and Auger transitions. Note that each element has a unique elemental spectrum. With the most commonly used excitation sources, the kinetic energy of photoelectrons is typically ranged between 0 and

1400 eV. Since inelastic mean free path of photoelectrons, λ, in solids is small [19], chemical information is obtained from the surface and few subsurface atomic layers. Quantitative information can be derived from the peaks areas, whereas chemical states can often be identified from the exact positions of the peaks and separations between them. The presence of chemical bonding causes binding energy shifts, which can be used to infer the chemical nature (such as atomic oxidation state) from the sample surface. Here, we limit ourselves to study elastomer samples. A complete description of XPS technique can be found in specialized literature [20, 21].

2.2. Advantages and shortcomings of XPS technique for characterization of elastomer surfaces. Operating conditions, measurements and semi-quantitative analysis

The standard XPS measurements are carried out under vacuum conditions by retarding-fields techniques. The most commonly used X-ray sources are Al Kα (1486.6 eV) and Mg Kα (1253.6 eV). The X-ray lines from these sources are narrow (less than 0.9 eV) and provide good energy resolution for many applications. Initially, a survey scan or wide energy range scan, typically from 1000 to 0 eV, should be obtained in order to identify the elements present on the surface. As each element emits electrons at characteristic energies, it is possible to identify all the elements present in the sample surface, except hydrogen and helium which are not detectable by this technique. Elastomers usually contain a small number of elements, of which the most common are C, O, N, F and Cl. Other elements like sulphur and zinc can be detected in small quantities. Sulphur is a typical curing agent, whereas zinc is usually employed as a curing activator [22]. In most of the cases, these elements will not be taken into account as they have no real influence on the surfaces properties. Normally, the elements are uniformly distributed in the bulk; however, under certain circumstances surface segregation may take place.

It should be stressed that XPS is a semi-quantitative technique. In order to quantify the amount of each element the integrated area of a particular peak should be divided by the corresponding relative sensitivity factor. The following is a generalized expression for determination of atom fraction, C_x, of a constituent x in a sample:

$$C_x = \left(I_x / S_x \right) / \left(\sum I_i / S_i \right),\qquad(2)$$

where I_x is the peak area and S_x is the atomic sensitivity factor of the x-th element. The denominator corresponds to the atomic fraction of other elements in the sample. Assuming a homogeneous distribution of elements, a strong line for each element in the spectrum should be analyzed. In case the requirement of homogeneity is not fulfilled, the assumption of homogeneity can be used as a starting point for further calculations. Reference published data on elemental sensitivity factors could be used for determination of S, although the type of instrument and analysis conditions should be considered. With this technique it is also possible to identify chemical states of a given element by measuring the high resolution or core level peaks.

Depth distribution of elements can also be obtained using XPS in destructive or non-destructive modes. In the first one, ion sputtering is used to remove surface layers. Sputtering and XPS can be applied consecutively or simultaneously. In the non-destructive mode, depth profiling is obtained by varying the detection angle of the emitted electrons. In this case the probed depth is limited to 3λ. More detailed information about both methods can be found elsewhere [20].

Another important problem in XPS analysis is related with sample degradation due to X-ray radiation. In fact, this degradation comes from the secondary electrons emitted during the X-ray exposure [23, 24]. In most of the cases this degradation is slow enough as compared with time required for XPS analysis, thus the changes in composition due to X-ray can be neglected. Notwithstanding, this problem should be considered when analysing chemically unstable materials. In our work, no sample degradation due to X-ray radiation has been observed for all studied elastomers.

XPS measurements were performed in ultrahigh vacuum with base pressure of 2×10^{-10} mbar using a Phoibos 100 ESCA/Auger spectrometer with Mg Kα anode (1253.6 eV). To avoid X-ray damage on the samples low X-ray power of 150 W was used. The core level narrow spectra were recorded using pass energy of 15 eV. For the data analysis, the contributions of the Mg K$_{\alpha}$ satellite lines were subtracted and the spectra were subjected to a Shirley background subtraction formalism [25]. The binding energy, E_B, scale was calibrated with respect to the C 1s core level peak at 285 eV. The surface area subject to XPS analysis was around 5.6 mm^2 that is large enough to obtain an average surface chemical composition. When modified samples were analyzed, the surface area subject to XPS analysis was smaller than the treated area, thus the contribution from untreated surfaces was negligible. The shape of C 1s core level peak (HR C1s) measured with high energy resolution was analysed using peak fitting in order to identify functional groups. Depending on the chemical environment of the carbon atoms, important chemical shift of C 1s peak can be observed. Decomposition of the experimental peak in components allows identification of the contribution from each component. For the analysis of HR C1s, the spectrum recorded from the untreated sample was used as a reference. The HR C1s was fitted leaving the full width at half maximum (FWHM) of the C–C/C–H component to vary freely while the other components were forced to adopt the same value. The fit of the treated samples was performed using the same values of FWHM and the binding energies (with uncertainty of ± 0.1 eV) as for untreated elastomer. The only remaining free parameter in the fit procedure was the area of the peaks. By doing so, new carbon species derived from the treatment processes could be identified. An example of the analysis of HR C1s is presented in Figure 2 where some spectra of untreated and modified elastomers are compared. Presence of new carbon species (in this case C-O and C-F bonds) can be identified from the shape of the HR C1s. The contribution from these groups varied depending on the surface treatment. However, identification of chemical groups can be difficult when different species produce similar chemical shifts (see Table 1 for E_B of main carbon bonds identified in the present study). For example, C=C bond was included into the group of C-C/ C-H components since the shift between these two groups is only 0.3 eV [26] that is below the resolution limit of the experimental system used in this study. Detailed analysis of XPS spectra is presented in section 3.3.

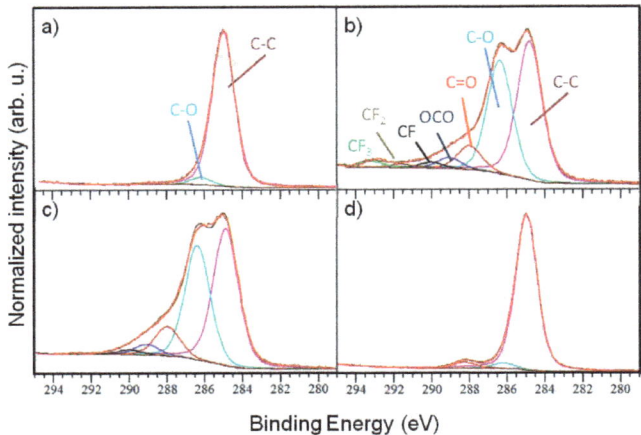

Figure 2. High resolution C1s core level spectra of elastomer samples: a) untreated EPDM, b) EPDM after fluorination with CF4, c) EPDM after fluorination with SF6, and d) HNBR after the same fluorination as c)

Bonds	E_B (±0.1 eV)	Ref.
C-C, C-H, C=C	285	[26, 27]
C-O, -CH-CF$_2$	286.3	[26-30]
C=O, -CH-CF	288.1	[26-30]
O-C=O, -CFH-CF$_2$	289	[26-30]
CF	289.8	[30, 31]
CF$_2$	291.8	[30, 31]
CF$_3$	293.3	[30-32]

Table 1. Components employed for the analysis of the C 1s core levels

2.3. Complementary techniques for the interpretation of the results of XPS

As we mentioned in the introduction section, surface and bulk properties of elastomers depend on the way the main constituents (C, H, O, N, etc.) are combined rather than on the presence of other chemical elements. Therefore, in some cases and depending on the light source employed for XPS analysis, it is difficult to distinguish between the presence of different functional groups, as occurs for HR C1s with C-O and C-N groups. Complementary information on surface chemistry of elastomers can be obtained from spectroscopy of inelastic scattering of light, e.g. Fourier Transformed Infra-Red spectroscopy (FTIR), Raman spectroscopy and others. Measurements of SFE of elastomer using sessile drop method is another very simple but powerful method which can provide valuable information on the type of the surface groups. The method is based on measuring the CA between a droplet of a certain liquid and an elastomer surface under well-controlled conditions. The CA is obtained from a balance of interfacial tensions between three phases: solid (S), liquid (L) and vapour (V) (Figure 3) and is defined from Young-Dupré equation:

$$\gamma_{SV} - \gamma_{SL} - \gamma_{LV} \cos\theta = 0 .$$ (3)

When a droplet contacts a rough surface, the measured or apparent contact angle, may differ from the intrinsic one, i.e. the CA of the same liquid on an ideally smooth surface of the same material. Wenzel [33] proposed to introduce a roughness factor, r, which is the real contact area divided by the geometrical, or projected, area. Homogeneous wetting regime of a liquid on a rough surface is described by:

$$r\cos\theta_a = \cos\theta .$$ (4)

The roughness factor can be determined numerically from 3D surface measurements obtained using appropriate technique, e.g. confocal microscopy, laser scanning profilometry, etc. [1]. Recently there were many criticisms on the Wenzel's approach. In [34] it was demonstrated that CA behaviour is determined by interactions of the liquid and the solid at the three-phase contact line alone and that the interfacial area within the perimeter is irrelevant. They suggested that Wenzel's equation is valid only to the extent that the structure of the contact area reflects the ground-state energies of contact lines and the transition states between them.

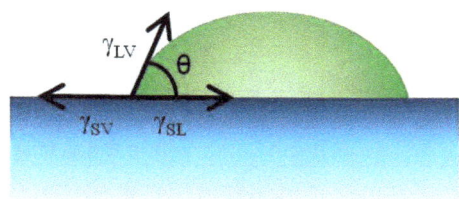

Figure 3. Schematic representation of contact angle

Depending on the specific method, the CA measurement allows determining total SFE, the polar and dispersive components of SFE (Fowke's approach), apolar Lifshitz – Van der Waals (LW) and polar acid - base components (van Oss's approach). According to van Oss's approach, the surface tension could be resolved into components due to dispersion, induction and dipole-dipole forces, and hydrogen bonding [35]. For non-metallic solid surfaces, in addition to apolar LW interactions, electron acceptor – electron donor interactions, or Lewis acid-base (AB) interactions may often occur. In this case the total surface tension is the sum of two components: γ^{LW} and γ^{AB} [36]. Unlike LW interactions, polar interactions are essentially asymmetrical. The polar component of the free energy of interaction between solid and liquid can be expressed as [35]:

$$\Delta F_{SL}^{AB} = -2\left(\sqrt{\gamma_S^+ \gamma_L^-} + \sqrt{\gamma_S^- \gamma_L^+}\right), \tag{5}$$

where γ^+ is the electron acceptor, or Lewis acid component, and γ^- is the electron donor, or Lewis base component of the surface tension. Then, the expanded form of Young-Dupré equation can be obtained by combining (3) and (5):

$$0.5\gamma_L^t\left(1 + \cos\theta\right) = \sqrt{\gamma_S^{LW}\gamma_L^{LW}} + \sqrt{\gamma_S^- \gamma_L^+} + \sqrt{\gamma_S^+ \gamma_L^-}. \tag{6}$$

If for a given liquid the components of surface tension γ_L^t, γ_L^{LW}, γ_L^+, and γ_L^- are known, (6) is a linear function of three unknown parameters corresponding to the components of surface tension of the solid surface $\sqrt{\gamma_S^{LW}}$, $\sqrt{\gamma_S^-}$, and $\sqrt{\gamma_S^+}$. As this equation is underdetermined, the components of surface tension for the solid can be found by measuring CAs using at least three different liquids with known and different components of surface tension. If the values of the components of surface tension for the three liquids are close together, the calculated values for three parameters for the solid will be "unduly sensitive" [36] to small errors in the values of the parameters of surface tension of the liquids, and in the measured CAs. To overcome this problem, CA measurements should be performed with more than three liquids. These will constitute an overdetermined system of linear equations which can be solved by least-square method. In order to reduce the measurement error, each measurement of the CA should be repeated several times. Mean value, $\overline{\theta_i}$, and standard error of mean, seθ_i, should be determined for each liquid from these measurements.

The resulting set of simultaneous equations is the following:

$$0.5\gamma_{Li}^t\left(1 + \cos\overline{\theta_i}\right) = \sqrt{\gamma_S^{LW}\gamma_{Li}^{LW}} + \sqrt{\gamma_S^- \gamma_{Li}^+} + \sqrt{\gamma_S^+ \gamma_{Li}^-}, \tag{7}$$

where subscript i indicates the liquid. It can be written in the matrix form:

$$\mathbf{Y} = \mathbf{Ab}, \tag{8}$$

where $\mathbf{Y}$ is the matrix of independent variable (left side of eq. (7)), $\mathbf{A}$ is the $(n\times3)$ matrix of known coefficients, n is the number of liquids used for CA measurements, and $\mathbf{b}$ is the vector of unknown parameters:

$$\mathbf{Y} = 0.5\left[\left(1 + \cos\overline{\theta_1}\right)\gamma_{L1}^t \cdots \left(1 + \cos\overline{\theta_n}\right)\gamma_{Ln}^t\right]^T, \tag{9}$$

$$\mathbf{A} = \begin{pmatrix} \sqrt{\gamma_{L1}^{LW}} & \sqrt{\gamma_{L1}^+} & \sqrt{\gamma_{L1}^-} \\ \cdot & \cdot & \cdot \\ \sqrt{\gamma_{Ln}^{LW}} & \sqrt{\gamma_{Ln}^+} & \sqrt{\gamma_{Ln}^-} \end{pmatrix}, \tag{10}$$

$$\mathbf{b} = \left[\sqrt{\gamma_S^{LW}} \ \sqrt{\gamma_S^-} \ \sqrt{\gamma_S^+} \right]^T . \tag{11}$$

Then, mean values of the surface tension components can be determined from the matrix equation:

$$\mathbf{b} = \left(\mathbf{A}^T \mathbf{D}^{-1} \mathbf{A} \right)^{-1} \mathbf{A}^T \mathbf{D}^{-1} \mathbf{Y} , \tag{12}$$

where $\mathbf{D} = \begin{pmatrix} (se\theta_1)^2 & 0 & 0 \\ 0 & \ddots & 0 \\ 0 & 0 & (se\theta_n)^2 \end{pmatrix}$

is the covariance matrix of errors of CA measurements.

The standard error of mean of the unknown parameters can be found from the main diagonal of the covariance matrix:

$$\mathbf{K} = \left(\mathbf{A}^T \mathbf{D}^{-1} \mathbf{A} \right)^{-1} . \tag{13}$$

The calculated values of the parameters of the surface tension should be tested for statistical significance using t-test. In case some of the parameters are not statistically significant, it can be zero set and removed from $\mathbf{b}$. Then, the calculation should be repeated using modified matrix $\mathbf{A}$. By doing so, the standard error of the parameters of solid can be reduced.

Matrix method is also very useful for the analysis of surface tension variation in time, e.g. due to ageing. In this case, CA measurements are performed at different periods of time using a set of several liquids as described above. This constitutes a set of simultaneous equations at the selected points of time.

$$0.5\gamma_{Li}^t \left(1 + \cos\overline{\theta}_i(t_j) \right) = \sqrt{\gamma_S^{LW}(t_j)\gamma_{Li}^{LW}} + \sqrt{\gamma_S^-(t_j)\gamma_{Li}^+} + \sqrt{\gamma_S^+(t_j)\gamma_{Li}^-} . \tag{14}$$

Therefore, $\mathbf{Y}$ and $\mathbf{b}$ change to $(n \times p)$ matrixes, where p is the number of time points:

$$\mathbf{Y} = 0.5 \begin{pmatrix} (1+\cos\theta_{11})\gamma_{L1}^t & \cdots & (1+\cos\theta_{1j})\gamma_{L1}^t & \cdots & (1+\cos\theta_{1p})\gamma_{L1}^t \\ \vdots & \vdots & \vdots & \vdots & \vdots \\ (1+\cos\theta_{n1})\gamma_{Ln}^t & \cdots & (1+\cos\theta_{nj})\gamma_{Ln}^t & \cdots & (1+\cos\theta_{np})\gamma_{Ln}^t \end{pmatrix} , \tag{15}$$

$$\mathbf{b} = \begin{pmatrix} \sqrt{\gamma_{S1}^{LW}} & \cdots & \sqrt{\gamma_{Sj}^{LW}} & \cdots & \sqrt{\gamma_{Sp}^{LW}} \\ \sqrt{\gamma_{S1}^-} & \cdots & \sqrt{\gamma_{Sj}^-} & \cdots & \sqrt{\gamma_{Sp}^-} \\ \sqrt{\gamma_{S1}^+} & \cdots & \sqrt{\gamma_{Sj}^+} & \cdots & \sqrt{\gamma_{Sp}^+} \end{pmatrix} . \tag{16}$$

Assuming that all measurements have the same error, the matrix of parameters of solid surface can be found from the following equation:

$$b = \left(A^T A\right)^{-1} A^T Y,$$ (17)

and standard errors of the unknown parameters can be found from the main diagonal of the covariance matrix:

$$K = s^2 \left(A^T A\right)^{-1},$$ (18)

where s^2 is the sample variance determined as:

$$s^2 = \frac{1}{n-p-1}(Y - Ab)^T (Y - Ab).$$ (19)

Although different substances can be used as probe liquids, the following five liquids are the most widely used: water, glycerol, diiodomethane, formamide, and ethylene glycol [1, 29, 37, 38]. The values of the components of surface tension for these liquids are listed in Table 2.

Liquid	γ^t	γ^{LW}	γ^{AB}	γ^-	γ^+
Water	72.80	21.80	51.00	25.50	25.50
Glycerol	64.00	34.00	30.00	57.40	3.92
Formamide	58.00	39.00	19.00	39.60	2.28
Ethylene glycol	48.00	29.00	19.00	30.10	3.00
Diiodomethane	50.80	50.80	0.00	0.00	0.00

Table 2. The components of surface tension for different probe liquids (from [29, 37])

Additionally, the Fowke's model can be used to determine the polar and dispersive components of surface energy. The following is the set of simultaneous Young-Dupré equations corresponding to the measurements of the CA for p liquids at time t_j:

$$0.5\gamma^t_{Li}\left(1 + \cos\overline{\theta}_i(t_j)\right) = \sqrt{\gamma^d_S(t_j)\gamma^d_{Li}} + \sqrt{\gamma^p_S(t_j)\gamma^p_{Li}}, i = 1\ldots p.$$ (20)

Since there are two unknown parameters in this model, the number of liquids used for the CA measurements can be smaller than for van Oss's model. After corresponding modification of A and b, solution of (20) can be found by the matrix method described above.

3. Case studies

3.1. Characterization of elastomer surface subject to ageing

Though synthetic elastomers like EPDM are very attractive to industry due to their high chemical stability and low permeability for water, they are sensitive to oxidation at elevated temperatures. Understanding of the chemical mechanisms of elastomer degradation is a key for designing advanced elastomers with higher resistance to oxidative degradation. Therefore, XPS and SFE analysis were employed to elucidate chemical changes produced on EPDM elastomer surfaces due to ageing.

XPS wide energy range scans were obtained for EPDM samples aged at 80 °C and 120 °C during up to 100 days. Surface chemical composition of the samples determined from these spectra as a function of ageing duration is shown in Table 3. The high carbon content in all samples arises from the contribution from the backbone structure of the elastomer. Oxygen, nitrogen, silicon and zinc are generally attributed to curing agents, amine-based accelerators and additives [1, 14].

With the increasing ageing duration, the O/C ratio also increases (last column of Table 3). In addition, for both temperatures there was certain increase in nitrogen and silicon concentrations for the 100 days ageing. As XPS is a superficial analysis technique, the variation of these elements present in small amounts on the surface could be related to diffusion processes during ageing and segregation of impurities on the surface. Being a thermally activated process, migration of additives is faster at higher temperatures, thus surface concentration of silicon after ageing at 120 °C is higher than at 80 °C.

Thermal ageing		Composition (% at.)					O/C ratio
T (°C)	t_a (days)	C	O	N	Si	Zn	
As received		93	5	0	2	-*	0.054
	5	90	7	0	2	1	0.078
80	50	88	9	1	1	1	0.102
	100	82	14	3	1	-*	0.171
120	100	80	13	2	4	1	0.163

Table 3. Chemical composition of some EPDM samples obtained from XPS wide energy range scan (with permission from [1])
*traces

The results of curve fitting procedure of the HR C1s are shown in Figure 4. The broad carbon peak in the range of E_B from 283 eV to 289 eV can be attributed to different carbon-based surface functional groups. C 1s peak was fitted with four Gaussian/ Lorenzian components with the maximum intensity at E_B of 285 eV, 286.3 eV, 288.1 eV and 289 eV. According to the literature, these energies can be assigned to C-C or C-H, hydroxyl (C-O/ C-OH), carbonyl (C=O) and carboxyl (O-C=O), respectively (see Table 1). Most part of carbon was in form of C-C / C-H. For the samples aged at 80 °C the amount of carbon bonded to oxygen, especially in form of hydroxyl, increased with the increase of ageing duration. After 100 days at 80 °C, carbon-oxygen bonds were composed of hydroxyl (20% with respect to carbon), small portion of carbonyl (4%) and traces of carboxyl (1%). This effect was similar to the evolution of the oxygen content registered in the wide energy range scan (Table 3). Similar behaviour was also observed by [39] and [7]. When comparing the samples aged during 100 days at 80 ºC and 120 ºC the portion of carbon-oxygen functional groups was lower at higher temperature, though both samples had similar surface contents of oxygen. According to [39] C-OH bonds are the main product of EPDM ageing as inferred from the C 1s core level peak. However, the results obtained in our work suggested that ageing above 100 ºC could cause hydroxyl desorption. This could explain the lower C-O content registered after the treatment at higher temperature.

Variations in the components of the SFE for the elastomer as a function of the ageing parameters were determined using acid-base regression method with the five liquids listed in Table 2. The results are shown in Figure 5. For both ageing temperatures γ^+ was statistically insignificant, so this term was omitted from the model. Since the component γ^+ is null, EPDM surface is mainly γ_S^- monopolar. In the absence of a parameter of the opposite sign, energy parameters of a monopolar surface do not contribute to the total surface energy (energy of cohesion) since the polar component $\gamma^{AB} = 2\sqrt{\gamma^+\gamma^-} = 0$ [35]. Therefore, the total SFE is controlled solely by LW interaction. However, monopolar surfaces can strongly interact with bipolar liquids.

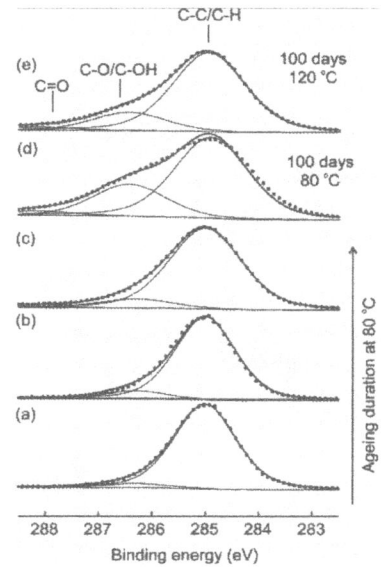

Figure 4. C 1s spectra of EPDM: a) as received, b) aged 5 days at 80 ºC, c) aged 50 days at 80 ºC, d) aged 100 days at 80 ºC, e) aged 100 days at 120 ºC. Dots – experimental data, solid lines – fitting (with permission from [1])

At 80 ºC, γ_S^{LW} increased exponentially with ageing duration reaching almost stable values after 60 days. Parameter γ_S^- had an induction period of approximately 5 days. These results agree with previous works in which the induction period during thermal oxidation of EPDM was determined as 130 h at 80 ºC [40] and 150 h at 150 ºC [7]. Variations of the induction period in different works can be due to differences in the EPDM composition, more specifically, in the carbon black and antioxidants content. After induction period, γ_S^- increased rapidly and reached the maximum in 30 days. Then, it remained almost constant with a slightly decreasing tendency, which, however, was within a standard error. The solid line connecting the filled circles in Figure 5a was obtained by fitting the experimental data with an exponential function having a time constant of 17.7±0.4 days.

At 120 °C (Figure 5b), γ_S^{LW} raised up at the beginning of ageing and then followed almost linear increasing behaviour with low rate. Surprisingly, after 100 days ageing at 120 °C γ_S^{LW} was approximately 5% smaller than for ageing at 80 °C. However, after 100 days γ_S^{LW} still maintained the linear growth, while at 80 °C it stabilized.

The evolution of γ_S^- was similar to that of γ_S^{LW}, although the initial increase was not as steep as for γ_S^{LW} component. There is a large difference in the behaviour of γ_S^- for both ageing temperatures. On short ageing periods γ_S^- was notably smaller at the higher temperature, but this difference vanished on large ageing periods. In addition, at 120 °C the induction period was not observed. Probably, the induction period at higher temperature was less than one day, so it could not be measured in these tests. This finding is consistent with [40] who reported shortening of the induction period to 10 h at 120 °C.

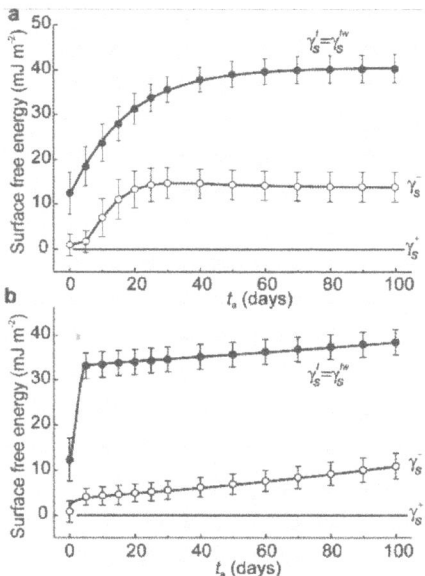

Figure 5. Components of SFE as function of ageing duration: a) ageing temperature 80 °C, b) ageing temperature 120 °C (with permission from [1])

The initial value of γ_S^t, which is equal to γ_S^{LW} in our case, is consistent with the findings of [39] for EPDM before weathering test at ambient temperature. They observed that γ_S^t first increased and then stabilized at 23.8 - 25.4 mJ m^{-2}. These values are almost two-fold smaller than in our thermal ageing experiments. This fact supports the hypothesis of a thermally activated nature of the processes responsible for the increase in the surface energy [1].

During ageing of EPDM, two competitive processes typically occur: (i) oxidation of the elastomer chains and (ii) crosslinking between the chains. The oxidation process resides in chain scission and recombination accompanied by formation of oxygen functional groups and radicals. Since double bonds are more chemically active due to the presence of a π-bond, cross-linking and oxidation at the initial stage of ageing mainly involves rupture of double bonds. Characteristic times for cross-linking of EPDM at 80 °C and 120 °C are 100 h and 12.5 h, respectively [40]. After these periods, the material is considered fully cross-linked (at given temperature) that implies significant reduction of the concentration of double bonds. Also, it is reasonable to expect that with the increasing temperature the degree of cross-linking increases and the residual concentration of double bonds decreases. During the induction period, cross-linking is the dominating process as can be inferred from the behaviour of γ_S^-, O/C ratio and very high activation energy for oxidation of EPDM, which ranges between 143.4 and 171.4 kJ mol^{-1} [41]. Further ageing of cross-linked elastomer is accompanied with slower oxidation of carbon chains. The higher reactivity of residual double bonds for EPDM aged at 80 °C can explain the steeper increase in SFE and higher concentration of oxygen after induction period. The evolution of SFE for ageing at 80° C is described by a first-order reaction with the activation energy between 63.5 and 83.7 kJ mol^{-1}[1]. These values are higher than those reported in [42], but similar to the activation energy for oxidation of long hydrocarbon chain alkanes and aromatics such as in heavy fuel oil [43]. For ageing at 120 °C the linear increase in SFE is described by a zero-order reaction. Zero-order reaction was reported also for surface degradation of fully cross-linked EPDM under artificial weathering conditions [39].

In conclusion, oxygen functional groups, mainly hydroxyl, were identified on EPDM surface after ageing. The presence of these groups was more pronounced after the treatment at 80 °C than at 120 °C. Higher ageing temperatures lead to faster cross-linking processes. At lower temperature C=C bonds are not fully consumed due to cross-linking [3], hence the oxidation processes at lower temperature is more intensive than at higher temperature. In addition, ageing at long durations promotes changes in the surface chemical composition of EPDM. These changes can be attributed to migration of additives towards the surface as reflected by the increase in Si and N concentrations after 100 days ageing at both temperatures.

3.2. Characterization of surface chemical composition of elastomer surfaces subject to sliding friction

Degradation of elastomer surfaces can be accompanied by formation of specific surface texture like smearing or microfibrill formation [44-46]. In [45, 46] it was speculated that these effects could be due to tribochemical reactions and thermooxidative degradation, however no cogent experimental evidences have been presented so far. In order to provide deeper insight into the mechanisms of elastomer failure, surface chemical composition and SFE were studied before and after friction as a function of the amount of carbon black (CB) filler in EPDM [14]. Carbon black is one of the most widely used reinforcing fillers [47-49] that improves the stiffness and the toughness of rubbers, while maintaining high flexibility and good physical and mechanical properties at low manufacturing costs. The amount of CB varied between 0 and 60 parts per hundred rubber (phr). The samples were subjected to roller-on-plate (ROP) friction tests under conditions detailed in [46]. Friction coefficient was not influenced significantly by the CB content, whereas wear rate decreased with increasing the CB content [14].

Surface chemical composition (at. %) outside and inside the contact zones was determined from the analysis of wide energy XPS spectra (Figure 6). The dominating carbon contribution (95 % and higher) was due to the elastomer backbone structure. The atomic concentrations of other elements including O, Si, S, N and Zn remained below 5 %.

In case some thermooxidative processes and/or tribochemical reactions occur at the contact zone, one can expect certain increase not only in the oxygen concentration, but also in oxygen bonding to carbon atoms in the friction zone. However, the observed behaviour of surface chemical composition was more complex. More specifically, two different tendencies were observed as far as the amount of oxygen in the friction zone is concerned. For unfilled EPDM, the amount of oxygen on the surface of friction zone increased, whereas for filled EPDM it decreased. Notwithstanding these variations, on the surfaces not subjected to friction and for all CB contents, no changes in the binding energy of the oxygen and carbon were observed in high-resolution O 1s and C 1s spectra (Figure 7). The single contribution of C at 285 eV (Figure 7b) from the C-C / C-H component implies absence of oxygen-containing functional groups (see Table 1).

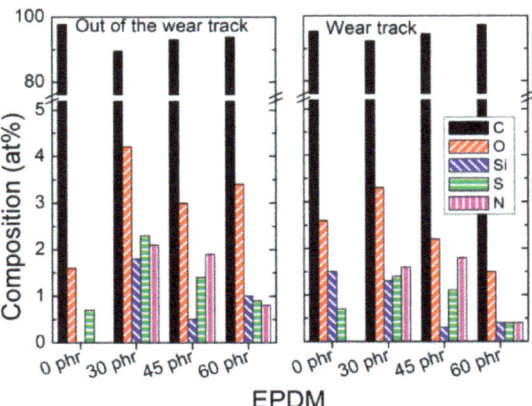

Figure 6. Surface composition of EPDM samples with different carbon black content determined from XPS analysis (with permission from [14])

Detailed analysis of the C 1s core level on the surface subjected to friction revealed that FWHM of the peaks were broader for EPDM 0 phr and 30 phr than for EPDM with higher

CB content (Figure 7b). The reason for this broadening could be initially attributed to higher surface roughness of these samples. However, an equivalent broadening did not occur for the O 1s peak. So, roughness could not explain the broadening of the C 1s core level peak. Moreover, the broadening was not completely symmetric and presented a shoulder at lower binding energies. The fitting of the spectra with a single component of the same FWHM for all samples evidenced this asymmetry (Fig. 7d). The shoulder corresponded to energies close to sp^2 carbon [26, 28, 29]. This finding suggests formation of carbon double bonds and/or graphitization in the elastomers with lower CB content. Since these elastomers have worse wear performance, peak broadening can be associated with higher wear rate and damage in the elastomer. A small mismatch between the fitting and the experimental data was observed at higher binding energies only for 0 phr EPDM. This mismatch could be attributed to the roughness effect since the contact surfaces of these samples were severely damaged. In the hypothetical case assuming that this mismatch was caused by C-O bonds, the amount of these species would be rather small. What is clear from these fittings is the absence of carbon-oxygen bonds in the wear track that could explain the different performance in response to friction of the EPDM samples.

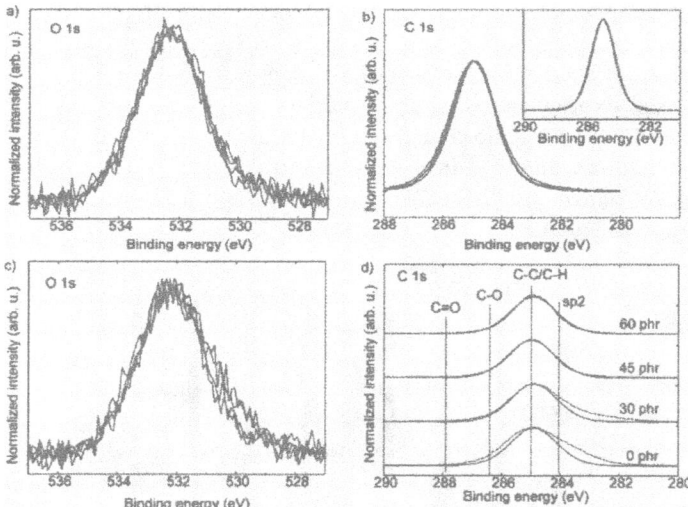

Figure 7. HR O 1s and C 1s core level spectra out of (a) and b) and in (c) and d) the wear track of the EPDM samples. Inset in b) represents the fitting of the EPDM 0 phr sample (with permission from [14])

Elastomer degradation is usually associated with bond scission and oxidation of the backbone structure [27, 39]. We argue that oxygen detected on the samples was not related with backbone structure oxidation since no C-O and C=O bonds were observed in XPS spectra. Migration of additives to the surface can be a plausible reason for the increase in oxygen. Actually, silicon was found at the characteristic binding energy of its oxide form (102 eV). Other authors have suggested that fracture of macromolecular chains is accompanied by generation of low molecular weight products as well as C=C structures [27, 50]. For the samples with larger amounts of CB (45 and 60 phr), the bond scission was smaller as can be inferred from their better wear performance, so the formation of C=C could not be appreciated with the given resolution of the XPS using a non-monochromatic light source.

On the wear track the amount of elements coming from additives (those different from carbon) as well as the amount of oxygen progressively decreased with the increasing amount of carbon black filler. Similarly to the unworn region, oxygen on the worn surface was associated mainly with silicon. Some changes in the sulphur spectrum also occurred. The as-received samples presented two peaks at about 162 eV and 168.5 eV. The first one is related to the S^{2-} sulphur state, while the second one is related to higher oxidation states. The peak at 168.5 eV significantly decreased after the removal of airborne contamination indicating superficial localisation of these oxides and the predominant S^{2-} state in the wear track. These findings suggested that the sulphur chemical state at the surface of the EPDM samples was altered in the ROP tests, and a part of the oxides located in the outer surface of the elastomer was removed.

Water CAs on worn, $\theta_{fr,a}$, and unworn, θ_{nfr}, surfaces were measured to study the changes in wettability caused by the presence of new superficial functional groups (mainly oxygen functional groups due to degradation). Subscript a denotes the apparent CA. The values of the roughness factor, r, and intrinsic CA for water, θ_{fr}, are shown in Table 4. For unworn samples, mean value of the CA was around 84° with no significant variations with different carbon black content. However, intrinsic CA was larger on the friction zone than on the unworn surfaces for all samples. The increase in the CA was statistically significant at the significance level 0.05. These findings imply that the worn surfaces were more hydrophobic than the initial ones. This behaviour is opposite to the tendency observed during ageing of a commercial EPDM with 52.6 phr of carbon black, when surface became more hydrophilic with 50% decrease in water CA [1]. The decrease in water CA in [1] was caused by thermooxidation of initially hydrophobic methyl-terminated surface of EPDM. This process was accompanied by an arrangement of polar oxygen functional groups (-C-OH, -C=O) on the outer surface layer [1, 2]. In case of frictional surfaces, no oxidation of the elastomer backbone could be found from XPS spectra. Furthermore, the increase in water CA could be associated with changes in the amount of additives present on the surface. Detailed analysis of the surface chemical composition, scanning electron microscopy and energy-dispersive X-ray spectroscopy revealed increase in zinc oxide and silica at the surface [14]. Both of these oxides have hydrophobic and superhydrophobic properties [51] that can explain the increase in water CA in the friction zone.

CB content (phr)	r	θ_{fr} (deg)	seθ_{fr} (deg)	θ_{nfr}(deg)	seθ_{nfr}(deg)
0	1.654	107	4.98	81	3.9
30	1.102	90.6	2.11	84	2.1
45	1.207	99.3	2.46	85	1.1
60	1.129	104	4.39	84	2.9

Table 4. Roughness factor (r), mean intrinsic contact angles on worn, θ_{fr}, and unworn, θ_{nfr}, surfaces with corresponding standard errors of mean (with permission from [14])

From the results of XPS and CA measurements we concluded that no thermooxidation processes were observed on friction zone under given experimental conditions for all EPDM samples and irrespective of the CB content. Chemical modification of the EPDM surface was due to mechanochemical effects rather than a thermooxidative effect [52-55]. Softer EPDMs with lower carbon black content were severely damaged during ROP test. The increase in C=C bonds for these samples can be attributed to bonds breaking accompanied by different radical reactions [56-58].

3.3. Characterization of surface chemical composition after atmospheric plasma treatments and thin coating of amorphous diamond-like carbon (DLC)

Surface modification is aimed at changing the characteristics of the surface and thin subsurface layer [59] or generation of active centres for further attachment of compounds [60]. By surface modification desired surface properties such as adhesion or wettability [61] can be obtained leaving the underlying bulk unchanged. By doing so, both the surface and bulk properties can be independently tailored and optimized. In this section we present some case studies with the purpose of demonstration the capabilities of XPS technique.

Plasma processing of materials is a crucial industrial technology in many areas including electronics, aerospace, automotive, and biomedical industries [62] due to its versatility [63]. Though nowadays plasma processing is performed mostly at low pressures, atmospheric plasma systems provide an appealing alternative to vacuum plasma systems because continuous processing can be performed at a lower cost [62, 64-66]. Operation under atmospheric pressure provides high flexibility and portability to this technique [32] and allows it expansion to processing of a larger number of materials [62]. One of the possibilities of atmospheric plasma treatments is the use of a plasma torch [67]. In this case, a reactive gas is added to the primary feed gas of the plasma torch in order to generate a flux of chemically active species, e.g. fluorine, toward treated surface. In this way, the surface being fluorinated should not be immersed in reagents and is not directly exposed to the plasma [62]. This treatment can be performed at room temperature and is faster than other fluorination methods [68]. All these advantages are important for industrial application. In our studies surface modification was carried out in two ways: using only ions of inert gases or using also chemically active gases. The reactive gases used for fluorination (commonly SF_6 and CF_4) should be thoroughly diluted with a carrier gas on order to create a stable plasma at atmospheric pressure. Nitrogen, argon or helium are typically used as a carrier gas as they can be easily ionized [65]. Due to the high reactivity of ionized fluorine-containing gases, surface reactions, etching, and plasma polymerization can occur simultaneously. The predominance of one or another will depend on the gas feed, the operating parameters and the chemical nature of the polymer substrates.

Table 5 shows the chemical composition of EPDM elastomers modified by atmospheric plasma using N_2, Ar and He carrier gases in combination with SF_6 and CF_4 fluorination precursors. The composition of untreated elastomers and those activated only by He plasma are included for comparison. The chemical composition was determined from XPS analysis.

All plasma-treated samples had fluorine (0.3 to 13 at. %), oxygen (13 to 22 at. %) and nitrogen (< 2 at. %) on their surfaces. It should be mentioned that plasma treatments promoted a significant increase in the oxygen content. This oxygen incorporation into the surface was similar when plasma activation only with He gas was used. Therefore, oxygen content on the treated surfaces does not significantly depend on the presence or absence of reactive gases during plasma treatments. Oxygen incorporation to the elastomer takes place through interactions between free radicals, O_2 and H_2O molecules from ambient air after the plasma treatment [64]. Traces of sulphur were detected on the EPDM surface after plasma processing using SF_6 [69]. The degree of fluorination for each treatment can be evaluated from O/C and F/C atomic ratios shown in Table 5. From these data it is evident that for fluorination purposes CF_4 is more effective than SF_6. Previously, these results were explained in terms of the dissociation products formed in the plasma [32, 66]. No significant differences in F/C ratio were found for the combination of inert carrier gases (Ar and He) and CF_4, whereas He was more effective for fluorination in combination with SF_6. At the same time, nitrogen carrier gas yielded lower F/C ratio. This is because radicals produced by dissociation of SF_6 and CF_4 could not react with N_2 due to the strong bond between nitrogen atoms. So far, only few information is available in the literature on these reactions [70, 71] to allow definite conclusion. The analysis of the HR C 1s (Figure 2 a, b) revealed the changes in the shape of the core level peak that evidenced not only significant increase in the C-O bonds after the fluorination treatments, but also the apparition of new functional groups involving carbon, oxygen and fluorine.

Process		Chemical composition (at. %)					F/C	O/C
		C	O	N	F	S		
Untreated		93	5	0	0	0	0	0.005
SF_6	N_2	80.4	16.3	1.5	1.3	0.5	0.02	0.20
	Ar	81.6	14.1	1.9	1.7	0.6	0.02	0.17
	He	71.1	19.6	1.6	6.8	0.9	0.09	0.27
CF_4	N_2	84.4	13.4	1.9	0.3	0	0.003	0.16
	Ar	65.9	19.3	1.8	13.0	0	0.20	0.29
	He	63.4	21.9	1.8	12.0	0	0.19	0.34
Activation	He	79	16	traces	-	-	-	0.20

Table 5. Surface composition of EPDM samples after atmospheric plasma treatments

	C1s core level components						
	C-C C-H	C-O -C-H-CF$_2$	C=O CH-CF	O-C=O CFH-CF$_2$	CF	CF$_2$	CF$_3$
Untreated	96	4	0	0	0	0	0
He activation	77	12	8	3	0	0	0
Fluorinated	45	36	7	5	3	3	1

Table 6. Analysis of the components of the C1s core level of EPDM elastomers: untreated, He activated and fluorinated

The presence of CF and CF$_2$ can be explained by H substitution and chain scissions [32]. CF$_2$- represents the main chain of the polymer,-CF- component could indicate cross-link sites [72] and -CF$_3$ component indicates end groups of polymer chains [72] and grafting [32]. These findings imply enhancing of cross-linking in the elastomer due to plasma processing with fluorine-containing gas.

Figure 2b and 2c show a comparison of HR C1s for elastomers treated with CF$_4$ and SF$_6$ using the same carrier gas. From the shape of the peaks one can observe again that CF$_4$ is more effective than SF$_6$ in fluorinating because of the presence of additional fluorine containing groups together with the increase in the total fluorine content already mentioned in Table 6. SF$_6$ molecules mainly dissociate into fluorine atoms and SF$_5$ radicals [66], whereas CF$_4$ produces fluorine, CF, CF$_2$ and CF$_3$ radicals that react with the elastomer surface leading to a substantial incorporation of fluorine [32]. Though fluorine atoms were considered as the main responsible of the fluorination process, CF$_x$ radicals could contribute to formation of highly fluorinated components (in particular CF$_3$ groups) [65]. Thus, higher content of CF$_x$ groups after treatment with CF$_4$ in comparison to SF$_6$ could be attributed to the larger number of fluorine-containing radicals. Our results demonstrated that the main effect of SF$_6$ resided in oxidation of the polymer surface, while CF$_4$, in addition to surface oxidation also induced incorporation of CF$_x$ radicals enhancing the efficiency of the fluorination process [4]. SFE measurements of fluorinated samples revealed that increase in the surface free energy, γ, was noticeable only for the samples which chemical composition was significantly modified, especially by incorporation of oxygen and fluorine polar groups. The increase in SFE related mainly with the electron donor, γ^-, contribution of the polar component, γ^{AB}, [73].

It should be noted that the modification of elastomer surface is not always as evident as in the above examples. The extent of surface modification induced by a particular surface treatment is highly dependent on the type of the elastomer. For instance, when the same treatment is carried out for HNBR, the extent of the surface modification was much less as compared with EPDM (Figure 2d). Surface chemical composition of treated HNBR was almost the same as for untreated one with no presence of fluorine containing groups and no increase in oxygen content [15]. A saturated backbone structure of elastomer such as HNBR makes it less reactive to the plasma treatments.

From systematic studies of fluorinating process under different conditions we concluded that high concentration of fluorine on the elastomer surface, typically higher than 7%, is associated with formation of fluorine-containing functional groups in form of CF, CF$_2$ or CF$_3$.The presence of CF and CF$_2$ can be explained by H substitution and chain scissions [32], where -CF$_2$- represents the main chain of the polymer, -CF- component could indicate cross-link sites [72], and-CF$_3$ component indicates end groups of polymer chains [72] and grafting [32]. These findings imply enhancing of cross-linking in the elastomer due to plasma processing with fluorine-containing gas. On the contrary, in processing with lower fluorinating efficiency (< 7%) no evidence of C-F bonds was found [4].

The analysis of other core levels such as O 1s or F 1s can provide complementary information on surface chemical groups. From the analysis of O 1s core level peak of fluorinated samples having one symmetric peak at about 532.6 eV (not shown here, see [4]), we concluded that hydroxyl or ether species were predominant after all fluorinating treatments [62]. This finding is in agreement with the C 1s analysis presented above, as can be inferred from small contributions in the C 1s peak from carboxyl and carbonyl species at 288.1 and 289 eV, respectively [26]. On the other hand, F 1s core level spectra had certain differences when using SF_6 and CF_4 with the same carrier gas, e.g. He (Figure 8). When SF_6 was used, the peak was symmetric and centred at about 687.2 eV, whereas for CF_4 it was centred at 688.1 eV. These energies are close to those reported in the literature for fluorine covalently bonded to carbon [62, 64]. These findings clearly indicated an incorporation of fluorine in the polymer chains. The behaviour observed in the F 1s peaks presented a good correlation with the C 1s analysis given above. Higher fluorine concentration in the sample is associated with the presence of CF_2 bonds according to the C 1s analysis and displacement of the main F 1s peak towards higher binding energies [26].

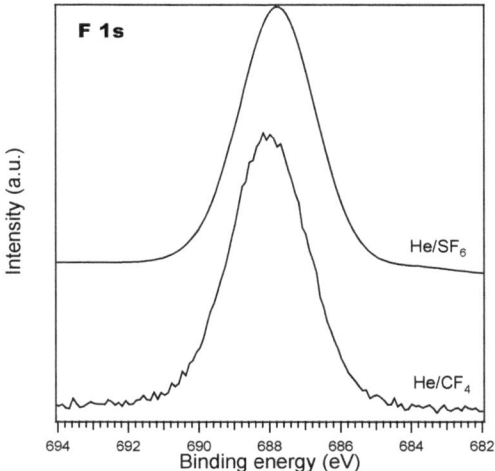

Figure 8. Comparison of the F 1s core level after fluorination treatments with different precursors

Another way to modify elastomer surfaces and improve their tribological properties consists in application of different coatings. Elastomer surfaces with low friction can be obtained by deposition of (3-aminopropyl)-triethoxysilane (APEO) or (3-glycidoxypropyl)trimethoxy-silane (GLYMO) coatings using siloxane precursors [74, 75] or polytetrafluoroethylene (PTFE). PTFE is commonly used as a coating on metallic substrates and it has been recently optimised for use on elastomers yielding low friction coefficient and enhancing other properties [75, 76]. Diamond like carbon (DLC) is another very promising candidate for coatings due to its excellent tribological properties and chemical inertness [6]. DLC coatings have been widely used on different substrates including metals, ceramics and other inorganic materials. Recently, elastic DLC coating on elastomers was developed [5]. Initially, the idea of applying a hard DLC film on soft elastomer materials with low elastic modulus received much scepticism because of the risk of loss of adherence and interfacial delamination. However, after successful demonstration of the efficiency of DLC coatings on elastomer substrates their application has been widely spread, especially in automotive industry. Further advances in application of DLC on elastomer surfaces can be made on the basis of deeper understanding of surface chemistry of coated systems. In this work we studied surface chemical composition of uncoated and DLC-coated NBR and HNBR elastomers using XPS and CA methods. HNBR is a hydrogenated elastomer with no double bonds in the elastomer backbone structure. Therefore, HNBR is less chemically reactive than NBR against the same treatment [4].

Surface chemical composition of elastomers obtained from the wide energy range scan XPS spectra are shown on Table 7. For all samples, carbon, oxygen and nitrogen were the main surface elements. Small amounts of other elements used in the elaboration processes of elastomers were also detected [22]. As we already mentioned before, these components have no important influence on the surface properties [13]. When DLC coating was applied, oxygen content slightly increased. In order to investigate changes in the main bonding of the elastomers, curve fitting procedure of XPS spectra of the C1s core level peaks was performed (see Figure 9 and Table 7). Four components: C-C/CH; C-O; C=O and O-C=O were used for fitting. These components were derived from the expected chemistry of the samples and taking into account natural oxidation process of the elastomers. The binding energies of these components are listed in Table 1. CN bond from the NBR structure was not considered due to the small contribution of nitrogen to the final composition. The results evidenced that C-C contribution corresponding to the backbone structure of the elastomers decreased for DLC-coated elastomers as a consequence of the formation of oxygen functional groups, mainly in form of C-O. HNBR presented larger variation of the carbon bond than NBR.

Despite the small variations observed in the surface chemical composition, DLC-coated elastomers presented better tribological performance reflected in a significant reduction in the coefficient of friction (COF) and friction noise [77]. Also, water CA increased for NBR after DLC deposition (Figure 10) indicating the increase in the hydrophobic character of NBR elastomer surfaces after DLC deposition. This finding is consistent with previous works where an increase in the hydrophobic properties of the DLC-coated elastomers was attributed to sp^2 and sp^3 hybridised carbon bonds in the DLC coating. One should bear in mind that higher hydrophobicity of the surface is usually related to a higher chemical stability. Actually, in our experiments SFE of NBR elastomers decreased by 9% after DLC deposition [77]. For hydrogenated HNBR elastomer, the variations in the hydrophobicity after DLC coatings were statistically insignificant. In contrast to NBR, SFE increased by 8% for DLC-coated HNBR. We suggested that hydrogenation of unsaturated bonds to form HNBR results in different reactivity of the elastomer towards the DLC coatings. Despite the fact that the same type of DLC coating was deposited on all the elastomers, the extent of the modifications was different depending on the substrate.

				C1s core level components			
	C	O	N	C-C C-H	C-O	C=O	O-C=O
NBR uncoated	98	1	0	97	3	0	0
NBR coated	98	1	0	93	7	0	0
HNBR uncoated	92	4	3	91	5	4	0
HNBR coated	89	6	3	86	11	2	1

Table 7. Surface composition and analysis of the components of the C1s core level of NBR and HNBR before and after DLC coating

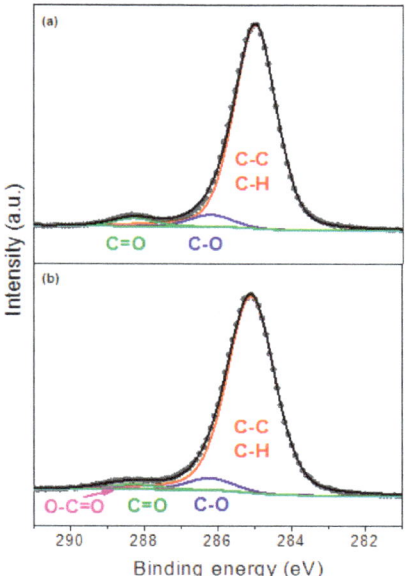

Figure 9. HRC 1s core level of HNBR elastomer before (a) and after DLC coating (b)

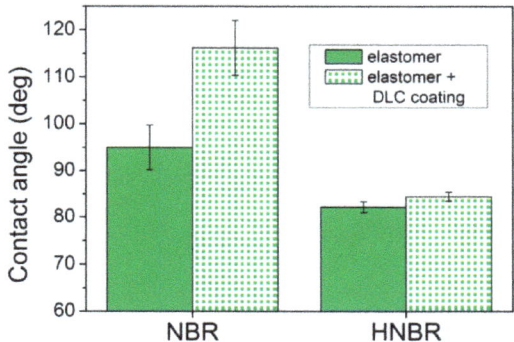

Figure 10. Water CA on elastomers before and after DLC coating

4. Concluding remarks

We have shown that XPS is a sensitive and versatile technique to characterize surface chemical composition of engineered elastomer surfaces. The combination of XPS with other techniques such as CA measurements or SFE calculations allows the evaluation of minute changes in surface chemical composition and structure of surface functional groups resulting from surface degradation or surface treatment. Therefore, factors like ageing processes, lubricant absorption or reaction of the elastomer chains under surface treatment have been analysed with this technique. As XPS binding energies are not only element-specific but also contain chemical information, it provides information about chemical states of a particular element. So, the degradation of the elastomer backbone structure during ageing and wear was evaluated in terms of C-O or C-OH bond formation. In general, the binding energy increases with increasing oxidation state and, for a fixed oxidation state, with the electronegativity of the ligands. The C 1s spectrum of fluorocarbon polymer is an example how the binding energy of carbon depends sensitively on the electronegativity of its neighbours. It was shown that an increase in the oxygen content is not necessarily related to oxidation of carbon bonds and, therefore, to degradation. The presence of new elements, e.g. fluorine due to plasma fluorination, is not always related to the formation of C-F bonds as well. This information cannot be obtained by many other analytical techniques, which makes XPS particularly interesting for this kind of studies. One limitation found was the lack of sensitivity to resolve the possible cross-linking after certain surface treatments. The energy gap between the C=C and C-C bonds are below 0.8 eV which is the range of resolution of nonmonocromated Mg Kα X-ray source. However, cross-linking effects were determined from the variation of SFE components and O/C and F/C ratios. When studying elastomer surfaces subjected to rubbing, no important oxidation indicative of thermochemical reactions was observed. Modifications of the elastomer surfaces were interpreted in terms of mechanochemical reactions and wear.

The final properties of elastomer components can be significantly modified even by small changes in chemical composition of thin surface layer. The extent of the surface modification is clearly influenced by the elastomer substrate and it is more significant for organic materials than other inorganic materials.

Author details

Lidia Martínez and Elisa Román
ICMM-CSIC, Dept. Surfaces and Coatings, Madrid, Spain

Roman Nevshupa*
CISDEM-CSIC, Spain

---

* Corresponding Author

Acknowledgement

Most of this study was developed under the framework of the EU project KRISTAL No. 515837-2 (6th FP). The authors acknowledge Dr. Yves Huttel for his fruitful discussion about the XPS results, IVW (Germany) and TRW (Spain) for the EPDM substrates, VITO (Belgium) for the plasma treatments, HEF (France) for the DLC coatings. The authors also acknowledge the Spanish National Research Council (CSIC) for grant PIE201160E085 and Spanish Ministry for Economy and Competitiveness for financial support provided in the projects MAT2011-29194-C02-02, BIA2011-25653 and "Ramón y Cajal" RYC-2009-04125.

5. References

[1] Nevshupa R, Martinez L, Alvarez L, Lopez MF, Huttel Y, Mendez J, et al. (2011) Influence of thermal ageing on surface degradation of ethylene-propylene-diene elastomer. Journal of Applied Polymer Science. 119:242-251.

[2] Laoharojanaphand P, Lin TJ, Stoffer JO (1990) Glow discharge polymerization of reactive functional silanes on poly(methyl methacrylate). Journal of Applied Polymer Science. 40:369-384.

[3] Desai SM, Bodas DS, Singh RP (2004) Fabrication of long-term hydrophilic elastomeric surfaces via plasma induced surface cross-linking of functional monomers. Surface and Coatings Technology. 184:6-12.

[4] Martínez L, Huttel Y, Verheyde B, Vanhulsel A, Román E (2010) Photoemission study of fluorination atmospheric pressure plasma processes on EPDM: Influence of the carrier and fluorinating gas. Applied Surface Science. 257:832-836.

[5] Nakahigashi T, Tanaka Y, Miyake K, Oohara H (2004) Properties of flexible DLC film deposited by amplitude-modulated rf p-cvd. Tribology International. 37:907-912.

[6] Lindholm P, Björklund S, Svahn F (2006) Method and surface roughness aspects for the design of DLC coatings. Wear. 261:107-111.

[7] Delor-Jestin F, Lacoste J, Barrois-Oudin N, Cardinet C, Lemaire J (2000) Photo-, thermal and natural ageing of ethylene–propylene–diene monomer (EPDM) rubber used in automotive applications. Influence of carbon black, crosslinking and stabilizing agents. Polymer Degradation and Stability. 67:469-477.

[8] Majumder PS, Bhowmick AK (1998) Friction behaviour of electron beam modified ethylene–propylene diene monomer rubber surface. Wear. 221:15-23.

[9] Lonkar SP, Kumar AP, Singh RP (2007) Photo-stabilization of EPDM–clay nanocomposites: Effect of antioxidant on the preparation and durability. Polymers for Advanced Technologies. 18:891-900.

[10] Farahani TD, Bakhshandeh GR (2005) The effect of curing on sorption and diffusion of a brake fluid in EPDM elastomer. e-Polymers. 47:1-10.

[11] Banik I, Bhowmick AK (2000) Electron beam modification of filled fluorocarbon rubber. Journal of Applied Polymer Science. 76:2016-2025.

[12] Tillier DL, Meuldijk J, Koning CE (2003) Production of colloidally stable latices from low molecular weight ethylene–propylene–diene copolymers. Polymer. 44:7883-7890.

[13] Degrange JM, Thomine M, Kapsa P, Pelletier JM, Chazeau L, Vigier G, et al. (2005) Influence of viscoelasticity on the tribological behaviour of carbon black filled nitrile rubber (NBR) for lip seal application. Wear. 259:684-692.

[14] Martinez L, Nevshupa R, Felhoes D, de Segovia JL, Roman E (2011) Influence of friction on the surface characteristics of EPDM elastomers with different carbon black contents. Tribology International. 44:996-1003.

[15] Martínez L, Álvarez L, Huttel Y, Méndez J, Román E, Vanhulsel A, et al. (2007) Surface analysis of NBR and HNBR elastomers modified with different plasma treatments. Vacuum. 81:1489-1492.

[16] Alisoy HZ, Baysar A, Alisoy GT (2005) Physicomathematical analysis of surface modification of polymers by glow discharge in medium. Physica A: Statistical Mechanics and its Applications. 351:347-357.

[17] Einstein A (1905) Über einen die erzeugung und verwandlung des lichtes betreffenden heuristischen gesichtspunkt. Annalen der Physik. 322:132-148.

[18] Hüfner S (2003) Photoelectron spectroscopy. Principles and applications, 3rd edition. Berlin: Springer, 662 p.

[19] Seah MP, Dench WA (1979) Quantitative electron spectroscopy of surfaces: A standard data base for electron inelastic mean free paths in solids. Surface and Interface Analysis. 1:2-11.

[20] Briggs D, Seah MP (1990) Practical surface analysis—auger and X-ray photoelectron spectroscopy. Chichester: Wiley, 657 p.

[21] Woodruff DP, Delchar TA (1986) Modern techniques of surface science. Cambridge: Cambridge University Press, 608 p.

[22] Mitra S, Ghanbari-Siahkali A, Kingshott P, Rehmeier HK, Abildgaard H, Almdal K (2006) Chemical degradation of crosslinked ethylene-propylene-diene rubber in an acidic environment. Part i. Effect on accelerated sulphur crosslinks. Polymer Degradation and Stability. 91:69-80.

[23] Akhter S, Allan K, Buchanan D, Cook JA, Campion A, White JM (1988) XPS and IR study of X-ray induced degradation of pva polymer film. Applied Surface Science. 35:241-258.

[24] Coullerez G, Chevolot Y, Léonard D, Xanthopoulos N, Mathieu HJ (1999) Degradation of polymers (PVC, PTFE, m-f) during X-ray photoelectron spectroscopy (ESCA) analysis. Journal of Surface Analysis. 5:235-239.

[25] Shirley DA (1972) High-resolution x-ray photoemission spectrum of the valence bands of gold. Physical Review B. 5:4709-4714.

[26] Beamson G, Briggs D (1992) High resolution XPS for organic polymers: The Scienta ESCA 300 database. Ney York: John Wiley & Sons, 306 p.

[27] Zhang SW (2004) Tribology of elastomers. Amsterdam: Elsevier, 282 p.

[28] Swaraj S, Oran U, Lippitz A, Friedrich JF, Unger WES (2005) Surface analysis of plasma-deposited polymer films, 6. Plasma Processes and Polymers. 2:572-580.

[29] Grythe KF, Hansen FK (2006) Surface modification of EPDM rubber by plasma treatment. Langmuir. 22:6109-6124.

[30] Nansé G, Papirer E, Fioux P, Moguet F, Tressaud A (1997) Fluorination of carbon blacks: An X-ray photoelectron spectroscopy study: I. A literature review of XPS studies of fluorinated carbons. XPS investigation of some reference compounds. Carbon. 35:175-194.

[31] Wen C-H, Chuang M-J, Hsiue G-H (2006) Plasma fluorination of polymers in glow discharge plasma with a continuous process. Thin Solid Films. 503:103-109.

[32] Hochart F, Levalois-Mitjaville J, De Jaeger R, Gengembre L, Grimblot J (1999) Plasma surface treatment of poly(acrylonitrile) films by fluorocarbon compounds. Applied Surface Science. 142:574-578.

[33] Wenzel RN (1936) Resistance of solid surfaces to wetting by water. Industrial & Engineering Chemistry. 28:988-994.

[34] Gao L, McCarthy TJ (2007) How Wenzel and Cassie were wrong. Langmuir. 23:3762-3765.

[35] Van Oss CJ, Chaudhury MK, Good RJ (1988) Interfacial Lifshitz-van der Waals and polar interactions in macroscopic systems. Chemical Reviews. 88:927-941.

[36] Wu W, Giese RF, Jr., van Oss CJ (1995) Evaluation of the Lifshitz-van der Waals/acid-base approach to determine surface tension components. Langmuir. 11:379-382.

[37] Navrátil Z, Buršíková V, St'ahel P, Šíra M, Zvěřina P (2004) On the analysis of surface free energy of DLC coatings deposited in low pressure rf discharge. Czechoslovak Journal of Physics. 54:C877-C882.

[38] Luner PE, Oh E (2001) Characterization of the surface free energy of cellulose ether films. Colloids and Surfaces A: Physicochemical and Engineering Aspects. 181:31-48.

[39] Zhao Q, Li X, Gao J (2008) Surface degradation of ethylene–propylene–diene monomer (EPDM) containing 5-ethylidene-2-norbornene (ENB) as diene in artificial weathering environment. Polymer Degradation and Stability. 93:692-699.

[40] Kumar A, Commereuc S, Verney V (2004) Ageing of elastomers: A molecular approach based on rheological characterization. Polymer Degradation and Stability. 85:751-757.

[41] Mason LR, Reynolds AB (1998) Comparison of oxidation induction time measurements with values derived from oxidation induction temperature measurements for EPDM and XLPE polymers. Polymer Engineering & Science. 38:1149-1153.

[42] Budrugeac P, Segal E (1994) On the kinetics of the thermal degradation of polymers with compensation effect and the dependence of activation energy on the degree of conversion. Polymer Degradation and Stability. 46:203-210.

[43] Ayala JA, Rincón ME (1981) The oxidation of fuel oil #6 studied by differential scanning calorimetry. ACS Fuel. 26:120-130.

[44] Schallamach A (1971) How does rubber slide? Wear. 17:301-312.

[45] Felhös D, Karger-Kocsis J (2008) Tribological testing of peroxide-cured EPDM rubbers with different carbon black contents under dry sliding conditions against steel. Tribology International. 41:404-415.

[46] Karger-Kocsis J, Mousa A, Major Z, Békési N (2008) Dry friction and sliding wear of EPDM rubbers against steel as a function of carbon black content. Wear. 264:359-367.

[47] Rathinasamy P, Balamurugan P, Balu S, Subrahmanian V (2004) Effect of adhesive-coated glass fiber in natural rubber (NR), acrylonitrile rubber (NBR), and ethylene–propylene–diene rubber (EPDM) formulations. I. Effect of adhesive-coated glass fiber on the curing and tensile properties of NR, NBR, and EPDM formulations. Journal of Applied Polymer Science. 91:1111-1123.

[48] Chou H-W, Huang J-S, Lin S-T (2007) Effects of thermal aging on fatigue of carbon black–reinforced EPDM rubber. Journal of Applied Polymer Science. 103:1244-1251.

[49] Wang M-J, Wolff S. Surface energy of carbon black. In: Donnet J-B, editor. Carbon black: Science and technology, second edition. Ney York: CRC Press; 1993. p. 289-355.

[50] Rizk RAM, Abdul-Kader AM, Ali ZI, Ali M (2009) Effect of ion bombardment on the optical properties of ldpe/EPDM polymer blends. Vacuum. 83:805-808.

[51] Zhang J, Huang W, Han Y (2006) Wettability of zinc oxide surfaces with controllable structures. Langmuir. 22:2946-2950.

[52] Nevshupa R, Roman E, de Segovia JL (2010) Model of the effect of local frictional heating on the tribodesorbed gases from metals in ultra-high vacuum. International Journal of Materials & Product Technology. 38:57-65.

[53] Nevshupa RA (2009) The role of athermal mechanisms in the activation of tribodesorption and triboluminisence in miniature and lightly loaded friction units. Journal of Friction and Wear. 30:118-126.

[54] Heinike G (1984) Tribochemistry. Munchen: Carl Hanser Verlag, 495 p.

[55] Kostetsky BI (1992) The structural-energetic concept in the theory of friction and wear (synergism and self-organization). Wear. 159:1-15.

[56] Malhotra M, Kumar S (1997) Thermal gas effusion from diamond-like carbon films. Diamond and Related Materials. 6:1830-1835.

[57] Butyagin PY (1971) Kinetics and nature of mechanochemical reactions. Russian Chemical Reviews. 40:901-915.

[58] Butyagin PY (1984) Structural disorder and mechanochemical reactions in solids. Russian Chemical Reviews. 53:1025-1038.

[59] Orellana LM, Pérez FJ, Gómez C (2005) The effect of nitrogen ion implantation on the corrosion behaviour of stainless steels in chloride media. Surface and Coatings Technology. 200:1609-1615.

[60] Goddard JM, Hotchkiss JH (2007) Polymer surface modification for the attachment of bioactive compounds. Progress in Polymer Science. 32:698-725.

[61] Minko S, Müller M, Motornov M, Nitschke M, Grundke K, Stamm M (2003) Two-level structured self-adaptive surfaces with reversibly tunable properties. Journal of the American Chemical Society. 125:3896-3900.

[62] Ho KKC, Lee AF, Bismarck A (2007) Fluorination of carbon fibres in atmospheric plasma. Carbon. 45:775-784.

[63] Mitra S, Ghanbari-Siahkali A, Kingshott P, Rehmeier HK, Abildgaard H, Almdal K (2006) Chemical degradation of crosslinked ethylene-propylene-diene rubber in an acidic environment. Part ii. Effect of peroxide crosslinking in the presence of a coagent. Polymer Degradation and Stability. 91:81-93.

[64] Felten A, Ghijsen J, Pireaux JJ, Johnson RL, Whelan CM, Liang D, et al. (2008) Photoemission study of CF_4 rf-plasma treated multi-wall carbon nanotubes. Carbon. 46:1271-1275.

[65] Fanelli F, Fracassi F, d'Agostino R (2008) Fluorination of polymers by means of he/CF_4-fed atmospheric pressure glow dielectric barrier discharges. Plasma Processes and Polymers. 5:424-432.

[66] Borisov S, Khotimsky VS, Rebrov AI, Rykov SV, Slovetsky DI, Pashunin YM (1997) Plasma fluorination of organosilicon polymeric films for gas separation applications. Journal of Membrane Science. 125:319-329.

[67] http://www.vitoplasma.com/en/30, last access on March 2012.

[68] Yasuda H, Hsu TS (1977) Some aspects of plasma polymerization of fluorine-containing organic compounds. Journal of Polymer Science: Polymer Chemistry Edition. 15:2411-2425.

[69] Barni R, Riccardi C, Selli E, Massafra MR, Marcandalli B, Orsini F, et al. (2005) Wettability and dyeability modulation of poly(ethylene terephthalate) fibers through cold SF_6 plasma treatment. Plasma Processes and Polymers. 2:64-72.

[70] Kumar SVK, Sathyamurthy N, Manogaran S, Mitra SK (1994) Possible reaction of atomic nitrogen with SF_x (x = 1–5) and CF_x (x = 1–3) fragments from N_2-SF_6 and N_2-CF_4 discharges. Chemical Physics Letters. 222:465-470.

[71] Radoiu MT (2003) Studies of 2.45 GHz microwave induced plasma abatement of CF_4. Environmental Science & Technology. 37:3985-3988.

[72] Tran ND, Dutta NK, Roy Choudhury N (2006) Weatherability and wear resistance characteristics of plasma fluoropolymer coatings deposited on an elastomer substrate. Polymer Degradation and Stability. 91:1052-1063.

[73] Schlögl S, Kramer R, Lenko D, Schröttner H, Schaller R, Holzner A, et al. (2011) Fluorination of elastomer materials. European Polymer Journal. 47:2321-2330.

[74] Verheyde B, Havermans D, Vanhulsel A (2011) Characterization and tribological behaviour of siloxane-based plasma coatings on HNBR rubber. Plasma Processes and Polymers. 8:755-762.

[75] Verheyde B, Rombouts M, Vanhulsel A, Havermans D, Meneve J, Wangenheim M (2009) Influence of surface treatment of elastomers on their frictional behaviour in sliding contact. Wear. 266:468-475.

[76] http://www.fluoroprecision.co.uk/ptfe-coated-elastomers-rubbers.html, last access on March 2012.

[77] Martinez L, Nevshupa R, Alvarez L, Huttel Y, Mendez J, Roman E, et al. (2009) Application of diamond-like carbon coatings to elastomers frictional surfaces. Tribology International. 42:584-590.

Chapter 6

Use of Magnetic Induction Spectroscopy in the Characterization of the Impedance of the Material with Biological Characteristics

Jesús Rodarte Dávila, Jenaro C. Paz Gutierrez and Ricardo Perez Blanco

Additional information is available at the end of the chapter

1. Introduction

The basic electrolytic experiment consists of a homogeneous[1] electrolytic solution with two identical electrodes (Fig 1). We know that a homogeneous solution hasn't boundaries or membranes, except the electrodes and the solution receipt.

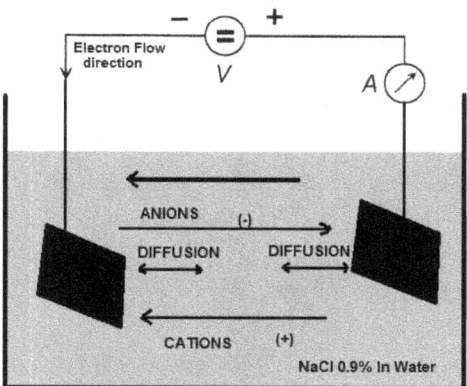

Figure 1. The basic electrolytic experiment

Electrolyte solution selected as the most important of the human body: NaCl aqueous solution with a concentration of 0.9% by weight [i].

---

[1] If NaCl is dissolved in water then NaCl is the solute (and the electrolyte), and water is the solvent, together form the solution

In a homogeneous conductive material the impedance (Z) is proportional to its length and inversely proportional to its cross sectional area (A) (Fig. 2).

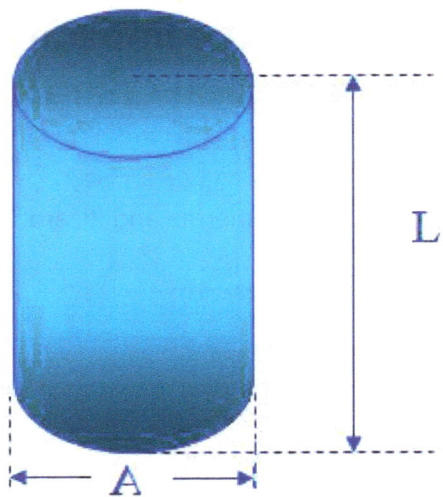

Figure 2. The impedance (Z) of homogeneous conductive material is directly proportional to its length and resistivity and inversely with its area

$$Z=\varrho L/A=\varrho L^2/V \tag{1}$$

According with the Figure 2 Z=impedance, L=length, A=area, V=volume ϱ=resistivity=$1/\sigma$ (conductivity). An empirical relationship can be established between the ratio (L^2 / V) and the impedance of the saline solution which contains electrolytes that conduct electricity through the sample. Therefore impedance (Z) = ρ L / A = ρ L^2 / V.

Hoffer et al. [2] and Nyboer [3] were the first to introduce the technique of four surface electrodes Bio-impedance analysis.

A disadvantage presented by this technique is the use of a high current (800 mA) and a high voltage to decrease the volatility of injected current associated with skin impedance (10 000 Ω/cm^2)[4]

Harris et al.[5], (1987) uses a four terminals device to measure impedance for the purpose of eliminating the effect of electrodes in an aqueous medium.

Asami et al.[6] , (1999) used a pair of coils submerged for monitoring the current induced in the coil pair, which he called electrode-less method, however still requires physical connections between the coils and electronic instruments.

To measure the complex spectrum of the permittivity of a biological culture solution Ong et al.[7], uses a remote sensor resonant circuit, to obtain the impedance of the environment by observing the resonant frequency and the frequency of zero reactance.

In another case for monitoring the fermentation process Hofmann et al.[8], Use a sensor based on a transceiver, as a way of overcoming the effect of having two metal electrodes to measure the impedance of the culture broth in a fermentation process, as in such processes the behavior of living cells is as small capacitors, then measure the impedance represented by these small capacitors correlates with the number and size of living cells in the system Hofmann et al., 2005).

2. Magnetic induction spectroscopy

"The Magnetic Induction Spectroscopy (MIS) aim is the non-contact measurement of passive electrical properties (PEP) σ, ε and μ of biological tissues via magnetic fields at various frequencies.[9]"

The basic requirements of this method are:

a. Creating a varying in time magnetic field, from an exciting coil to induce the field to the object under study.

b. Obtaining information generated from the disturbance or "reaction" of coils-environment system through the Receiver / Sensor coil.

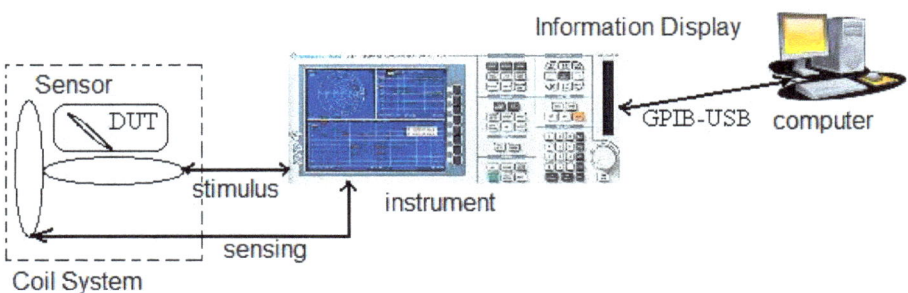

Figure 3. Measurement System: composed of a coil arrangement, network analyzer, and a pc

3. Measurement system

Following the protocol of the method used in Figure 3 we present our system Equipment-Interface into three sections:

a. Computer: Using the platform that provides National Instrument [10] - LabView V8.6, displays and processes the information obtained from the coil system from the Instrument.

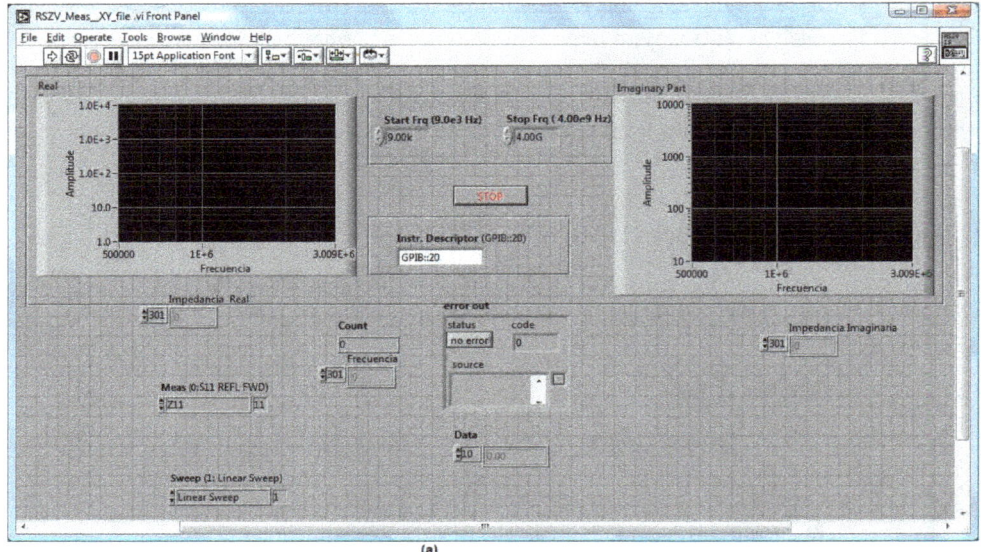

(a)

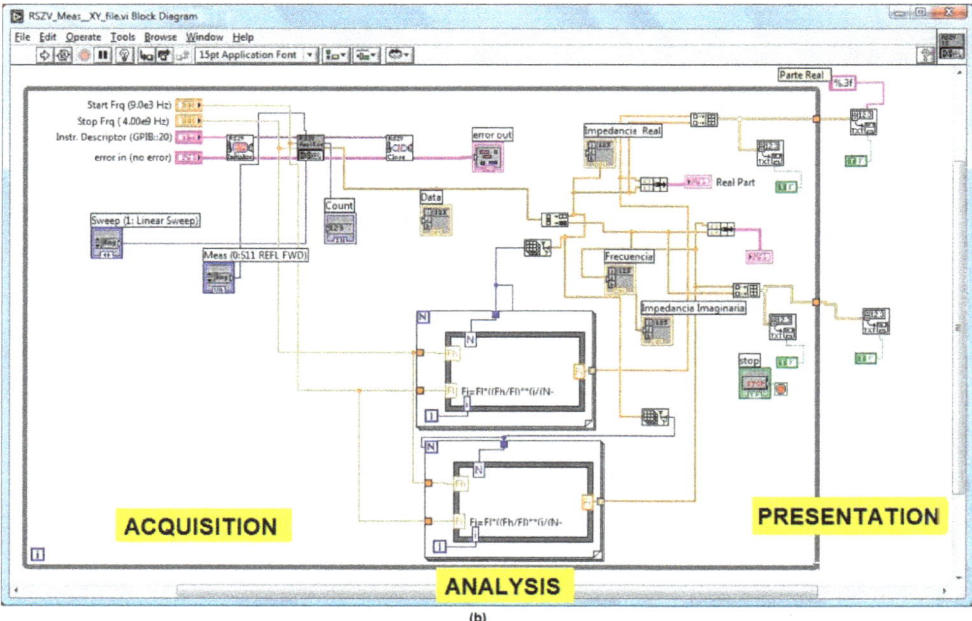

(b)

Figure 4. Virtual Instruments LabVIEW (National Instrument VI), a) Front panel (user), b) Block diagram panel (interconnections), uses them to automate the acquisition and management of information.

Use of Magnetic Induction Spectroscopy in the Characterization of the Impedance of the Material with Biological Characteristics

175

b. Instrument: Commercial Equipment, RS ZV Vector Network Analyzer [11] "Rohde & Schwartz" which performs the frequency sweep from 100 kHz to 4 MHz range, applying it to the exciting coil, the same equipment then captures the data or information from the receiver coil and sends to the computer via General-Purpose Interface Bus Universal Serial Bus or GPIB-USB.

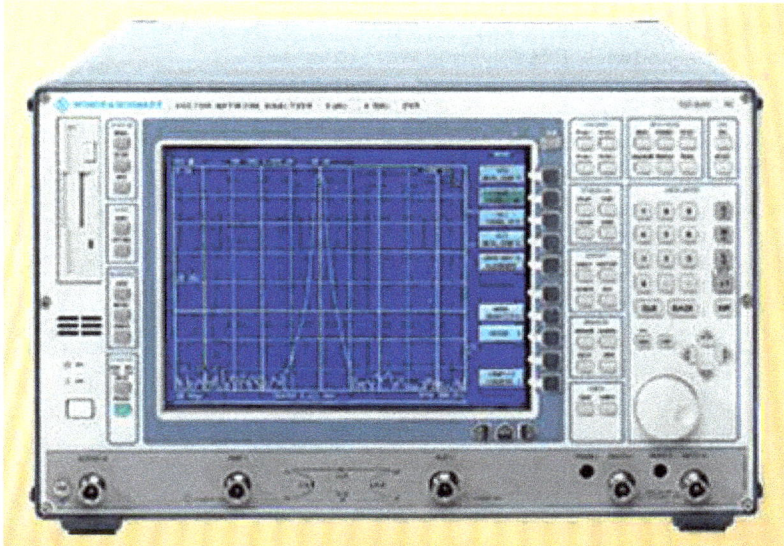

Figure 5. Network Analyzer "Rohde & Schwartz", as a frequency sweep source

c. Coil System: Fig. 6 show this system, consisting of three coils, an exciter, a receiver completely perpendicular to the exciting, adjusting it to the mutual inductance between two coils is minimal, and a third coil to function as "mirror-sensor" of the magnetic field generated by the exciting coil.

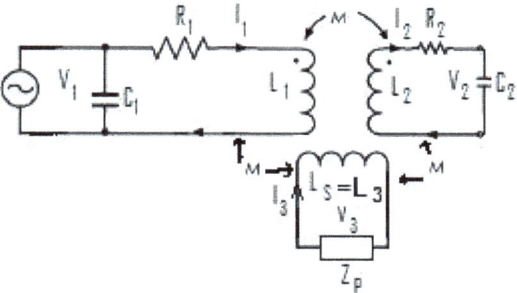

Figure 6. Representative schematic of the three coil arrangement

3.1. Drive coil L1

For generating a spectral magnetic field a flat coil was used as a transmitting antenna built on a phenol board as printed circuit with the following characteristics:

- Spiral coil [12] internal diameter of 2 cm and an outer diameter of 9.5 cm with 20 turns, with cooper tracks width of 500 μm , and an equal distance between them, see Figure 7
- 24 μH Inductance measured experimentally in a frequency range of 100 KHz to 5 MHz
- Shielding is used to minimize capacitive coupling at the top and bottom of the exciting coil, forming a sandwich, this shielding was "grounded".

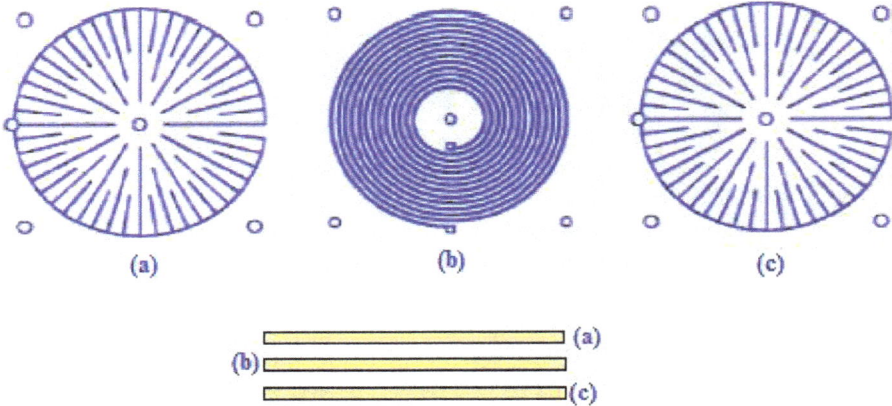

Figure 7. Coil and Shielding employee

The Network Analyzer employee served as a source of power in a frequency range of 100 KHz to 5 MHz applied. The combination coil and capacitance of the cables, that although were coaxial cables it presented a resonance frequency of 4.612 MHz This initial structure was proceeded with a series of measurements with actual physical capacitors with a dual role, first see the system answer at different values, and the second as reference calibration [13].

3.2. Receiver coil L2

For reception of the spectrum magnetic field, also use a flat coil as the receiving antenna, built on a phenol board as printed circuit with the following characteristics almost as similar to the transmitting antenna:

- Spiral coil internal diameter of 2 cm and an outer diameter of 9.5 cm with 20 turns, with cooper tracks width of 500 μm, and an equal distance between them, see Figure 8
- 24μH Inductance measured experimentally in a frequency range of 100 KHz to 5 MHz
- Shielding is used to minimize capacitive coupling in the bottom of the receiving coil, this shield was "grounded".

Use of Magnetic Induction Spectroscopy in the Characterization of the Impedance of the Material with Biological Characteristics

177

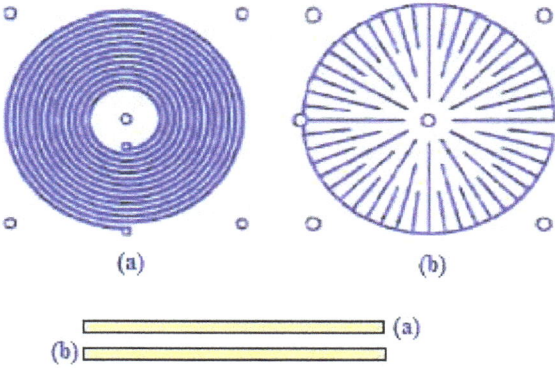

Figure 8. Coil and shield used

- In addition to the above conditions and in order to minimize inductive coupling both coils are placed in perpendicular way see Figure 9, so that the only magnetic field received were the projected by the coil sensor.

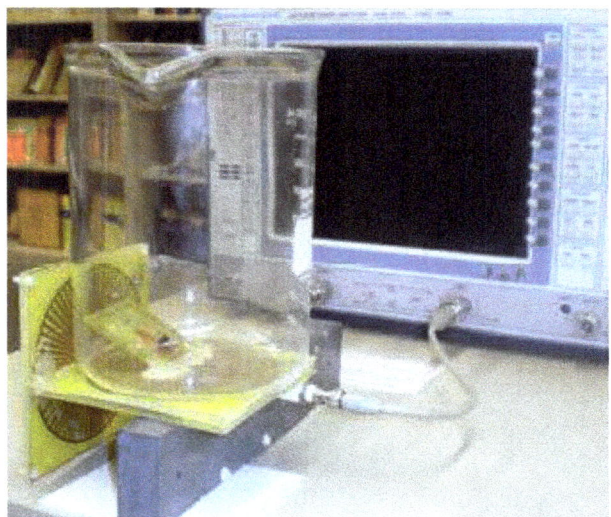

Figure 9. Mechanical available of the coils to minimize Inductive coupling between transmitter coil and receiver coils

Network Analyzer employee close the system, as seen in Figure 9, the application is through the coil L1, placed horizontally, which serves as basis for the deposition of both the saline and the samples biological tissue. Receiving and monitoring of the signal through the coil L2, vertically positioned and perpendicular to L1

3.3. Mirror-sensor coil

As a passive coil mirror-sensor used a flat square coil of about 7.92 µH, see Figure 10, with a measured value of 9.716 µH, at a frequency applied of 1 MHz, Coil with a capacitor of 330 nF added to make a 1 MHz resonant circuit, finally resulting a resonant frequency of 1.14 MHz, like the previous coils was constructed on phenol board as a printed circuit having the following characteristics:

- The inductance L, was determined by experimental measurements, confirming the calculations used in the approximation developed by Ong et al[14].

$$L = 1.39 * 10^{-6}(OD + ID) * N_L^{\frac{5}{3}} * log(4 * \frac{OD+IDD}{OD-ID}) \qquad (2)$$

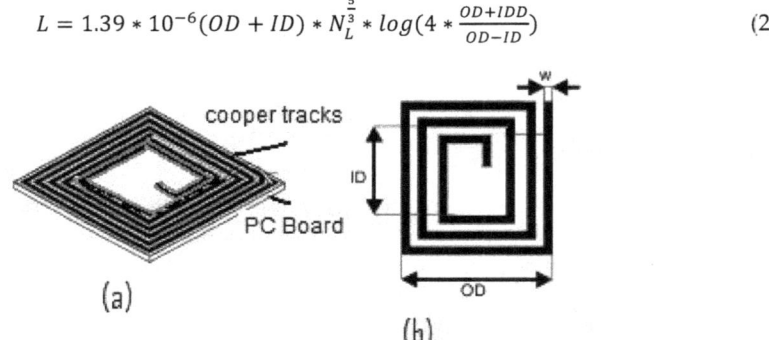

Figure 10. Layout of the Mirror-Sensor coil with interdigital capacitor (not shown)

- Square spiral coil internal diameter of 3.6 cm and an outer diameter of 4.7 cm with 10 turns, according to the expression [1] L = 7.92 µH
- To minimize resistive-capacitive effect between turns, when immersed in saline was used an insulating paint.

4. Characterization of the system

The technique of network development has in its whole with the element to explore interesting properties, which generates all possible parameters of the immittance associated with the two-port network, Figure 11.

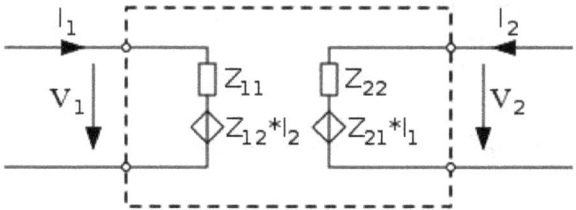

Figure 11. Electronic models of the transmitter coil and receiver coil

Lse of Magnetic Induction Spectroscopy in the Characterization of the Impedance of the Material with Eiological Characteristics

179

The formulation of network equations obtained by choosing voltages across the capacitors which are not physically in our arrangement, but it is necessary to include consideration coaxial cables within the coil system, which by their physical characteristics have a value of capacitance and current flows through the antenna-coils (inductors) as variables used in the description of the network, produces many equations as reactive elements present in a given network.

VI relations of our model, without loss, respecting the conventions of voltage and current shown in Figure 11, are[15]:

$$V_1 = Z_{11} * I_1 + Z_{12} * I_2 \tag{3}$$

$$V_2 = Z_{21} * I_1 + Z_{22} * I_2 \tag{4}$$

Since the primary reactive elements of this network are the inductors $Z_{11} = j\omega L_1$ y $Z_{12} = j\omega M$; $Z_{22} = j\omega L_2$ y $Z_{21} = j\omega M$ and since there is an inductive link between the coils, equations [2] and [3] are reconsidered:

$$V_1 = j\omega L_1 * I_1 + j\omega M * I_2 \tag{5}$$

$$V_2 = j\omega M * I_1 + j\omega L_2 * I_2 \tag{6}$$

in equations [4] and [5], M is the mutual inductance (dimension H) between the coils L_1 and L_2, defined by $M = k * \sqrt{L_1 * L_2}$ where k (dimensionless number, ≤ 1) is the coupling factor between the inductors (inductive link ratio), controllable with the relative position between them.

The model presented in Figure 6 allows the development of matrix impedance parameters of a bi-port, as expressed in equations [3-6], and involves the mutual inductance of the coil system, then presents the mathematical analysis of the inductive link.

The design and structure of the coil system aims to make the inductive link between the coils L1 and L2 is minimized by establishing a physical way the perpendicularity between L1 and L2, seeking with this strategy to establish a relationship of mutual inductance increased by engaging in two-port network a third coil that provides a "mirror" effect, the Figure 12 shows a representation of the this inductive link.

The mutual inductance M_{12} of the inductor L_2 in relation to L_1, which (according to the reciprocity theorem) it was found experimentally equal to the mutual inductance M_{21} of the inductor L_1 in relation to L_2, theoretically defined as the coefficient of mutual magnetic flux $\Phi_{12} = \Phi_{21}$ which spans both coils due to the current flowing in each coil respectively.

$$M_{12} = \frac{\Phi_{12}}{I_1} \tag{7}$$

$$M_{21} = \frac{\Phi_{21}}{I_2} \tag{8}$$

"The Biot-Savart law, states that if a small length of conductor δl carrying a current i, then the magnetic field strength at a distance r and angle θ is

$$\delta H = \frac{i\delta l sin\theta}{4\pi r^2} \tag{9}$$

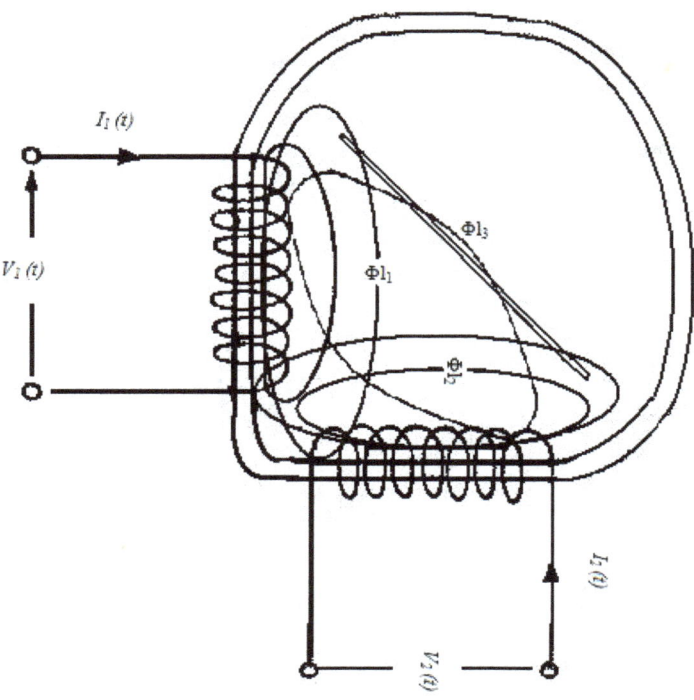

Figure 12. Approximate representation of variable in time magnetic induction in the coils system

(Sine θ merely states that if power is not in an optimal direction, then the field at that point decreases.) This is confirmed by Ampere's rule, $\int H dl = i$; The line integral of the magnetic field in a closed loop is equal to the electric current flow through the closed loop" [16]. In particular, the magnetic field strength H in a circular path of radius r around an electric current i in the center is

$$H \cdot 2\pi r = i, o \ B = \frac{\mu i}{2\pi r} \tag{10}$$

Because L_1 and L_2 have very similar characteristics, and considering the expression [9] may approximate the magnetic field concentric loops through both coils (see Figure 8a), B_1 is the magnetic field of the first ring, the flow magnetic Φ_2 through second ring we can determine from B_1.

$$\Phi_2 = B_1 A_2 = \left(\frac{\mu_0 I_1}{2R_1}\right)\pi R_2^2 = \frac{\mu_0 \pi I_1 R_2^2}{2R_1} \tag{11}$$

where μ_0 is the magnetic permeability of free space, R_1 and R_2 are the radii of the rings which form the coils.

The mutual inductance is then:

$$M = \frac{\Phi_2}{I_1} = \frac{\mu_0 \pi R_2^2}{2R_1} \tag{12}$$

This expression shows that M depends only on geometric factors, R_1 and R_2, and is independent of the current in the coil. As regards the expression $M = k * \sqrt{L_1 * L_2}$ to obtain an approximation of mutual inductance between coils, since low values of coupling coefficients "k" with air-core coils are obtained usually in the order of 0.001 to 0.15 [17] experimentally is considered as the value of $k_1 = 0.00167$ (caused by a misalignment both angular and lateral between coils L_1 and L_2), and a value of $k_2 = k_3 = 0.465$ (caused by an angular misalignment between coils L_1, L_2 over L_3), so we get the following values:

$$M_{31} = M_{13} = 7.1\mu H$$

$$M_{23} = M_{32} = 7.1\mu H$$

$$M_{21} = M_{12} = 0.04 \ \mu H$$

By obtaining these values, we proceeded to simulate the circuit in PSpice [18], see Figure 13, to verify that they were not far from the actual physical model, which was obtained following graphs.

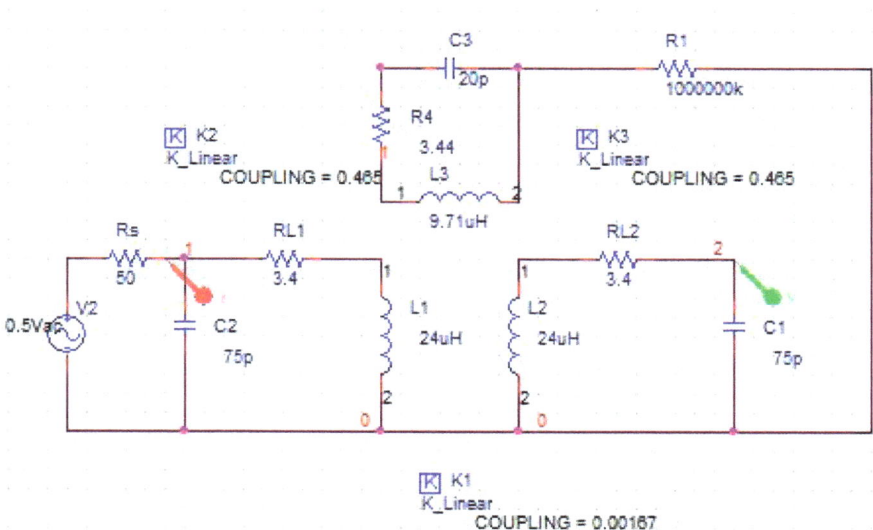

Figure 13. System coils with their respective coupling factors for achieving the simulation circuit had to be "grounded" L3 but with the highest permissible value, simulating a physical disconnection between the two coils.

In Figure 14 we can observe the system performance in terms of voltage induction refers to both L_2 and L_3.

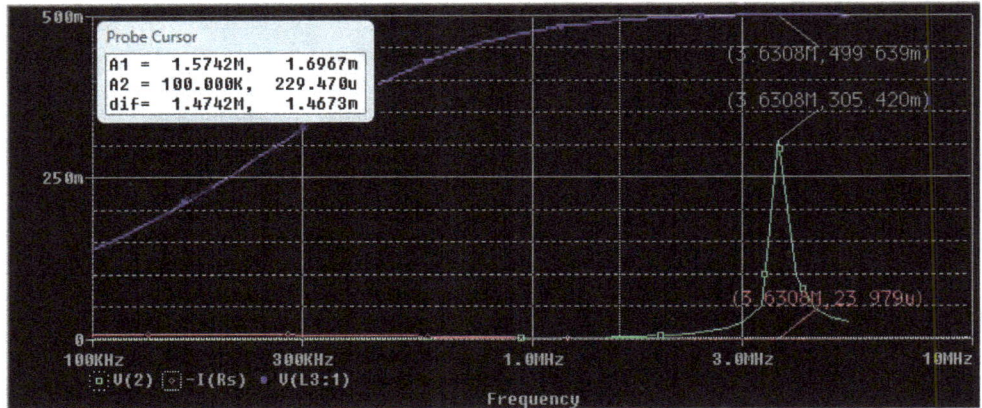

Figure 14. Response of the coil system to a frequency sweep from 100 kHz to 5 MHz, with estimated coupling factors.

The obtained simulation results permit the development equations which describe the model in Figure 6; these can be expressed in matrix form:

$$\begin{bmatrix} V_1 \\ V_2 \\ V_3 \end{bmatrix} = j\omega \begin{bmatrix} L_1 M_{12} M_{13} \\ M_{21} L_2 M_{23} \\ M_{31} M_{32} L_3 \end{bmatrix} \begin{bmatrix} i_1 \\ i_2 \\ i_3 \end{bmatrix} \tag{13}$$

M_{ij} are mutual inductances, and the proportionality factors of the currents are the impedances so that the matrix can be rewritten in terms of the Z parameters.

$$\begin{bmatrix} V_1 \\ V_2 \\ V_3 \end{bmatrix} = j\omega \begin{bmatrix} Z_{11} Z_{12} Z_{13} \\ Z_{21} Z_{22} Z_{23} \\ Z_{31} Z_{32} Z_{33} \end{bmatrix} \begin{bmatrix} i_1 \\ i_2 \\ i_3 \end{bmatrix} \tag{14}$$

As experimentally determined that $Z_{12} = Z_{21}$ be as small as possible to perceive impedance changes as the sensor coil L_3, so we have:

$$Z_{12} = \frac{V_1}{i_2}|i_1 = 0; Z_{21} = \frac{V_2}{i_1}|i_2 = 0$$

$$\frac{V_3}{i_3} = Z_3 \tag{15}$$

If we consider the equations [14] and [15] in combination with the matrix [12], expressing V2 in terms of Z3, we have:

$$V_2 = j\omega M_{21} i_1 + j\omega M_{23} \frac{V_3}{Z_3} \tag{16}$$

5. Making a reference

Considering the previous development, and articles concerning the use of systems like ours (Ong. 2000, Hofmann 2005) is therefore important to have baseline measurements

Use of Magnetic Induction Spectroscopy in the Characterization of the Impedance of the Material with Biological Characteristics

183

of physical elements such as capacitors, to be served at one time as models of reference and calibration, was determined to carry out the completion of the procedure detailed below.

Figure [16] the electrical circuit presented is the model of a capacitive type humidity sensor [19], the behavior of this sensor is comparable to the behavior of L_3 so we can take the development of the equation for Z_P that would be the Z_3 response, which could take the complex impedance as:

$$Z_P = R_s + \frac{R_p}{1+jwR_pC} \tag{17}$$

Misevich et al refers to the resistance changes predominantly in the range of 10-100 MΩ in a circuit as shown in Figure [15], fall about 0.5 Ω in a very humid environmental, is not our case so is taken as a criterion for considering a single value of R_P of 100Ω, likewise considering the impedance presented by R_P, varying w in the range of 12e^{+6} to 25e^{+6} rad / s the impedance C in parallel with R_P take R_P value in the range of C values from 10pF-100pF, and goes to take the R_s value in the range of C values from 1nF-100nF, see table 1, that added to the XL values presented by w is negligible this value so we can reduce our equation to:

$$Z_P = \frac{R_P}{1+jwR_PC} \tag{18}$$

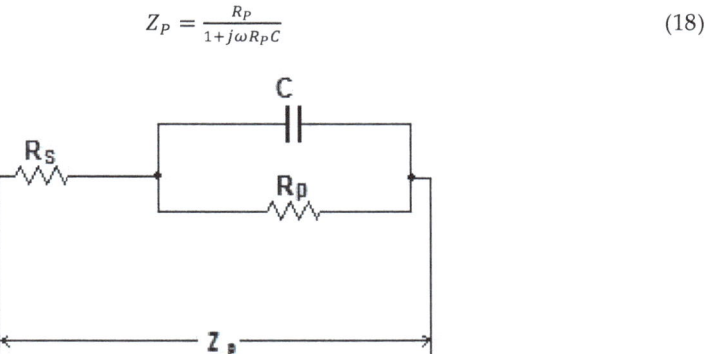

Figure 15. Misevich, K.W. shows the development of a capacitive humidity sensor element impedance equivalent to measuring the coil system, where $R_s = 3.44\Omega$ series resistance of the sensor, $R_P = 100\Omega$, C multivalued see table 2

Val. Cap./Freq.	2.5MHz	3MHz	3.5MHz	4MHz
10pF	98.4Ω	98.1Ω	97.8Ω	97.5Ω
100pF	86.4Ω	84.1Ω	82Ω	80.0Ω
1nF	38.9Ω	34.6Ω	31.25Ω	28.5Ω
10nF	5.98Ω	5.04Ω	4.35Ω	3.82Ω
100nF	0.63Ω	0.53Ω	0.45Ω	0.4Ω

Table 1. $Z_P = Z_3$ values obtained by calculation.

$$Z_3 := \frac{R_3}{1+(j \cdot R_3 \cdot C_3)} = 97.459 - 17.73i \qquad (19)$$

Expression [19] used for calculating the impedance of Table 1 using MathCad [20]; Therefore the values obtained with small capacitors values by calculating means as measured by the analyzer is in a range of 5% in 3.09 MHz, the resonance frequency, and not higher capacitor values that are very de-correlated, this is due to the length of the coaxial cables, overriding the capacitance introduced by these.

Val. Cap./Freq.	2.5MHz	3MHz	3.5MHz	4MHz
10pF	96.18Ω	102.7Ω	107.8Ω	112Ω
100pF	100.9Ω	107Ω	110.5Ω	112.6Ω
1nF	60.2Ω	41.4Ω	29.5Ω	24.5Ω
10nF	22.6Ω	29.6Ω	36.2Ω	42.7Ω
100nF	29.6Ω	35.7Ω	41.7Ω	47.63Ω

Table 2. Z_3 values obtained by direct measurement by network analyzer.

If we consider the matrix equation [12] and equation [18], V_3 can be expressed in terms of $Z_P = Z_3$, we have then:

$$V_3 = \frac{j\omega M_{23} i_2}{1 - \frac{j\omega L_S}{Z_3}} \qquad (20)$$

Considering the matrix equation [12] and equation [19], we have V_1 expressed in terms of Z_3:

$$V_1 = j\omega M_{12} i_2 + j\omega M_{13}\left(\frac{j\omega M_{23} i_2}{Z_3 - j\omega L_S}\right) \qquad (21)$$

If we consider the equation [14 and 20] we express Z_{12} as:

$$Z_{12} = j\omega M_{12} - \omega^2 \left(\frac{M_{23} M_{31}}{\sqrt{\left(\frac{1}{R_p}\right)^2 + (\omega C)^2} - j\omega L_S}\right) \qquad (22)$$

Figure 16 shows the values obtained by performing calculations using equation [22] implemented in Mathcad, it is appreciated also that around the two resonance frequencies in the system the magnitude of Z_{12} is the same.

Magnitude

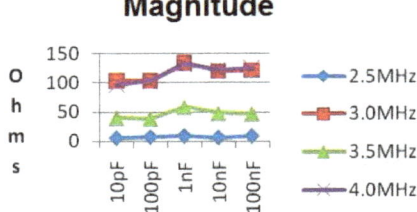

Figure 16. Comparison between Z_{12} Magnitude values of impedance calculation in Mathcad.

Use of Magnetic Induction Spectroscopy in the Characterization of the Impedance of the Material with
Biological Characteristics

185

6. Validation and calibration

Figure 17a shows the graphical model used for our experiment, and Figure 17b refers as a practical way to the interface between the network analyzer and sensor description given above in a comprehensive manner.

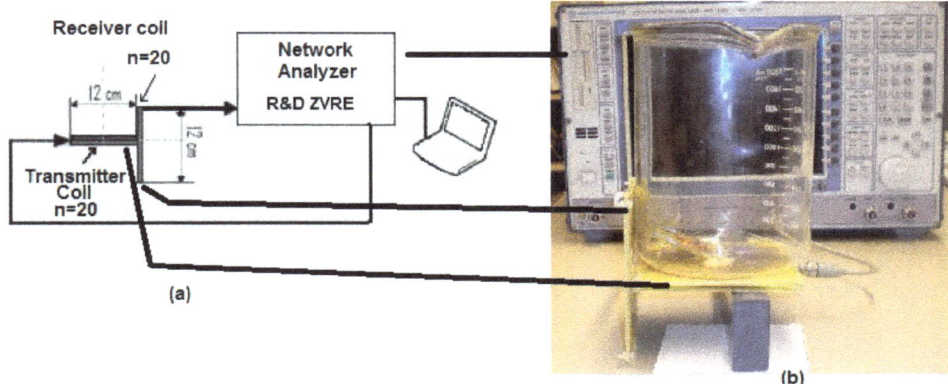

Figure 17. Magnetic Impedance Spectroscopy Method: a) block diagram, and b) practical implementation

Figure 18 a shows a block diagram of a three-port circuit, the voltages and currents are indicated, the upper terminal of the instance is positive with respect to the terminals of the bottom and the currents flowing inwards as indicated by KCL for each port, Figure 18b shows the block diagram of a real form.

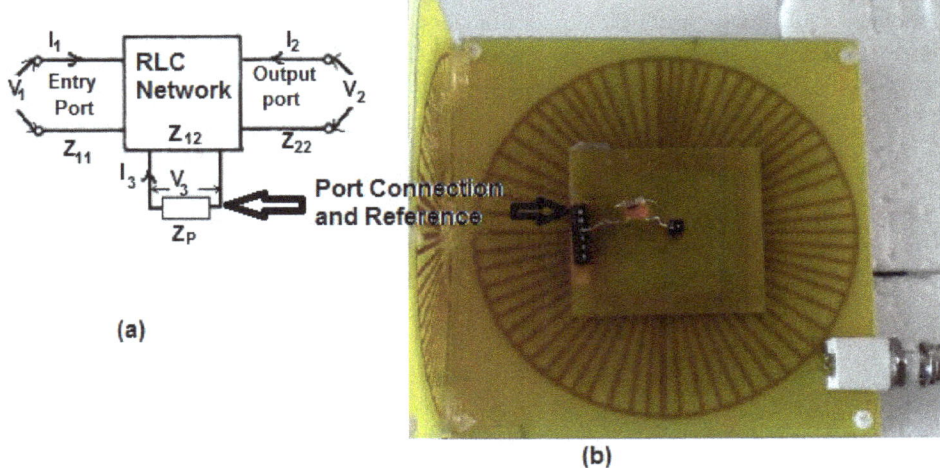

Figure 18. Three-port circuit in preparation for constant section cell measuring based on measurement of physical components.

The values obtained for both measurement and calculation as shown in Table 2 allows us to make the cell preparation of constant section for measuring conductivity. For this preparation had to consider the mechanical error sources such as physical dimensions of the saline cell, the effects of temperature on the cell, and error handling and positioning of it. It also ponder the discrepancy between the measurements of a real physical element such as a capacitor in parallel with a resistor, and the resistivity and permittivity expected of an electrolytic cell, it was therefore necessary to have a truer reference of known salt solution to evaluate deviation of our measurements and bring the measuring physical model to a virtual model and simulate their behavior in this way to "induce" their performance, come to perform these measurements with electrodes.

In Figure 19 shows a sample obtained from a series of measurements made under the scheme of Figure 18, changing the values of the capacitances in parallel with resistors first small values, reaching values used to 1 kΩ .

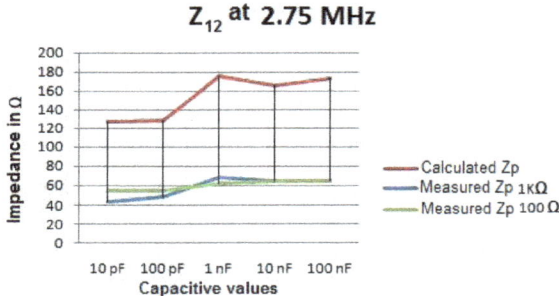

Figure 19. Comparisons between reference measurements with physical components and estimated values

Although performing multiple measurements and their subsequent acquisition of information due to the speed and flexibility of use of equipment to automate this process, include only the most representative of the behavior of a saline cell, as seen in Figure 20, with various degrees of salinity.

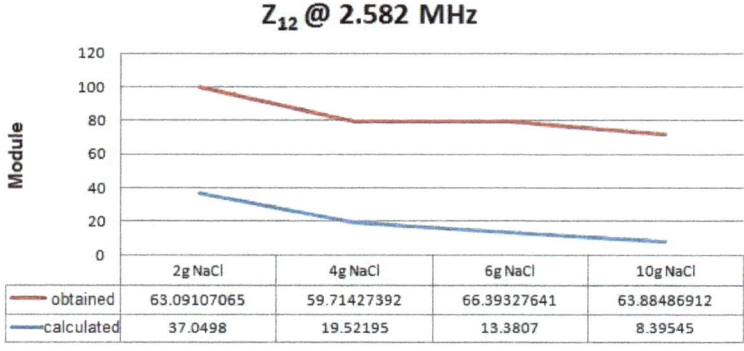

Figure 20. Measurement of Z_{12} of four saline cells

Values obtained with the saline cell, Figure 20, as well as the physical components are between 4% and 8% above the Z12 average obtained from the saline cell, provides that comparatively physical model is closer to cell model representing a saline cell of Figure 21, which is formed by a resistance of 100 Ω in parallel with a capacitor of 1 nF, with a response very "right" to the resonant frequency of the coil system which is approximately 3 MHz

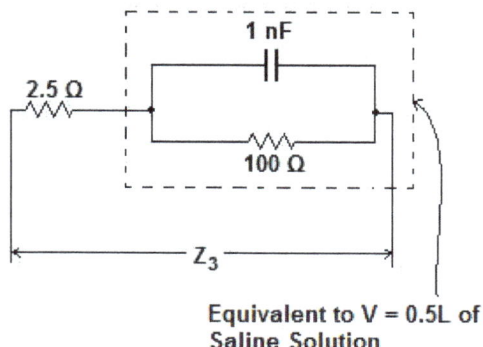

**Equivalent to V = 0.5L of
Saline Solution**

Figure 21. Physical model equivalents to a 500 mL saline cell at a frequency of ~ 3 MHz

The conductivity of a saline solution with 2 g of salt dissolved in 1L of deionized water having a conductivity measurement $5.8\mu S/cm$, presents a $\sigma = 3.9$ mS / cm at 25 ° C and $\sigma = 3.4$ mS / cm at 20 ° C, this represents a 0.2% concentration and 0.034 M.

If we use this solution as the first reference electrolytic cell and according to equation [1], with an equal volume to 500 mL and a length of 85 mm, we calculate a 37.05Ω impedance (Z) , with these values and data from our physical model equivalent gives a $\sigma = 2.29$ mS / cm which leads us to obtain a correction term 3.9/2.29 = 1.7.

7. Measuring saline cell permittivity with mis

The resistivity of an aqueous sodium chloride dissolution (NaCl) [21] is obtained from considering the current density (J), determined this by the ions types (Na^+ and Cl^- are the most abundant ions) of the solution, this is directly proportional to the factor "α" of dissociation of molecules, approximately equal to 1 if the electrolyte is strong, with a concentration "c "gram-equivalent, and a mobility" μ "of the ions, also the electric field" E "as shown in the following relationship: $J = F \alpha c \mu E$.

A solution consisting of 9 grams of sodium chloride dissolved in one liter of water in medicine[2] is called Normal Saline, since the concentration of 9 grams per liter divided by 58 grams per mole (approximate molecular weight of sodium chloride) provides 0.154 moles per liter, that is, contains 154 mEq / L of Na^+ and Cl^-. The fact contain more solute per liter,

---

[2] In chemistry, the normal concentration of sodium chloride is 0.5 mol of NaCl assuming complete dissociation. Physiological dissociation is approximately 1.7 ions per mole, so that a normal NaCl is 1/1.7 = 0.588 molar. This is approximately 4 times more concentrated than the medical term "normal saline" of 0.154 mol

makes this solution with a slightly higher osmolarity[3] than blood, on an average day the natremia[4] can range between 130-150 mEq / L (normonatremia) . However, the osmolarity of normal saline is very close to the osmolarity of the NaCl in the blood[22].

From Figure 22 and considering the dimensions of our saline cells, presents three values of resistivity, $\varrho = 2.65\Omega m$, $\varrho = 0.55\Omega m$ and $\varrho = 1.3\Omega m$, respectively.

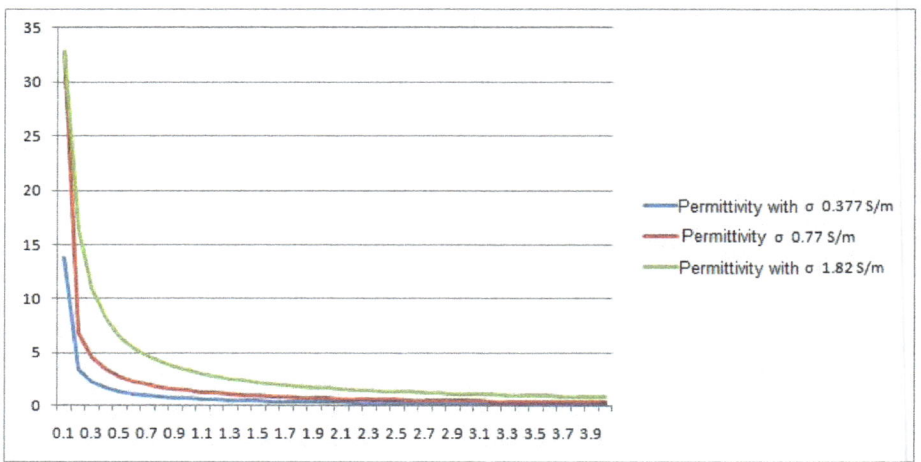

Figure 22. Measured frequency behavior of three electrolytic cells

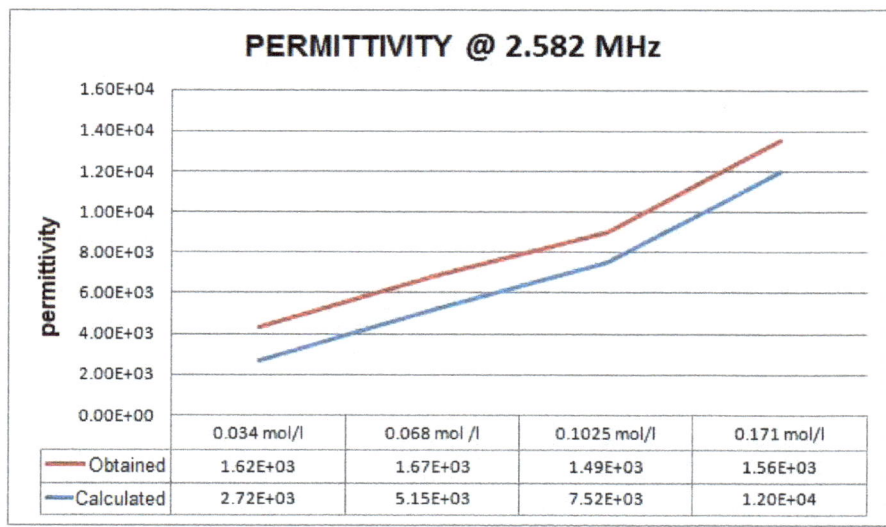

Figure 23. Saline solutions response at resonance frequency of the coil system

---

[3] Measuring solute concentration
[4] Concentration of sodium in the blood

Use of Magnetic Induction Spectroscopy in the Characterization of the Impedance of the Material with Biological Characteristics

189

Figure 23 presents a summary of values acquired with the network analyzer, Figure 24 represents the values obtained with the HP4192A impedance analyzer.

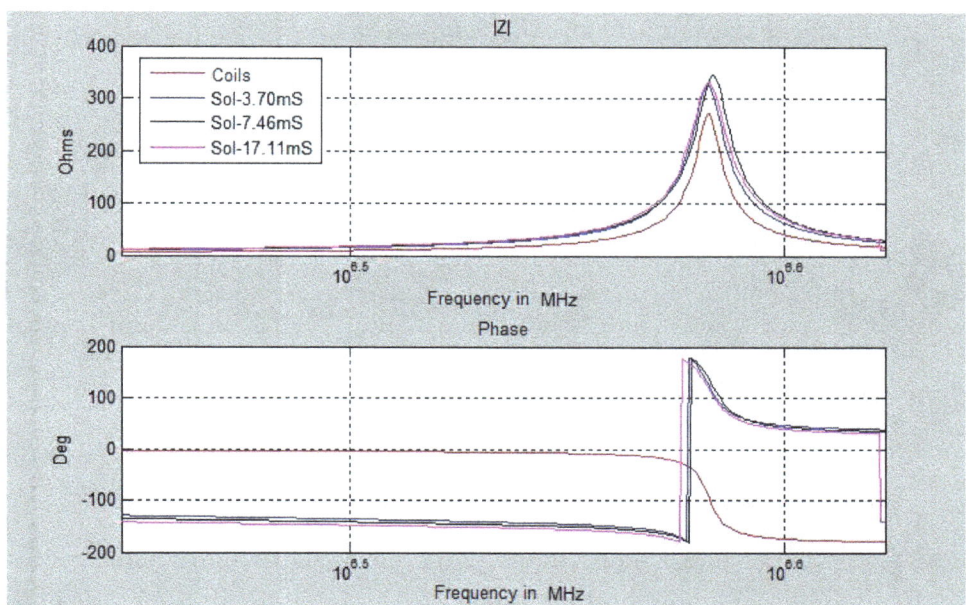

Figure 24. Frequency responses of Saline Solutions.

3. Biological suspensions

Typically, in vitro measures, is considered a sample of uniform section (A) and length (L), in the majority of cases it is more practical to consider the volume (V) in aqueous solutions. In this case the ratio between the conductivity and the permittivity of the sample with the resistance and reactance are:

$$C = \frac{\varepsilon\,\varepsilon_0\,V}{L^2}\ [F] \tag{22}$$

$$R = \frac{L^2}{\sigma\,V}\ [\Omega] \tag{23}$$

$$Z = \frac{L^2}{(\sigma + j\,\omega\,\varepsilon\,\varepsilon_0)\,V} \tag{24}$$

Following experiments with saline solutions, was considered to have the information needed to proceed to carry out experiments with biological solutions, hence the first use of coil system as a reference and according to the above expressions, and also with a 500 mL volume, and a tank length of 85 mm was obtained with a suspension of high concentration of biomass (80 g / l of yeast), a 8.305 dB attenuation, comparing this with a 13.71 dB attenuation of coil system, this means a representation of a 65% increase in impedance in the suspension.

It should be mentioned that the embodiment of the vast majority of measurements for suspension was made with the HP4192A impedance analyzer, considering the structure and existing coil system, the utilization ratio of the impedance analyzer is because the probes used to interconnect arrangement coils are constructed on purpose, greatly decreasing the amount of error attributable to the length of the coaxial cables.

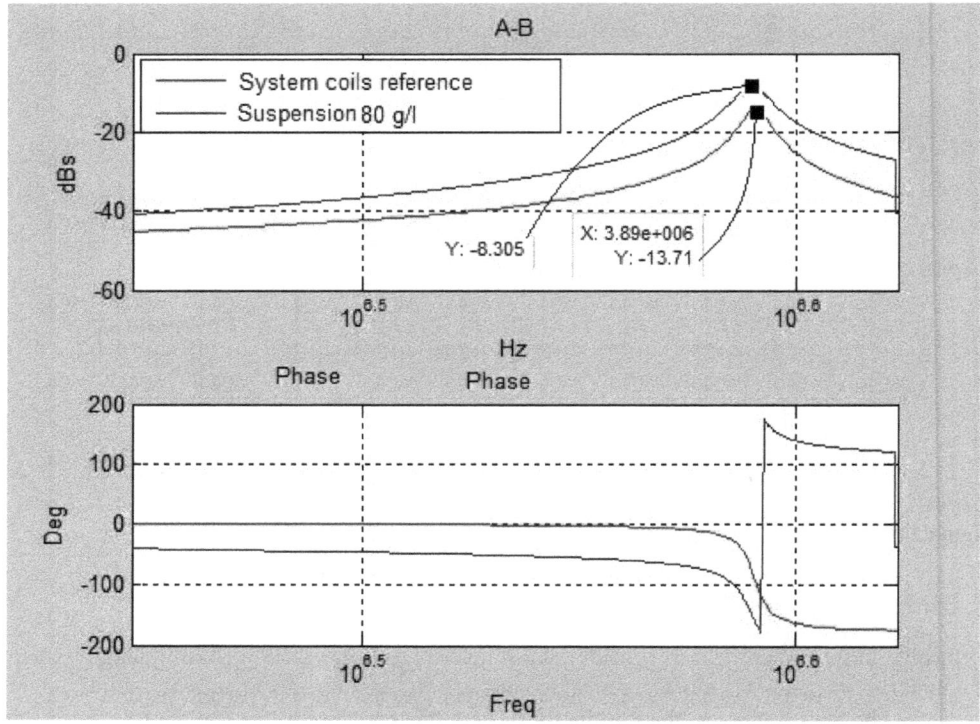

Figure 25. Impedance Representation of Biological Suspension, comparatively to the coil system.

After 35 minutes, the yeast was deposited at the bottom of the container; we proceeded to extract 350ml of water and recovered the same amount of water (350ml), but now with a salinity 8.78mS/cm.

The interesting thing in this experiment as shown in Figure 25, having increased conductivity, as a result of water replaced without electrolytes, with water with electrolytes (8.78mS/cm) decreased by 6% the impedance of the suspension by the lower proportion of yeast and major salinity, $Z_{lev} = 384.37\Omega$, $Z_{lev} + NaCl = 361.87\Omega$, $\Delta\Omega = 22.5$

9. Comparison of results

Measurements of saline cell and a biological suspension with HP impedance analyzer, allow us to characterize the results:

Lse of Magnetic Induction Spectroscopy in the Characterization of the Impedance of the Material with
Biological Characteristics

191

	Z (Ohms)	ε	μ (H/m)	σ (S/m)
2g NaCl	325.3	14.45	2.13E-04	0.00307409
4gNaCl	334.9	14.039	2.13E-04	0.00298597
10gNaCl	333.3	14.11	2.13E-04	0.0030003

Table 3. Characteristic values of a saline cell saline with varying degrees of salinity at resonance
requency of the coils system @ 3.825 MHz

	Z (Ohms)	ε	μ (H/m)	σ (S/m)
80g_yeast	395.64	11.68	2.16E-04	0.00252755
80g_yeast+6h	392.42	11.77	2.17E-04	0.00254829
80g_yeast+24h	385.08	11.99	2.17E-04	0.00259686

Table 4. Characteristic values of a biological suspension with various lengths of time at resonance
frequency of the coils system @ 3.892 MHz

10. Conclusion

The presentation of results, by their very short nature, could be interpreted as an activity which is not time consuming, but the opposite is true, because from the moment of preparation of the experiments, there are always a number of imponderables, such as materials or materials that are needed to carry out the measurements do not have them, at least, operating conditions, climate, lighting, etc..

For the characterization of substances, suspensions and / or solutions must take into account, and in a very particular, maintain the same amounts, and make measurements in a repetitive manner, so as to place on record its findings to conditions geometric, physical and mechanical properties, different and contrast with conditions similar to those obtained. The measurements carried out with the Network Analyzer, which was mostly used equipment, are quite contrasting with measurements made with the HP impedance analyzer. As we can see with the first, which strongly influences the distance at which measurements were made, not with the impedance analyzer as it had a "fixture" on purpose.

Speaking of the actual material used and specific the experiment with saline cell, it was found that one of the factors that influence in the admittance at low frequencies is the electrical permittivity of water. The values depend on the frequency of measurements, and the frequency sweep, this due to the response of the dielectric constant of the solution, which varies considerably with frequency.

Using a saline solution with a high degree of salinity, more than 9 g / l of salt, we see that in the graphs of the results if ε'' / ε' '<2, the ε'' accuracy degrades. This may be because the impedance of solutions with high salt is essentially resistive. In an opposite manner with a "window" at 999 KHz frequency on average, before the resonance frequency of the coil circuit, the impedance of the solution is essentially reactive, and since this type of impedance measurements of reflection / transmission were performed mainly with the network analyzer is a significant higher level of uncertainty in measuring the smaller capacitive component where ε'' is derived.

The HP impedance analyzer measurements, both saline solutions and suspensions, allowed us to compare results, specifically ε" values. In the suspensions could be evaluated especially an increase in the impedance, over time, possibly due to increased cell growth, hence an increase in the capacitive reactance of the suspension.

Considering the results obtained, we can consider the approach to the characterization as reasonably good since the relative permittivity is much greater than unity for the dielectric (both saline solutions and biological suspensions).

One of the properties of biological tissue, such as conductivity, is well reflected by the frequency dependence, particularly in the range of 10 KHz-10MHz (β dispersion range). The conductivity at low frequencies denotes the volume of extracellular fluid essence, the additional contribution of intracellular fluid volume with a significant increase in the applied frequency causes a significant increase in conductivity.

And finally and considering the substantial increase in interest in the development of magnetic induction spectroscopy (MIS) as a valid option for obtaining the conductivity of the human body without the need for direct contact with tissue (Korzhenevskii and Cherepenin 1997 Griffiths et al. 1999, Korjenevsky et al. 2000); In addition to other passive electrical properties of biological tissues (Hermann Scharfetter, Casañas Roberto and Javier Rosell, 2003).

Since MIS is based on measurements of small changes in magnetic fields, typically of the order of 1% or less at frequencies up to 10 MHz, and also because of their physical limitations is not recommended at frequencies below 10 kHz, this also represents a great challenge in electronic design, possibly one of its greatest disadvantages.

Author details

Jesús Rodarte Dávila, Jenaro C. Paz Gutierrez and Ricardo Perez Blanco
Department of Electrical and Computer Engineering, Juarez City Autonomous University,
North Charro Avenue, Juarez City, Chihuahua, México

Acknowledgement

My eternal gratitude to Dr. Ramon Bragos Badia all his efforts and attention afforded to this humble servant, without which there would have been possible to achieve this work.

Ramon the Humans are more humans for humans like you, Thanks. Also I would like to thank to Petra Salazar for his typing work of this paper, without your help this work may be stay in the abstract.

11. References

[1] Grimnes S. Bioimpedance and Bioelectricity Basics, Academic Press ISBN 0-12-303260-1;p7

2] Hoffer EC, Clifton KM, Simpson DC. Correlation of wholebody impedance with total body volume. J Appl Physiol 1969;27:531–4.

3] Nyboer J. Electrical impedance plethysmograph, 2nd ed. Springfield, IL: CC Thomas; 1970.

4] Boulier A, Fricker J, Thomasset A-L, Apfelbaum M. Fat-free mass estimation by the two-electrode impedance method. Am. J Clin. Nutr. 1990; 52:581–5.

5] Harris A. Ductal epithelial cells cultured from human fetal epididymis and vas deferens: relevance to sterility in cystic fibrosis, Journal of Cell Science, Vol 92, Issue 4 687-690, Copyright © 1989 by Company of Biologists

[6] Asami K. Real-Time Monitoring of Yeast Cell Division by Dielectric Spectroscopy, Biophys J, June 1999, p. 3345-3348, Vol. 76, No. 6

[7] Ong K.G., Monitoring of bacteria growth using a wireless, remote query resonant-circuit sensor: application to environmental sensing. Copyright © 2001 Elsevier Science B.V. All rights reserved.

[8] Hofmann M. C., Transponder Based Sensor for Monitoring Electrical Properties of Biological Cell Solutions. Copyright © 2005 The Society for Biotechnology, Japan Published by Elsevier B.V.

[9] Scharfetter H., Casañas R., and Rosell J., "Biological Tissue Characterization by Magnetic Induction Spectroscopy (MIS): Requirements and Limitations,"IEEE Transactions on Biomedical Engineering, vol. 50, NO. 7, JULY 2003.

[10] http://www.ni.com/support/labview/lvtool.htm

[11] Service Manual R&S®ZVx Vector Network Analyzer Family: http://www2.rohde-schwarz.com/en/products/test_and_measurement/network_analysis/ZVx-%7C-Key_Facts-%7C-4-%7C-658.html

[12] Riedel C.H., Keppelen M., Nani S., Merges R.D. y Dössel. (2004) "Planar System for Magnetic Induction Conductivity Measurement Using a Sensor Matrix". Institute of Physics Publishing., Physiol. Meas. 25

[13] Rodarte J. (2008) Método Experimental de Medida de Impedancias Inalámbricamente Usando un Analizador de Redes. Instituto Tecnológico De Chihuahua., Electro 2008.

[14] Ong, K.G. and Grimes, C. A., "A resonant printed-circuit sensor for remote query monitoring of environmental parameters". Smart Mater. Struct., 9, 421–428 (2000).

[15] Edminister J., Nahvi M., "CIRCUITOS ELECTRICOS Y ELECTRONICOS", Cap. 13. Cuadripolos (circuitos de dos puertas), ISBN: 8448145437, Ed: McGrawHill, Serie Schaum

[16] Calculation of Formulae - Self-inductance ; www.ivorcatt.com/6_2.htm

[17] Donaldson N.deN., Perkins T.A., "Analysis of resonant coupled coils in the design of radio frequency transcutaneous links", Med. & Biol. Eng. & Comput., 1983, 21, 612-627

[18] Commercial Software "PSpice" Ver 9.2 Copyright 1886-1999 by Cadence Design Systems

[19] Kenneth W. Misevich, Capacitive Humidity Transducer, IEEE Transactions on Industrial Electronics and Control Instrumentation, vol. IECI-16, NO. 1, JULY 1969.

[20] Mathcad, version 14.0.0.163 Copyright © 2007 Parametric Technology Corporation. All Rights Reserved.

[21] Pething, 1979 R. Pething. Dielectric and Electronic Properties of Biological Materials, Wiley (1979).

[22] Awad S., Allison S.P., Lobo D.N., "The history of 0.9% saline.", Clin. Nutr. 2008 Apr; 27 (2): 179-88. Epub 2008 Mar 3.

Chapter 7

Non-Destructive Surface Analysis by Low Energy Electron Loss Spectroscopy

Vitaliy Tinkov

Additional information is available at the end of the chapter

1. Introduction

The modern progress in such priority scientific directions as microelectronics, nanotechnology, material science, heterogeneous catalysis, etc., are impossible without obtaining quantitative information about physical–chemical properties from the nano-size near surface region of the materials.

It is known that the physical–chemical properties of the metallic alloy surfaces differ markedly from that of the bulk and, mainly, it is caused by segregation one of the alloy components on the surface [1]. It is related to the fact that the physical–chemical state of the surface substantially influences such surface processes as adsorption, catalysis, oxidation, friction and wear. Recently such phenomena as the thermo-induced surface segregation of alloy components as used for obtaining chemically-active surfaces have been widely used; being of great interest in terms of heterogeneous catalysis and the development of new nanotechnological processes. Study of the kinetics of surface segregation permits the determination of the bulk diffusion coefficients of the segregated elements; knowledge of which then permits the controlled change of surface structures under heat treatment and etc.

At the present for the investigation of the physical–chemical properties of the metallic alloy surfaces the nondestructive methods are widely used, such as an Rutherford Backscattering Spectrometry (RBS), X-ray Photoelectron Spectroscopy (XPS), Low Energy Ion Scattering (LEIS), Ultraviolet Photoelectron Spectroscopy (UPS) and other [2].

Physical phenomena such as secondary electron emission (SEE) can be used for investigation of the near surface region of a solid with a purpose to obtain quantitative information concerning its crystal structure, element composition and the electronic states of atoms [3,4]. On the Figure1 the total energy distribution of reflected SEE from a surface is shown which is irradiated by an electron beam of primary energy E_0. The shape is due to some types of interaction: elastic and inelastic scattering together with secondary electron emission. There are four ranges in N(E), in each if which one of these interactions predominates.

The elastic interaction produced a narrow peak on the right, where the electrons retain their energy E_0 and merely show altered momentum direction (Region I). The broadening is due to thermal spread in the beam energy and is also affected by the analyzer resolution. With standard equipment the broadening is usually 0.5-1 eV, so phonon excitations (energy loss 10-50 meV) can be detected only by special techniques involving highly monoenergetic primary beams and improved analyzer resolution, as realized in high resolution energy loss spectroscopy. With other methods, one can assume that the elastic peak is due to group electrons from the beam that have undergone elastic and quasi elastic interactions with the surface. Various methods are applied to the elastic backscattering, particularly diffraction ones such as Low Electron Energy Diffraction (LEED) and high energy diffraction with back scattering which have been applied to the spatial distributions of the backscattered beams. The methods have been applied to the atomic structures and dynamic characteristics in ordered surface layers.

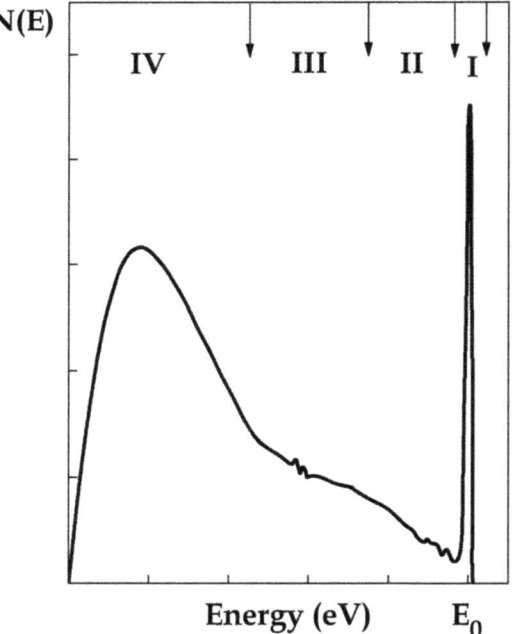

Figure 1. Total energy distribution of secondary electron emission from a surface which is irradiated by an electron beam of primary energy E_0

The broad low energy maximum (Region IV) is due to the true secondary electrons which have energies from zero up to some tens of eV and are formed by repeated inelastic electron-electron scattering in the cascade process. The true secondary electrons may constitute up to 70% of total energy distribution. Their energy distribution is related to the random filling of the final states and to the cascade multiplication.

In Region III there is also a fine structure due to electrons from the solid escaping in the vacuum by the Auger process (Auger Electron Spectroscopy (AES)). The Auger electron spectrum for a given element has a characteristic form and certain energies which have meant that AES is widely used in elemental analysis.

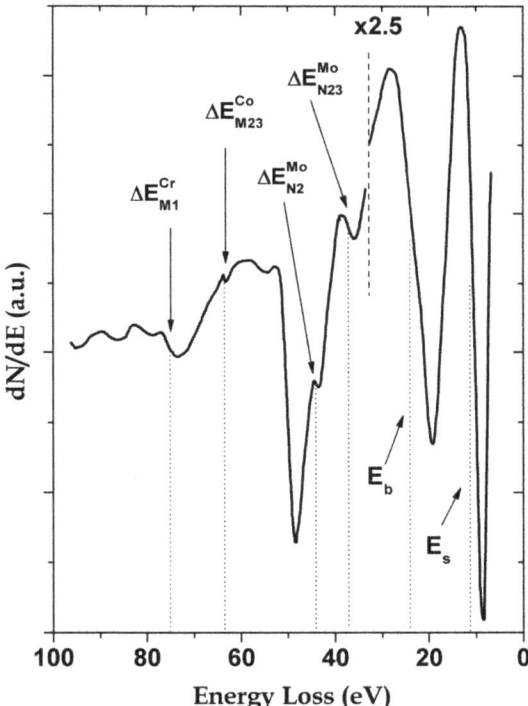

Figure 2. Example low EELS spectra obtained for the Co–Cr–Mo alloy surface at the primary energy E_0 = 350 eV with identification of energy losses

Region II is due manly to electrons that have lost some of their energy by inelastic scattering; directly by the elastic peak, one finds electrons that have suffered discrete energy losses from the excitation of inter- and intraband electronic transitions, surface and bulk plasmons, hybrid modes of plasmons and ionization losses (Ionization Spectroscopy). That range is usually 30-100 eV. Usually, the losses related to surface and bulk plasmon excitations are most intensive lines in the electron energy loss spectrum. The spectra of plasma oscillations are potential data carriers about composition and chemical state of elements on the surface of solid and in the adsorbed layers. The energy losses are called as characteristic losses because losses do not depend on the primary electron energy E_0 and its value is individual for the chemical element and compound. Region II is called as Electron Energy Loss Spectroscopy (EELS). At energy $E_0 < 1000$ eV it can be called as low EELS. On the Figure 2 really low EELS spectra is shown with interpretation of losses for the Co-Cr-Mo alloy surface which was measured at the primary electron beam energy $E_0 = 350$ eV in dN/dE mode [5].

Ionization Spectroscopy (IS) is a variant on EELS. Gerlach et al. [6,7] first applied this in terms of analysis of the surface composition analysis for V, Ni, Pd and Mo as impurities on the surface of polycrystalline metals without depth analysis. The IS method is based on measuring the energy spectra of electrons, which have lost a particular portion of the energy ΔE for the excitation of electrons from internal atomic levels into the empty states (conduction band) of the solid. Having lost energy ΔE, and after being inelastically scattered, the primary electrons escape into vacuum and are registered on the background secondary emission spectrum as individual monochromatic groups which form spectral lines. The advantages of IS as compared to other methods of electronic spectroscopy are (i) the position of ionization lines in the spectrum with respect to the lines of elastic scattered electrons is determined by the binding energy of electrons in the ground state and by the distribution of the density of empty states and does not depend on the value of the primary electron energy E_0, (this allows easy separation of IS lines from the AES lines) and, (ii) the possibility to vary the probing depth of the near-surface region because the change in primary energy E_0 induces a change in the mean free path λ of electrons. On the Figure 3 the different between Auger process and ionization process is shown which are generated by an electron beam of primary energy E_0.

Auger process   Ionization process

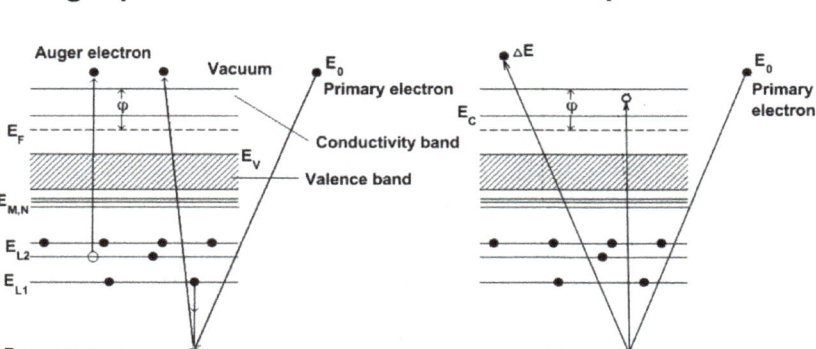

Figure 3. Example of different between Auger process and ionization process which are generated by an electron beam of primary energy E_0

The aim of the present chapter is to show the application of low Electron Energy Loss Spectroscopy as non-destructive method, namely Ionization Spectroscopy and surface and bulk plasmon excitations, at investigation of physical-chemical properties materials in the nano-size near surface region.

2. Low energy electron loss spectroscopy

2.1. Ionization spectroscopy

2.1.1. Physical model

Ionization Spectroscopy is based on the measurement of the energy spectra of electrons, which have lost a particular portion of the energy ΔE_β for the excitation of electronic transitions that are typical for a given kind of atom β. The position of an intensity line (IL) in the spectrum with respect to the primary electron energy E_0 is determined by the binding energy of electrons in the ground state and by the distribution of the density of empty states, but it does not depend on the value of E_0, on the work function or on the value of the surface charge.

The calculation of the contribution to the intensity of an IL by the electrons having lost an amount of energy ΔE_β at the depth Z from the sample surface by the ionization of the core states of the atoms β is simple when a traditional experimental configuration is used (an incident beam of the primary electrons is directed perpendicularly to the sample surface (θ_0 = 0) and the secondary electrons are registered at the angle θ with respect to the normal). In this case calculations within the framework of a two-stage model allow us to obtain the following expression for the intensity of an IL [8]:

$$I_\beta\left(Z, E_{0j}\right) = K\sigma_\beta \tilde{r}_\beta n_\beta(Z)\exp(-Z/\Lambda_\beta),$$ (1)

where K is an instrumental factor, σ_β is the ionization cross-section of the core level, $n_\beta(Z)$ is concentration of atoms β at depth Z from surface, $\tilde{r}_\beta$ is the elastic scattering factor of electrons. Λ_β is the effective free-path of electrons in a sample with respect to inelastic collisions, which is determined by the equation

$$\Lambda_\beta^{-1} = \lambda_0^{-1} + \left(\lambda_\beta \cos\theta\right)^{-1}.$$ (2)

For the Pt-Me (Me: Fe, Co, Ni, Cu) alloys [9]

$$\lambda_0 = \frac{1194}{E_{0j}^2} + 0.429 E_{0j}^{1/2} \; ; \; \lambda_\beta = \lambda(E_{0j} - \Delta E_\beta).$$

An effective probing depth in IS amounts to ~ $3\Lambda_\beta$ because the secondary electrons created in the near-surface region of this thickness contribute for 95% to the total intensity of an IL. An increase of the effective probing depth upon increasing the energy E_0 also results in an increased contribution from the deeper layers of the concentration profile into the IL intensity. This enables us to carry out a layer-by-layer reconstruction of the concentration profiles of the elements using the energy dependencies of the IL.

After integration of Eq. (1) with respect to depth and spatial angle of the four-grid energy analyzer, an expression for the total IL intensity has the following form

$$I_\beta(E_0) = 2\pi K\sigma_\beta \int\limits_{0}^{\infty} \int\limits_{\Theta_{min}}^{\Theta_{max}} \tilde{r}_\beta n_\beta(Z)\exp(-Z/\Lambda_\beta)\sin\Theta dZ d\Theta \,, \tag{3}$$

where $\Theta_{min} = 4°$ and $\Theta_{max} = 70°$ are respectively the minimum and maximum values of polar angle for the standard quasi-spherical four-grid energy analyzer.

As pointed out above, the offered method is essentially not sensitive to the type of the energy analyzer used. Only the values of Θ_{min} and Θ_{max} that correspond to the concrete conditions of an experiment should be substituted in Eq. (3). In the case of a binary A-B alloy, usually the ratio of intensities of A to B

$$R_A(E_{0j}) = \frac{I_A(E_{0j},\Delta E_A)}{I_B(E_{0j},\Delta E_B)} \tag{4}$$

is measured experimentally in order to eliminate the instrumental factor K, which is often unknown.

Let us consider as new variables the relative concentrations of the elements in a layer with number i:

$$N_\beta(i) = n_\beta(i)v_\beta, \ \beta = \overline{A,B} \,, \tag{5}$$

where v_β is the atomic volume of a pure component of an alloy. After replacing the integral in Eq. (3) by a summation over N and substituting the expression for I_β into formula (4), integration with respect to the width of isolated layer d leads to:

$$R_A(E_{0j}) = \frac{\sigma_A \lambda_i^A v_B}{\sigma_B \lambda_i^B v_A} \cdot \frac{\sum\limits_{i=1}^{N} N_A(i)P_A(i,E_{0j})}{\sum\limits_{i=1}^{N} N_B(i)P_B(i,E_{0j})} \,, \tag{6}$$

where

$$P_\beta(i,E_{0j}) = \exp\left[-\frac{(i-1)d}{\lambda_0}\right] \int\limits_{\Theta_{min}}^{\Theta_{max}} D_i(E_{0j},\Theta)\tilde{r}_\beta\left[1-\exp\left(-\frac{d}{\Lambda_\beta}\right)\right]d\Theta$$

$$D_i(E_{0j},\Theta) = \exp\left[-\frac{(i-1)d}{\lambda_0\cos\Theta}\right]\frac{\cos\Theta\sin\Theta}{\lambda_0 + \lambda_i^\beta\cos\Theta}\,, \tag{7}$$

i = 1, 2, ..., N - 1, and i = N

$$P_\beta(i,E_{0j}) = \exp\left[-\frac{(N-1)d}{\lambda_0}\right]\int\limits_{\Theta_{min}}^{\Theta_{max}} D_N(E_{0j},\Theta)\tilde{r}_\beta d\Theta, \tag{8}$$

Following the approach offered in [10], the expression (6) is transformed into a system of Linear equations (SLE) with respect to $N_A(i)$ using the relation $N_A(i) + N_B(i) = 1$. As a result, we obtain ($j = 1, 2, ..., M$)

$$\frac{\sum\limits_{i=1}^{N} N_A(i) \left\{ \frac{\sigma_A \lambda_i^A v_B}{\sigma_B \lambda_i^B v_A} P_A(i, E_{0j}) + R_A(E_{0j}) P_B(i, E_{0j}) \right\}}{\sum\limits_{i=1}^{N} P_B(i, E_{0j})} = R_A(E_{0j}), \tag{9}$$

where E_{0j} is the energy of primary electrons for which we have measured the ratio of IL intensities $R_A(E_{0j})$. Assuming that all interlayer distances in the near-surface region of a single-crystal alloy are identical and equal to d, for summations in Eqs. (6) and (9) the number of terms is selected that corresponds to the selection of N monolayers parallel to a free surface, using the relationship $(N-1) \cdot d = 3\Lambda_{max}$ (where $\Lambda_{max} = \left(\Lambda_A(E_0^{max}) + \Lambda_B(E_0^{max}) \right)/2$). The system of linear equations (9) can be solved only when it is determined or overdetermined, i.e. if the inequality $M \geq N$ is true for this system.

In the following sections, methods are presented for building the solution of Eq. (9) and for the numerical calculation of the concentration profiles within the framework of the described model.

2.1.2. Layer-by-layer reconstruction methods

A system of equations, describing the deviations of the concentrations in a monolayer i ($N_A(i)$) from their bulk value N_A, can be represented in matrix form by the expression

$$\sum\limits_{i=1}^{N} Q_{ji} \delta N_A(i) = \tilde{R}_A(E_{0j}), \tag{10}$$

where $\delta N_A(i) = N_A(i) - N_A$, $\tilde{R}_A(E_{0j}) = R_A(E_{0j}) - N_A \sum\limits_{i=1}^{N} Q_{ji}$, and an explicit form of the

matrix elements Q_{ji} is evident from the expression (9). However, the practical solution of the Eq. (10) presents particular difficulties because the matrix elements Q_{ji} correspond to close energy intervals that do not differ sufficiently. As a result, the determinant of the matrix Q is close to zero and the system (10) is ill-conditioned. As a consequence, the errors in the matrix elements Q_{ji} and in the $\tilde{R}_A$ values can result in an incorrect solution.

To construct a stable approximation for the solution of system (10), the condition–gradients projection method, the conjugate gradients projection method, the method of conjugate gradient projected on the Π^+–space and also the regularization method [11] were used in the present work. More detailed information on techniques for solving ill-posed problems can be found in [10,11]. The described regularization algorithm for the reconstruction of the elemental concentration profiles in a binary alloy on the basis of energy dependencies of the ratio of IL intensities is implemented in FORTRAN codes.

2.1.3. Results of the layer-by-layer reconstruction

Approbation of method of nondestructive layer-by-layer analysis was performed for the single crystal Pt80Co20 alloy with (100) and (111) surface orientations [12]. Initially, Pt80Co20(111) alloy surface was in a disordered state. First, we measured the spectra of the ionization losses for the clean (100) and (111) surfaces of Pt80Co20 alloy and polycrystals of platinum and cobalt in the dN/dE mode. The energy losses $\Delta E^{Pt}_{O_{2,3}} = 54 eV$, $\Delta E^{Co}_{M_{2,3}} = 62 eV$ and their IL were recorded in the range of primary electron energy $E_0 = 200 - 500$ eV.

In order to ignore in calculations an instrumental factor K (it is often unknown), the ionization cross-section σ_β and elastic scattering factor $\tilde{r}_\beta$ and the possible influence of matrix effects (for example, due to difference of atomic radii for platinum and cobalt $r_{Pt}/r_{Co} = 1.104$), usually the IL ratio of elements is measured, with normalization on standards by following equation:

$$R(E_0) = \frac{R_{alloy}(E_0)}{R_{st}(E_0)}, \qquad (11)$$

where $R_{alloy}(E_0) = I^{alloy}_{Co}(E_0)/I^{alloy}_{Pt}(E_0)$; $R_{st}(E_0) = I^{st}_{Co}(E_0)/I^{st}_{Pt}(E_0)$; I^{alloy}_{Co}, I^{alloy}_{Pt} and I^{st}_{Co}, I^{st}_{Pt} are intensity lines of the ionization losses of the alloy components and standards (pure metals), respectively. Figure 4 shows the ratio of ionization peaks of Co to Pt as a function of the primary electron energy for (111) and (100) faces of Pt80Co20 alloy at room temperature before and after normalization on the standards.

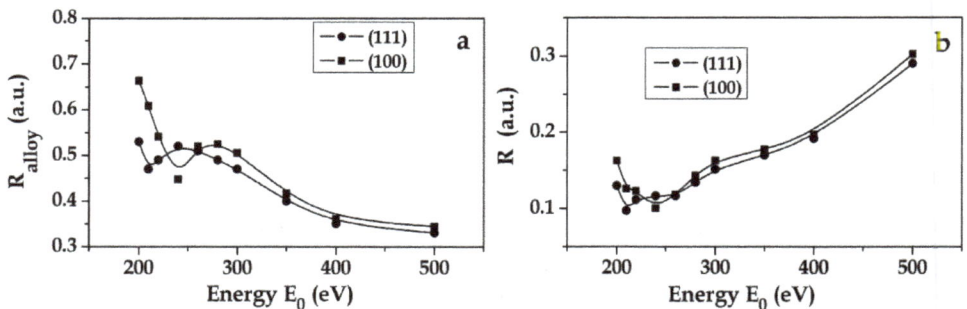

Figure 4. The ratio of ionization peaks of Co to Pt as a function of the primary electron energy for (111) and (100) faces of Pt80Co20 alloy at room temperature: (a) – before and (b) – after normalization on the standards

Based on experimental data R(E0), we calculated the layer-by-layer Pt concentration profiles for (100) and (111) faces of alloy Pt80Co20 by means of the condition–gradients projection method, the conjugate gradients projection method, the method of conjugate gradient projected on the Π^+–space with total level of experimental errors less than 3%. On Figure 5 the averaged Pt concentration are shows to all three methods by histograms.

It can be seen that the upper layer contains only platinum atoms for both faces and practically does not contain cobalt atoms. Moreover, there are strong orientation effects that affect on length of the platinum concentration oscillations for the (100) and (111) faces. The deeper oscillation is observed for a more "loose" (100) face, which affects the depth composition up to eight atomic layers. Whereas for the close-packed (111) face these changes are damped on the ifth level. The presented results are in good agreement with experimental data of concentration profiles which were obtained by means of LEED and LEIS [13,14].

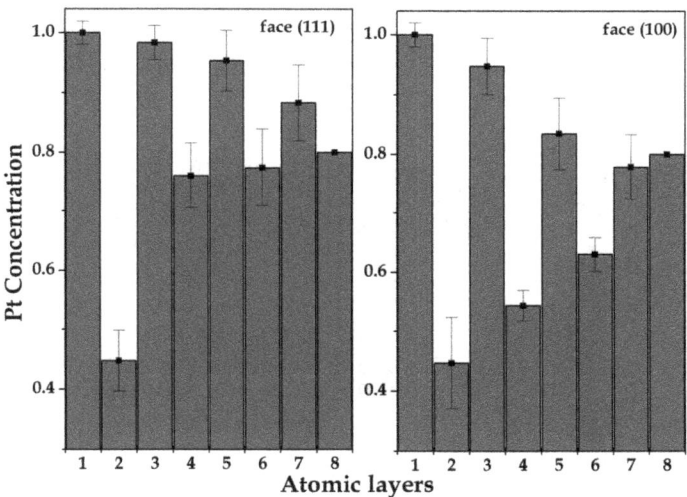

Figure 5. Layer-by-layer Pt concentration profiles reconstructed from the ionization spectra for (111) and (100) faces of Pt$_{80}$Co$_{20}$ alloy at room temperature

Non-destructive method of layer-by-layer analysis by IS can be effective at study of temperature concentration profiles. Authors [8] investigated influence of heating on concentration profile of Pt$_{80}$Co$_{20}$(111) (see Figure 6). Heating the sample to 613 K leads to a depletion of Pt atoms in the 2nd layer ($C_{Pt}^{(2)} = 24\%$) and to an insignificant enrichment of Co atoms in layers 3-6 in comparison with the profile at room temperature. Increasing the temperature further to 673 K is accompanied by a negligible segregation of Co from the second layer ($C_{Pt}^{(2)} = 31\%$) to the first ($C_{Pt}^{(1)} = 97\%$), while deeper layers remain practically unchanged. At 823K, a sandwich-like structure of the type Pt/Co/Pt was found in the first three atomic layers. As is obvious from Figure 6, heating the sample causes a smoothing of the oscillations in deeper layers towards the bulk concentration of the alloy. However, the first layer still consists of pure Pt up to 873 K. Further increasing the temperature gradually results in completely smoothed oscillations.

Consequently, the sample was slowly cooled during 10 hours from 1123 K to room temperature. As a result of this procedure, a chemically ordered alloy surface of the L1$_2$ type was obtained. LEED shows super-structural reflections in a diffraction pattern at E$_0$ = 112 eV. The result of the layer-by-layer reconstruction for the ordered state shows that the 1st

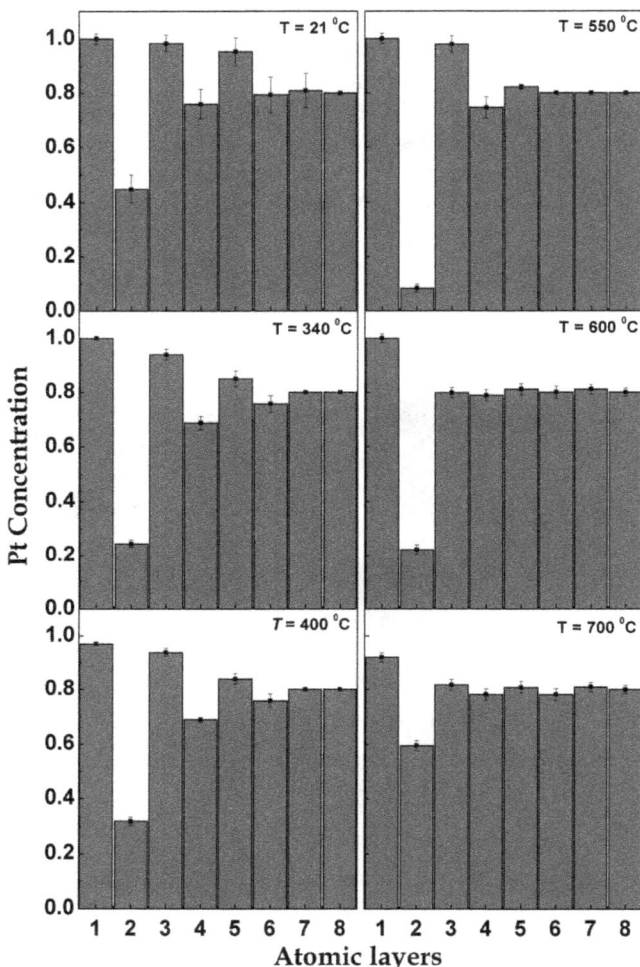

Figure 6. Layer-by-layer Pt concentration profiles reconstructed from the ionization spectra for the Pt80Co20(111) alloy at the different heating

atomic layer consists of pure platinum, and that the other atomic layers have concentrations near the bulk value of the alloy. Probing the surface with primary electrons of 58 eV (corresponding to a probing depth of two atomic monolayers [9]), a p(2x2) structure was found. The appearance of these additional super-structural reflections in a diffraction pattern can be caused by two possible phenomena: chemical ordering at the surface of the alloy and/or a reconstruction of the surface [14].

In work [15] Electron Energy Loss Spectroscopy has been employed for investigation of the effect of 600 eV Ar+-ion irradiation in the dose range $7 \cdot 10^{16} - 4 \cdot 10^{17}$ ions/cm^2 on the atomic structure and surface composition of Pt80Co20(111) alloy. Using the ionization energy loss spectra, a layer-by-layer concentration profile of the alloy components was reconstructed for different doses of ion irradiation of the surface. The Ar+-ion bombardment of the alloy was found to result in the preferential sputtering of Co and in the enrichment of the near-surface region by Pt atoms with formation of an altered layer, which is characterized by a non-monotonic concentration profile dependent on the irradiation dose. The results obtained are discussed in the framework of the models of preferential sputtering and radiation-induced segregation.

Application of IS for the investigation of composition changes on the depth is not limited to the study by single crystal alloys. In references [8,16] IS was used to study the surface segregation in the ternary Co–Cr–Mo system. Since it was polycrystalline alloy, there can't be applied layer-by layer analysis with profile reconstruction. Nevertheless, the integral distribution of elements on the probing depth can be investigated by means of IS.

According to reference [16] the concentration of the Co–Cr–Mo alloy components on E_0 can be calculated by following expression

$$C_i = \frac{I_i(E_0) / I_i^{st}(E_0)}{\sum_{i=Co,Cr,Mo} I_i(E_0) / I_i^{st}(E_0)}, \tag{12}$$

where i = Co, Cr and Mo metals, $I_i(E_0)$ and $I_i^{st}(E_0)$ are intensity lines of the ionization losses of the alloy components and standards, respectively. For the thermodynamic equilibrium state the ionization spectra of the alloy components at different temperatures were measured. The condition of the thermodynamical steady–state of the alloy depended on the prolonged heating of the sample at every preset temperature for 15 hours. Figure 7 shows the concentration dependences $C_{Co,Cr,Mo}$ on E_0 for the polycrystalline alloy at a different heating temperatures.

At first, we estimated the thickness of the probing layers for Co-Cr-Mo alloy at change of E_0 from 200 eV to 800 eV. For estimation of the probing depth we used experimental data for the inelastic mean free path (IMFP) λ which are collected in reviews [9] for pure Co, Cr, Mo metals. After, these data was approximated by following equation

$$\lambda = kE_0^n, \tag{13}$$

where k, n are fitting parameters. As result k = 0.36 and n = 0.5 and variation $\lambda(E_0)$ is from 5Å to 10Å.

For the non–annealed Co–Cr–Mo alloy the Mo atoms showed preferred segregation in the outermost layers at a room temperature. Gradual increase of the probing depth by changing the primary electron energy E_0 to 600 eV shows that the Mo concentration in the near-surface region decreases and the Cr concentration greatly increases, while the Co concentration does not exceed 5–7 at.%. However, at the energy E_0 = 200 eV it was detected that $C_{Co} \approx 42$ at%, $C_{Cr} \approx 20$ at% and $C_{Mo} \approx 38$ at% were present in the near-surface layers.

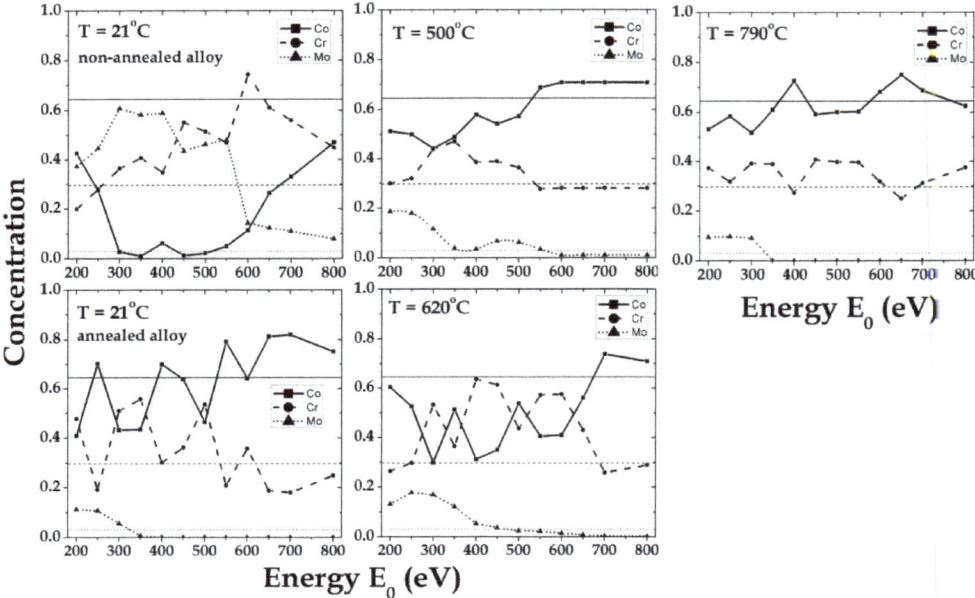

Figure 7. Concentration profiles of Co, Cr, Mo for the Co-Cr-Mo alloy at the different heating by means IS. Horizontal lines are bulk concentration for Co, Cr and Mo, respectively.

On the other hand, for an increase of probing depth at $E_0 > 250$ eV , the concentration of Co atoms sharply decreases to 1–6 at% in the below-surface region at $E_0 = 250 – 550$ eV. With the increase of primary electron energy $E_0 > 600$ eV the Co concentration rises but that of Cr and Mo atoms falls. Only an approximate tendency $C_{Co,Cr,Mo}$ towards the bulk concentration of the alloy is observed at $E_0 = 800$ eV. Heating of the alloy to temperature T = 500ºC essentially induced a change of the surface concentration in the Co–Cr–Mo alloy as compared to the surface concentration for the non–annealed state. Thus, the near surface layers contain $C_{Co} \approx$ 51 at%, $C_{Cr} \approx 30$ at% and $C_{Mo} \approx 19$ at% at the energy $E_0 = 200$ eV. With an increase of the probing depth the Mo concentration is lowered and the concentration of Co atoms is increased. Whereas in the interval of the energies $E_0 = 250 – 400$ eV the sharp growth of Cr concentration is observed and at $E_0 > 550$ eV the alloy composition is close to the bulk value. Further heating of alloy to T = 620ºC promotes an increase in concentration of Co atoms in the near surface region of Co–Cr–Mo alloy (at the $E_0 = 200$ eV). At the primary electron energy $E_0 = 250 – 650$ eV in the deeper layers growth in Cr concentration is detected as compared to the Cr bulk value and only at energy $E_0 > 700$ eV the composition of the alloy comes towards that of the bulk. Further increase of the alloy heating temperature to T = 790ºC is accompanied by smoothing of the alloy composition to the bulk. Nevertheless, an insignificant Mo segregation was still detected in the outermost layers of the alloy.

After prolonged annealing at T = 790ºC the ternary Co–Cr–Mo alloy was slowly cooled to room temperature over 12 hours. The concentration profile for the annealed alloy is shown on Figure 7. Also for the annealed state the preferred segregation of the Mo and Cr atoms is observed. At the energy E_0 = 200 eV the outermost layers contain $C_{Co} \approx$ 40 at%, $C_{Cr} \approx$ 50 at% and $C_{Mo} \approx$ 10 at%. At increasing primary electron energy E_0 the Mo concentration sharply diminishes and at E_0 > 400 eV Mo atoms are not detected anymore, though insignificant oscillations of the composition for the Cr and Co atoms are found near to the bulk concentration at varying E_0. We suggest that the thermodynamic steady–state of the alloy corresponds to that of the annealed alloy at room temperature, but not for the non–annealed alloy.

2.2. Kinetics of surface segregation by IS

IS can be effective at investigation of kinetic processes in the thin layers of a solid. In work [17] for studying kinetics of surface segregation of the $Pt_{80}Co_{20}$(111) alloy, the temperature interval T = 613 - 973K (T = 340 - 700 °C) is chosen at which the bulk alloy is in the ordered state. A special device allowed heating the sample to predetermined temperature, keeping it constant and changing with an accuracy ±2°C. Platinum - Pt alloy and 1%Rh thermo-couple was welded to the investigated sample for control temperature. Spectra of ionization losses were measured at every chosen temperature with a fixed time interval for platinum ($\Delta E_{O_{2,3}}^{Pt} = 54eV$) and cobalt ($\Delta E_{M_{2,3}}^{Co} = 62eV$). Primary electron beam with energy E_0 = 250 eV was taken for surface probing which by converting into monatomic layers corresponds to the 3rd monatomic layer over the depth [8]. Figure 8 shows the kinetics of segregation for Pt and Co atoms in the near-surface region at different temperatures of $Pt_{80}Co_{20}$(111) alloy.

Note, that diffusion processes (internal diffusion) for single crystal alloys course mainly according to the vacancy mechanism [1]. Under heating up to 613K the kinetics of segregation atom $C_β(t)$ has a classical dependence which may be provisionally divided into two regions: I is the region of fast diffusion when strong segregation of Co atoms is observed; II is the saturation region when the steady-state equilibrium of segregating atoms is set in the near surface alloy region. The character of kinetic curve $C_β(t)$ dependence changes substantially at higher temperatures. Thus, when heating is up to 673K the fast diffusion region I has more gentle appearance and region II acquires two characteristic sites: IIa is the region of changing the direction of Pt and Co atoms segregation (temporary S-shaped fold), smoothly transient into IIb, which is region of steady-state equilibrium of segregating elements. We consider that such S-shaped fold is associated with eventual formation of the ordered phase in the near surface region. One of the reasons for nucleation of the composition close to the ordering is a decrease of interatomic interaction constants and as the consequence, an increase of the amplitude of thermal atomic oscillations. We suppose that, most probably, an ordered phase is formed between the 3-5th atomic layers. Since at probing of the alloy surface by electrons with the energy E_0 = 200 eV (1-2nd monolayer [8]), the integral concentration of Pt and Co atoms was C_{Pt} = 0.9 at.% and C_{Co}= 0.1 at.% within the whole time and temperature interval. This confirms the preferred segregation of Pt atoms in topmost layers. The further increase temperature for $Pt_{80}Co_{20}$(111) alloy will lead to growth of the vacancies number due to thermal oscillation expansion of atoms and hence it will accelerate diffusion processes. Heating of alloy up to 823K, especially at 873K, leads to increase Co atoms enrichment in the near-surface alloy region as compared to other heating temperatures, and to decrease time which needed for the possible alloy ordering (IIa region). At heating of sample higher ordering temperature the rapid segregation of Co atoms was observed in region I which soon will be replaced by the segregation of atoms Pt (region II) and tend to bulk concentration. Such character of temporal diffusion of cobalt and subsequent segregation of platinum we suppose with redistribution of atoms in the near surface region of alloy.

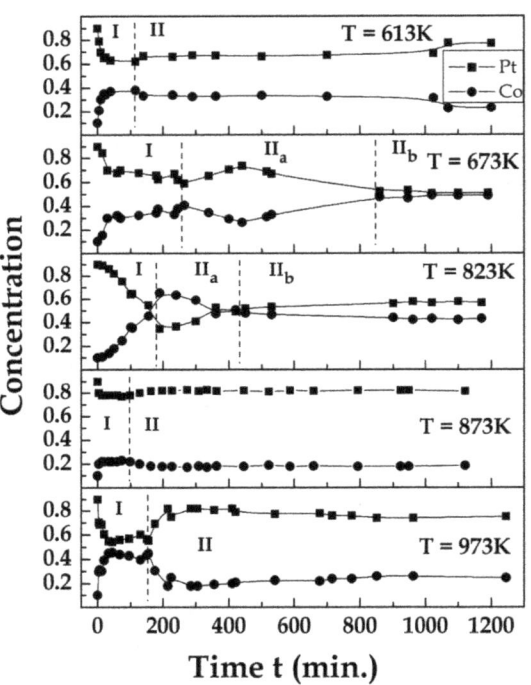

Figure 8. Kinetics of surface segregation in the near surface region for $Pt_{80}Co_{20}(111)$ alloy at different heating

At studying kinetics of segregation of the binary alloys in the work [18] was established that the concentration C_x^t of segregating atoms to the surface from the bulk for the time t out of the depth x may be given by the following ratio:

$$C_x^t = C_\infty - C_\infty \left(1 - \frac{1}{\alpha}\right) \exp\left(\frac{x}{\alpha d} + \frac{Dt}{\alpha^2 d^2}\right) erfc\left[\frac{x}{\sqrt{4Dt}} + \left(\frac{Dt}{\alpha^2 d^2}\right)^{1/2}\right], \tag{14}$$

where C_∞ is the bulk concentration of diffusing atoms; D is the diffusion coefficient; d is the thickness of the surface layer; α is the degree of surface enrichment defined as

$$C_s^t = \alpha C_0^t, \tag{15}$$

where C_s^t is the concentration of segregating atoms in the surface region; C_0^t is the atom concentration at the depth d at the initial time. By comparing formula (14) and (15) provided that $\alpha \gg 1$, we get

$$C_x^t = \alpha C_\infty \left[1 - \exp\left(\frac{Dt}{\alpha^2 d^2}\right) erfc\left(\frac{Dt}{\alpha^2 d^2}\right)^{1/2} \right]. \tag{16}$$

Data approximation of the kinetics of segregation atoms cobalt by Eq. (16) allowed to determine the mean coefficient values of cobalt diffusion at different temperatures, the order of which corresponds to diffusion bulk values. According to these results, the temperature dependence of Co diffusion coefficient in $Pt_{80}Co_{20}(111)$ alloy was plotted (Figure 9), by which pre-exponential factor $D_0 = 5.1$ m^2 s^{-1} and energy activation $E = (327\pm22)$ kJ/mol were determined. The value of energy activation is close to sublimation heat of pure cobalt $E = 309.73$ kJ/mol.

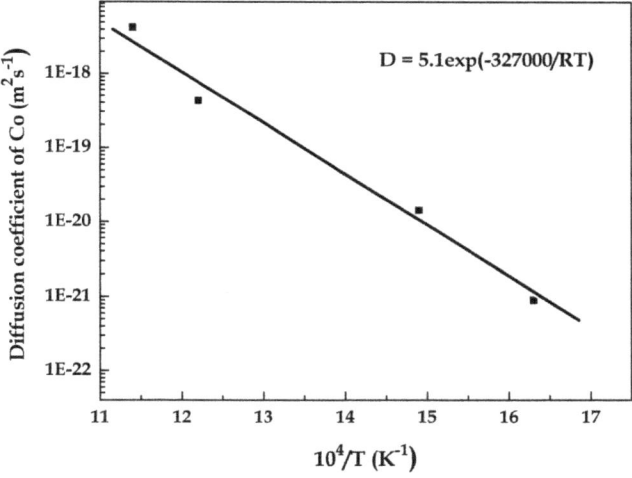

Figure 9. Diffusion coefficient of Co atoms in $Pt_{80}Co_{20}(111)$ alloy surface

2.3. Plasmon excitation

Plasmon excitations are potential data carriers about composition and chemical state of elements on the surface and bulk of solid and in the adsorbed layers.

2.3.1. Plasmon energy

A longitudinal plasma wave along the crystal produces long–range Coulomb forces between positive and negative charges and excites collective oscillations. These are called plasmons in the case of a free–electron gas model. The plasmon energy is obtained from the Fourier modes of the electron density $\rho(r) = \sum_k \rho_k e^{-ik \cdot r}$ and the ϱ_k are amplitudes of harmonic density fluctuations obeying [3]:

$$\ddot{\rho}_k + \omega_p \rho_k = 0 \tag{17}$$

in which ω_p is the Langmuir frequency. Then plasmon energy can be determined by the following expression:

$$E_b = \hbar\omega_b = \hbar\sqrt{\frac{e^2 N}{m\varepsilon_0}}, \tag{18}$$

where $\hbar$ is Plank's constant; ω_b is the cycle frequency of the bulk plasmon; e and m are the electronic charge and mass, respectively; n is the number of valence electrons per unit volume and ε_0 is the permittivity of the free space. The surface plasmon energy E_s is related to bulk plasmon energy by the following equation [19]:

$$E_s = E_b / \sqrt{1 + \varepsilon_s}, \tag{19}$$

where ε_s is the dielectric constant. In the framework of the model under consideration, $\varepsilon_s = 1$, i.e. $E_s = E_b / \sqrt{2}$.

In references [5, 20-24] the surface and bulk plasmon excitations were investigated for the $Pt_{80}Co_{20}(111)$ and $Cu_{75}Pd_{25}(100)$ single crystal alloys, ternary Co–Cr–Mo alloy and amorphous and crystalline $Fe_{73.6}Cu_1Nb_{2.4}Si_{15.8}B_{7.2}$ (FINEMET) alloy surface and their alloy components in range primary electron energy E_0 from 150 eV to 800 eV. It was found that the experimental values of plasma oscillation energy for all pure elements differ from the theoretical calculations but the data obtained in the given works are in good agreement with the results obtained by other authors.

Actually, the difference between experimental data and the free–electron gas model has been observed repeatedly for a lot of chemical elements. This may be a result of: (i) incomplete participation of valence electrons in the collective excitations; (ii) the involvement of filled d-band states and the appearance of inter– or intra-band transitions in characteristic spectra for the transition metals; (iii) cleanness and roughness of the surface region of specimens [3].

For example, on Figure 10 the bulk plasmon energies are shown for the range of primary electron beam energy 150 – 650 eV for pure Fe, Si, B, Nb, Cu and $Fe_{73.6}Cu_1Nb_{2.4}Si_{15.8}B_{7.2}$ alloy. It is known that, for silicon, the surface and bulk plasmon energy are 12 eV and 17 eV, respectively [4]. In our experimental data the plasmon energies are ~ 9 eV and ~15 eV. Most probably the shift of plasmon energies toward lower energy is related to the surface effects when comparing with other work because the probing depth is not deep and varied from 5.4 Å – 5 Å for silicon in the chosen range of E_0. Appearance of silicon oxides on the surface there can be eliminated since forming of oxides would lead to considerable increase in plasmon energy. For preparation of an atomically clean surface the amorphised silicon surface was first bombarded by argon ions and, subsequently, the sample was annealed at a high temperature. Consequently, both amorphous and crystalline phases can exist in the surface layers of Si. Also we cannot eliminate the fact that residual defects and implanted ions of argon may exist in the near-surface region, which is caused by ion irradiation. In the case of silicon we suggest that the total contribution of the above-mentioned surface effects will influence the shift of energy of plasma excitations to a lower energy. The experiments showed that, for the pure Fe, Si, B, Nb and Cu, the plasmon energy relation E_b/E_s exceeds the theoretical values and is equal to 1.79, 1.67, 2.13, 1.78 and 1.27, respectively. This discrepancy between theory and experiment has been observed repeatedly for many metals [3]. It should be noted that the theory supposes a perfectly flat surface in the vacuum–solid region and does not take into consideration the real physical–chemical state of the metallic surfaces.

According to reference [24] the experimental number of the valence electrons per unit volume n_{alloy} for the amorphous $Fe_{73.6}Cu_1Nb_{2.4}Si_{15.8}B_{7.2}$ alloy on E_0 can be calculated by the following expression

$$n_{alloy}(E_0) = \sum_j N_j \, n_j (E_0) , \tag{20}$$

where j = Fe, Si, B, Nb and Cu metals; N_j is number of the j–atoms per unit bulk (in our approach for the amorphous state, it is a bulk atomic concentration of the alloy components); $n_j (E_0)$ is the experimental number of the valence electrons per unit volume of the pure j–elements at fixed energy E_0. Substituting Eq.(20) into Eq.(18) and using experimental data we calculated the surface and bulk plasmon energy depending on primary electron energy E_0. The results of the calculations for the bulk plasmon energy of FINEMET are shown in Figure 10.

The obtained results are in good agreement with experimental data. For the surface plasmon the design function $E_s(E_0)$ is localized between values for the amorphous and crystalline alloy whereas for $E_b(E_0)$ there is a different situation. At low average primary electron energy $E_0 < 200$ eV the calculated function $E_b(E_0)$ is absolutely identical to energies of the bulk plasmon for the amorphous alloy and at energy $E_0 > 250$ eV the function $E_b(E_0)$ is close to the experimental data E_b for the crystalline state of the alloy. It was observed that, for the crystalline alloy, the energy of plasmon excitations is localized at lower loss energies as compared to those for the amorphous state.

It is known that the electronic state densities in the surface layers can be induced by such an effect as surface segregation. These phenomena are typical for major complex alloys when the composition in the near surface region differs from the bulk composition and it is caused by minimization of the free surface energy of the alloy [1]. Therefore the crystalline and phase structure and altered surface layers will also influence the dispersion of the surface and bulk plasmons when changing the primary E_0 or probing depth on amorphous or crystalline states of the alloy.

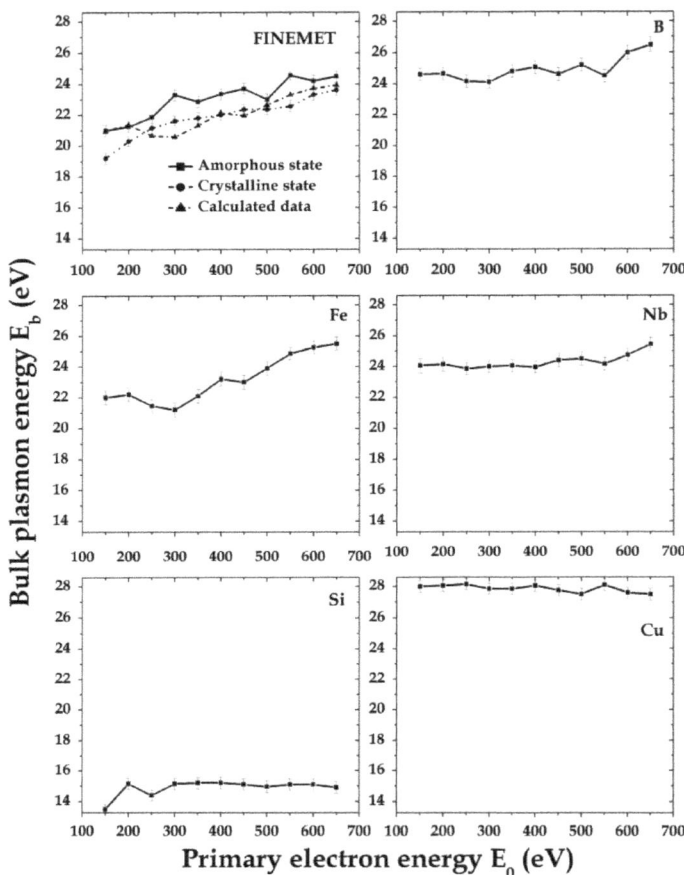

Figure 10. Dependence of the bulk plasmon energy E_b on the primary electron energy E_0 for the $Fe_{73.6}Cu_1Nb_{2.4}Si_{15.8}B_{7.2}$ alloy ribbons surface and pure alloy components

Similarly situation was observed at study plasmon energies for the disordered and ordered states of $Pt_{80}Co_{20}(111)$ alloy where are $E_s = 10.57$ eV, $E_b = 22.17$ eV and $E_s = 15.8$eV, $E_b = 25.31$ eV, respectively [21,22]. The plasma oscillations for the disordered state are localized at lower loss energies than it was established for ordered state. For the ordered alloy the bulk plasmon energy is 2–3eV more than that of the disordered alloy, whereas the difference for the surface plasmon energy is 4–7 eV in the whole range E_0. Probably it is related to changes of the DOS of valence electrons at the ordering alloy and surface segregation in the atomic layers.

Surface and bulk plasmon energy is sensitive not only to surface segregation, phase state etc but to heating too. EELS has been employed for investigation of the surface and bulk plasmon excitations versus heating in the Co–Cr–Mo alloy surface for the primary electron beam energies E_0 ranging from 150 to 800 eV (see Figure 11) [23].

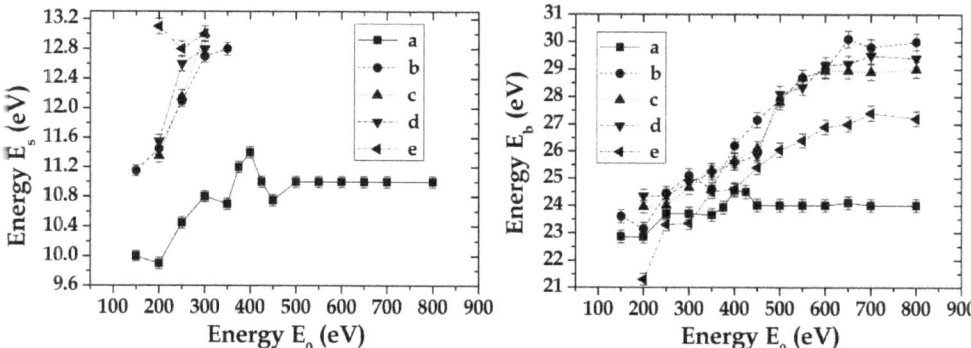

Figure 11. Dependence of surface and bulk plasmons energy from the primary electron energy E_0 for the Co-Cr-Mo alloy at different heating: (a) non-annealed state at $T = 21$ °C, (b) annealed state at $T = 21$ °C, (c) $T = 500$ °C, (d) $T = 620$ °C, (e) $T = 790$ °C.

As shown on Figure 11 for the annealed alloy the energies of surface plasmon E_s and bulk plasmon E_b are localized at greater energies than for the non–annealed alloy. In the range of the primary electron energy $E_0 = 150 - 800$ eV for the surface plasmon E_s this difference is $1 - 2$ eV. For the (non–)annealed alloy at room temperature the surface plasmon energy E_s have a linearly growth with a increase of the probing depth of alloy. Significant changes of bulk plasmon energy was observed for the annealed alloy in the range of the primary electron energy $E_0 = 150 - 800$ eV. For the non–annealed alloy in this energy range of E_0 the bulk plasmon is varied in small region of energy $E_b = 22.8$–24.5 eV, whereas it strongly changes for $E_b = 23.1$–30.1 eV in case of annealed alloy. For the (non–)annealed alloy the plasmon energies E_b are close in the energy range $E_0 = 150 - 200$ eV. With an increase of primary electron energy E_0 for the annealed alloy the bulk plasmon energy linearly increases and remains unchanging at $E_0 > 650$ eV.

At heating the surface plasmon energy E_s is shifted with an increase of the energy, and than more temperature of sample the more shift of energy E_s. However, the energy shift of surface plasmon, which is induced by heating there strongly differs against to annealed and non–annealed states of alloy. In all region of heating of the ternary Co–Cr–Mo alloy the energy of bulk oscillation E_b increases linearly with an increase of the primary electron energy E_0. With respect to dependence E_b from E_0 for the annealed state the alloy heating to temperatures 500°C and 620°C it is accompanied by growth of plasma energy E_b in the range of the energy $E_0 = 150 - 350$ eV (the near surface region) and decrease this value at the $E_0 > 400$ eV. Further heating of alloy to $T = 790$°C promotes to an insignificant shift of long wavelength plasmon oscillations E_b to sideways decrease of their energy in all region of E_0 as compared to other temperatures of Co–Cr–Mo alloy. Thus, for example, at the temperatures 500°C and 620°C the difference of bulk plasmon energy from 0.1 eV to 1.2 eV modulo with respect to annealed alloy, whereas at $T = 790$°C it changes from 0 eV to 2.7 eV at corresponding energies E_0. In the range of the primary electron energy $E_0 > 650$ eV the bulk plasmon energy E_b has a linearly dependence in all temperature regions. We suppose that this value will correspond to the real bulk plasmon energy crystal at the given temperature of heating.

The authors [25–34] investigated the influence of heating on EEL spectra from the surface of pure elements: C, Al, Ni, Mo, Ta, Pb, Nb, W and Ag. It was observed that owing to heating the surface and bulk plasmon energy suffers shifts in the characteristic spectra. After leaded systematic analysis of this effect by means of Transmission Electron Microscopy with EELS detector, a method of definition of the linear expansion coefficient was proposed using data to thermo–induced shifts of long wavelength plasmons [25–29]. This approach was based on the supposition that at heating of metal in consequence of the expansion/compression of crystal lattice the conductive electron density will lower/raise as a result it must lead to decrease/increase plasmon energy. In this case, number of valence electrons per unit $n(T)$ changes due to the thermal expansion of the crystal, Eq. (18) is rewritten using the linear thermal expansion coefficient $\alpha(T)$ of the crystal as follows:

$$E_b(T) = E_b(T_0)\left\{1 - \frac{2}{3}\int_{T_0}^{T} \alpha(T')dT'\right\},$$

$$E_b(T_0) = \hbar\sqrt{\frac{e^2 n(T_0)}{m\varepsilon_0}}$$

(21)

The obtained results for the thermal expansion coefficient are in a good agreement with tabular data for the clean Al, Ag and Pb. Also angular–resolved high resolution EELS was applied to study the plasmon excitations in the spectra of poly and single crystals as a function of temperature T. For example, in Ref. [33] a particular attention was devoted to silver because of the presence of an extremely sharp surface plasmon as observed for thin films and for all low Miller index surfaces. It was established that energy displacement of surface plasmon depends on temperature because of thermal expansion of the solid. Though Jensen et al. [34] observed with EELS strong temperature effects on the surface plasmon energy on graphite, which have been explained as a consequence of the unusual semimetallic band structure. Therefore this approach does not give us an ambiguous explanation of the reason for the plasmon shift in the ternary Co–Cr–Mo alloy. As noted above, the heating of alloy to T = 620ºC promotes to an increase of the bulk plasmon energy with respect to the annealed alloy and only at T = 790ºC the bulk plasmon suffers shift with a decrease of energy. More over, this approach doesn't take into account changes of the surface plasmon energy and their coupling with bulk plasmon in the near surface region.

It is known that the electrons in metals, which are neutralized by the fixed positive ions tightly sufficiently coupled between themselves and disposed in the lattice site, it is possible to consider as the special type of plasma [3]. From the classical point view the plasma oscillations in metals are oscillations of valence electrons with respect to positive ions which formed the lattice. These oscillations are conditioned owing to long–range Coulomb forces. Therefore, besides of the crystal lattice parameter at heating of ternary Co–Cr–Mo alloy it is necessary to take into account the change of Coulomb interaction force between the plasmons and atomic core that can not be calculated within the framework of the classic approach. The chemical elements Co and Cr are metals for which either the valence electrons are strongly bond s, d–electrons and the core electrons are weakly bond. Probably that at change of heating of these metals and those alloys the shift of plasmon energy with an increase or decrease of energy also will be determined by the change of Coulomb interaction force between the valence electrons and the core. It can lead to change of free s, d–electron concentration and effective electronic mass m (in Eq.(18)), which do participate in plasma excitations and, as a result, to shift the plasmon loss line relative to initial state of Co–Cr–Mo alloy.

Fact, for the pure chemical elements the surface and bulk plasmon energy can substantially differ from the plasmon energies of their alloys or compounds. In reference [16] was observed that in the near surface region the profile concentration versus temperature is differs to bulk of ternary Co–Cr–Mo alloy. Therefore, it is necessary to expect a displacement of long wavelength plasmon oscillations in the range of the energies $E_0 = 150 – 650$ eV as a result of segregation of the alloy components. However as far as the changed composition of the near surface region of alloys can strongly influence on plasma excitations at different temperatures at the present time is not clean.

The experimental data obtained in Figure 11 are indicated about the complex nature of the plasmon shifts in the near surface region of ternary Co–Cr–Mo alloy. Although most authors meet an opinion, that the energy plasmon shift mainly can be related to lattice parameter of solids, we suppose that in case of the complicated Co–Cr–Mo system the shift of plasmon energy will be defined by the summary balance of above mentioned possible causes at heating.

2.3.2. Intensity lines of plasmons

The nature of the surface plasmon appearance in the EELS spectra is related to the physical and chemical state of the surface layer nanosize thickness. It is also known that probability of the surface plasmon excitation by primary electrons will be directly related to their probing depth of the solid. Growth of the primary electron energy will lead to the increasing of the bulk plasmon excitation probability and, on the other hand, to the decreasing of the surface plasmon excitation probability and to damping of the surface plasmon intensity line in the EELS spectra. Consequently, for every chemical element and their alloys it is possible to define the range of the primary electron energy, in which the line of the surface plasmon will be detected in the characteristic loss spectra. Based on this concept, in references [20, 21] it was proposed to determine of the ratio R_s (in a.u.) of IL surface and bulk plasmons from the energy E_0 by the following equation:

$$R_{s,b}(E_0) = \frac{I_{pl}^{s,b}(E_0)}{I_{pl}^{s}(E_0) + I_{pl}^{b}(E_0)}, \tag{22}$$

where $I_{pl}^{s,b}$ is IL of the surface and bulk plasmons from primary electron energy E_0.

In works [20 - 24] the changes of IL for surface and bulk plasmon were studied for the $Pt_{80}Co_{20}(111)$ and $Cu_{75}Pd_{25}(100)$ single crystal alloys, ternary Co–Cr–Mo alloy and amorphous and crystalline $Fe_{73.6}Cu_1Nb_{2.4}Si_{15.8}B_{7.2}$ (FINEMET) alloy surface and their alloy components in range primary electron energy E_0 from 150 eV to 800 eV. There was found that damping of the function $R_s(E_0)$ is different for all specimen and different value of the primary electron energy E_0 for which the intensity line of the surface and bulk plasmons are equal. In case of pure elements the damping of function $R_s(E_0)$ is related to a decrease in probability of their excitation dependant on respective probing depth of the near surface layer and contrariwise this increases the probability excitation of the bulk plasmon and with altered near surface layers. In case of alloy, there was advanced a assumption that decay of intensity line of surface plasmon relative to bulk plasmon can be associated with changing of surface composition on the depth for the alloys and it confirms an assumption as to possibility of establishing the range of primary electron energy E_0, at which the electron beam will probe only the near surface region for the different materials.

Good correlation between the damping of surface plasmon $R_s(E_0)$ and concentration profile was established for the $Cu_{75}Pd_{25}(100)$ alloy surface at room temperature [20] and for the $Pt_{80}Co_{20}(111)$ alloy surface and Co–Cr–Mo alloy at a different heating [8, 23].

The results of measurement for $Pt_{80}Co_{20}(111)$ alloy are shown on Figure 12 at different heating. For the disordered alloy the damping of surface plasmon R_s have a more prolonged dependence compared to the ordered alloy at room temperature. If we will estimate a probing depth at primary electron energies E_0 = 550 eV and E_0 = 350 eV when surface plasmon does not appear ($R_s \approx 0$) in the EELS spectra, then we founds approximately 6-7th and 2-3rd atomic layers (bulk concentration) for the disordered and ordered states of $Pt_{80}Co_{20}(111)$ alloy, respectively (see Figure 6).

As in case of thermo–induced shift of plasmon excitations the changes in IL of surface plasmon relative to bulk plasmon were observed. Heating of alloy induces decreasing intensity line of surface plasmon and then higher temperature that more damping of surface plasmon R_s at variation of the primary electron energy E_0. In case of the $Pt_{80}Co_{20}(111)$ alloy surface with increasing of heating the damping of oscillating concentration depth profile is decreases [8]. More over, there is observes correlation between damping of surface plasmon R_s relative to bulk plasmon and damping of oscillating concentration depth profile at every given temperature.

The results of measurement for the Co-Cr-Mo alloy are shown on Figure 13 at different heating. For the non–annealed alloy the damping of surface plasmon R_s has a prolonged dependence and only at E_0 > 800 eV the surface plasmon peak disappears in EELS spectra. For the annealed state of alloy with an increase of the energy E_0 the dependence R_s decays quickly compared to the non–annealed alloy and at energy E_0 > 350 – 400 eV the surface plasmon does not appears in EELS spectra. As in case of shifts of the surface and bulk plasmon energy at heating of alloy the essential changes on intensity lines of surface plasmon relative to bulk plasmon were observed.

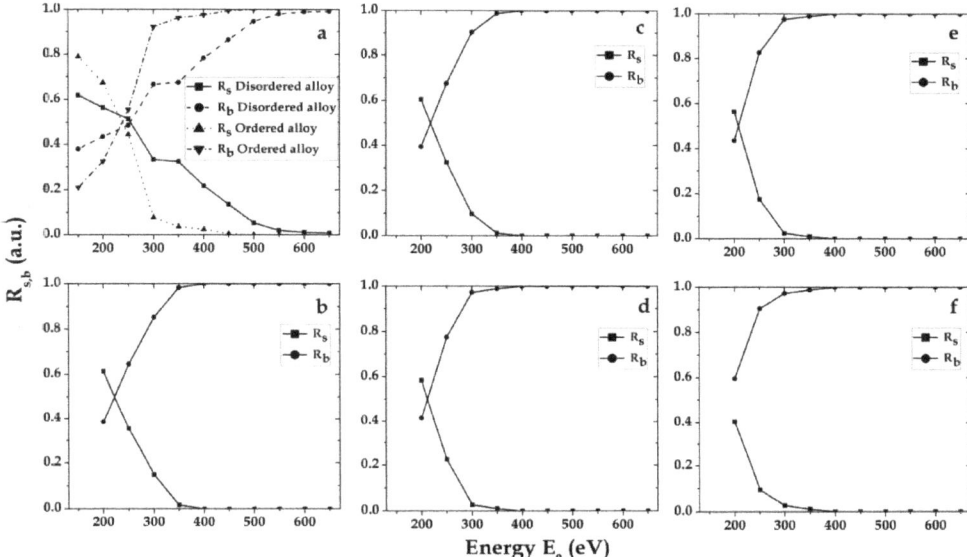

Figure 12. Dependence $R_{s,b}(E_0)$ from the primary electron energy E_0 for the Pt80Co20(111) alloy at different heating: (a) T = 21 °C, (b) T = 340 °C, (c) T = 400 °C, (d) T = 550 °C, (e) T = 600 °C, (f) T = 700 °C

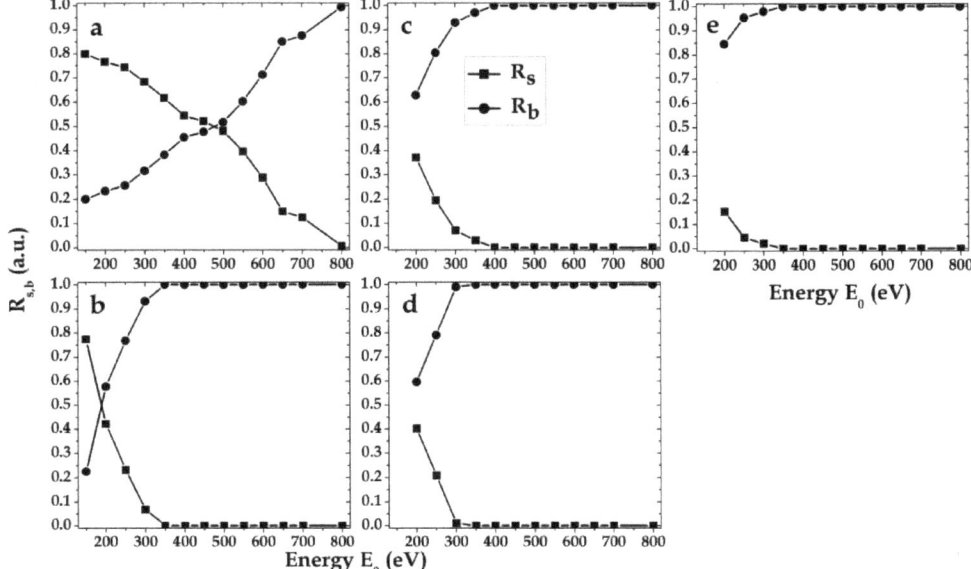

Figure 13. Dependence $R_{s,b}(E_0)$ from the primary electron energy E_0 for the Co-Cr-Mo alloy at different heating: (a) non-annealed state at T = 21 °C, (b) annealed state at T = 21 °C, (c) T = 500 °C, (d) T = 620 °C, (e) T = 790 °C.

Heating of Co-Cr-Mo alloy induces decreasing IL of surface plasmon as well as probability of their appearance in EELS spectra dependence on temperature at variation of the energy E_0. High–temperature heating of Co–Cr–Mo alloy promotes to increase the emission and background of secondary electrons in characteristic spectra, that did not allow us exactly to separate the peaks of plasma oscillations at small primary electron energy $E_0 = 150$ eV. As shown on Figure 13 for temperatures 500°C and 620°C the line of surface plasmon disappearances at $E_0 > 350$ eV and $E_0 > 300$ eV, respectively. At heating of alloy to T = 790°C the surface plasmon is detected only in range of the energy $E_0 = 200 - 350$ eV, however dependence R_s from E_0 decays quickly compared to other temperatures. Besides increasing the intensity line of plasmons with the increase of heating of Co–Cr–Mo alloy the observed a broadening of bulk plasmon line. The correlation between the damping of function R_s from E_0 and formation of concentration profile in the near surface region of alloy was established for the non–annealed Co–Cr–Mo alloy (see Figure 7). For the annealed alloy the surface plasmon detects in EELS spectra in the range of $E_0 = 150 - 350$ eV. This range of energy E_0 corresponding to the near surface region where was observed the largest variation of alloy composition relative to bulk concentration. The similar situation occurs at heating to T = 790°C for which the alloy concentration in the near surface region comes towards to the bulk at energy $E_0 > 300$ eV. Only the qualitative the correlation between the damping function R_s from E_0 and comes towards to the bulk concentration are observes for the Co–Cr–Mo alloy at the temperatures 500°C and 620°C.

3. Conclusion

Low Electron Energy Loss Spectroscopy can be used as effective non-destructive method at investigation of physical-chemical properties materials in the nano-size near surface region.

Ionization energy losses allows to investigation layer-by-layer concentration profile for the singe crystal alloys with monolayer resolution, element distribution on the depth for the polycrystalline alloys and study of kinetics of surface processes at thermo-induced treatment or after ion irradiation of the surface.

Plasmon excitations are very sensitive to structural and chemical state of surface and bulk and it can be used for study of electronic states of free electrons in the near surface region and influence of different kinetic processes on changing of electronic structure of materials.

Analysis of intensity line of surface and bulk plasmons depending on primary electron energy E_0 allows to define a surface-bulk interface when electron beam probes just near surface region with different physical-chemical properties as compared to the bulk material. These results have good correlation with data of surface composition on depth which obtained by IS and AES.

Author details

Vitaliy Tinkov
Department of the Surface Atomic Structure and Dynamic, Institute for Metal Physics of NAS of Ukraine, Ukraine

Acknowledgement

The author would like to acknowledge professor M.A. Vasylyev for his help in discussion this paper.

6. References

[1]   Vasiliev, M A. Surface effects of ordering in binary alloys. Journal of Physics D: Applied Physics 1997;30(22) 3037-3070.

[2]   O'Connor, D.J., Sexton, B.A. & Smart, R.St.C. Surface Analysis Methods in Materials Science, 2nd edit, Berlin: Springer; 2003.

[3]   Raether, H. Excitation of Plasmon and Interband Transitions by Electrons. Springer Tracts Modern Physics 1980;88 1-196.

[4]   Lüth, H. Surface and Interfaces of Solids. Springer Series in Surface Science 1993;15 1-356.

[5]   Vasylyev, M.A., Tinkov, V.A. & Gurin, P.A.. Electron Energy Loss Spectroscopy study of the near surface region of dental Co–Cr–Mo alloy. Applied Surface Science 2008;254 4671–4680.

[6]   Gerlach, R.L., Houston, J.E. & Park, R.L. Ionization spectroscopy of surfaces. Applied Physics Letters 1970;16(4) 147-188.

[7]   Gerlach, R.L. Ionization Spectroscopy of Contaminated Metal Surfaces. Journal of Vacuum Science & Technology 1971;8 599-604.

[8]   Vasylyev, M.A., Tinkov, V.A., Blaschuk, A.G., Luyten, J. & Creemers, C.. Thermo–stimulated surface segregation in the ordering alloy $Pt_{80}Co_{20}(111)$: Experiment and Modeling. Applied Surface Science 2006;253 1081–1089.

[9]   Seah, M.P. & Dench, W.A. Quantitative electron spectroscopy of surfaces: A standard data base for electron inelastic mean free paths in solids. Surface and Interface Analysis 1979;1(1) 2-11.

[10] Baschenko, O.A. & Nefedov, V.I. Depth profiling of elements in surface layers of solids based on angular resolved X-ray Photoelectron Spectroscopy. Journal of Electron Spectroscopy and Related Phenomena 1990;53 1 – 18.

[11] Cherkashin, G.Yu. Inverse problem: the concentration depth profile of elements from ARXPS data. Journal of Electron Spectroscopy and Related Phenomena 1995;74 67 – 75.

[12] Vasylyev, M.A., Blaschuk, A.G. & Tinkov, V.A. Reconstruction of concentration profiles in the surface region of $Pt_{80}Co_{20}$ alloy for (100) and (111) faces by means of ionization spectroscopy. Metal Physics and Advanced Technologies 2003;25(12) 1617–1632.

[13] Bardi, U., Atrei, A., Zanazzi, E., Rovida, G. & Ross, P.N. Study of the reconstructed (001) surface of the $Pt_{80}Co_{20}$ alloy. Vacuum 1990;41(1-3).

[14] Gauthier, Y., Baudoing-Savois, R., Bugnard, J.M., Bardi, U. & Atrei, A. Influence of the transition metal and of order on the composition profile of $Pt_{80}M_{20}(111)$ (M = Ni, Co, Fe) alloy surfaces: LEED study of $Pt_{80}Co_{20}(111)$. Surface Science 1992;276(1-3) 1 – 11.

[15] Vasylyev, M.A., Chenakin, S.P. & Tinkov, V.A. Electron Energy Loss Spectroscopy study of the effect of low–energy Ar^+–ion bombardment on the surface structure and composition of $Pt_{80}Co_{20}(111)$ alloy. Vacuum 2005;78 19–26.

[16] Tinkov, V.A., Vasylyev, M.A. & Gurin, A.P. Investigation of the thermo–stimulated surface segregation in the ternary Co–Cr–Mo alloy by means of Ionization Spectroscopy. Vacuum 2009;83 1014–1017.

[17] Vasylyev, M.A., Tinkov, V.A., Sidorenko, S. & Voloshko, S. The temperature dependence of the atoms Co diffusion coefficient in $Pt_{80}Co_{20}(111)$ alloy. Defect and Diffusion Forum 2007;265 19–23.

[18] Lea, C. & Seah, M.P. Kinetics of surface segregation. Philosophical Magazine 1977;35(1) 213-228.

[19] Maier, S.A. Plasmonics: Fundamentals and Applications, Springer Science+Business Media LLC; 2007.

[20] Vasylyev, M.A., Tinkov V.A. & Nieuwenhuys, B.E. Electron energy–loss spectroscopy of the metals Pd, Cu and the ordered $Cu_{75}Pd_{25}(100)$ alloy. Journal of Electron Spectroscopy and Related Phenomena 2007;159 53–61.

[21] Vasylyev, M.A. & Tinkov, V.A. Low energy electron induced plasmon excitations in the ordering $Pt_{80}Co_{20}(111)$ alloy surface. Surface Review and Letters 2008;15(5) 635–640.

[22] Tinkov, V.A. & Vasylyev, M.A. Thermo–induced shift of plasmon energy in electron loss spectra for the ordering $Pt_{80}Co_{20}(111)$ alloy surface. Surface Review and Letters 2009;16(2) 249–258.

[23] Tinkov, V.A. & Vasylyev, M.A. Thermo–induced plasmon excitations in the near surface region of ternary Co–Cr–Mo alloy. Vacuum 2011;85(8) 787-791.

[24] Tinkov, V.A., Vasylyev, M.A. & Galstyan, G.G. Low energy electron induced characteristic losses in the $Fe_{73.6}Cu_1Nb_{2.4}Si_{15.8}B_{7.2}$ (FINEMET) alloy surface. Vacuum 2011;85(6) 677-686.

[25] Watanabe, H. Experimental Evidence for the Collective Nature of the Characteristic Energy Loss of Electrons in Solids – Studies on the Dispersion Relation of Plasma Frequency. Journal of the Physical Society of Japan 1956;11(2) 112-119.

[26] Abe, H., Terauchi, M., Kuzuo, R. & Tanaka, M. Temperature Dependence of the Volume-Plasmon Energy in Aluminum. Journal of Electron Microscopy 1992;41(6) 465-468.

[27] Abe, H., Terauchi, M. & Tanaka, M. Temperature Dependence of the Volume-plasmon Energy in Silver. Journal of Electron Microscopy 1995;44(1) 45-48.

[28] Leder, L.B. & Marton, L. Temperature Dependence of the Characteristic Energy Loss of Electrons in Aluminum. Physical Review 1958;122 341–343.

[29] Imbuch, A. & Niedrig, H. Temperature effect on energy loss spectrum of fast electrons in aluminium and lead foils between 3 K and 295 K. Physics Letters A 1970;32(6) 375-376.

[30] Apholte, H.R. & Ulmer, K. Temperaturabhängigkeit der charakterischen energieverluste in niob, molybdän, tantal und wolfram. Physics Letters 1966;22(5) 552-553.

[31] Heimann, B. & Hölzl, J. Variation of Characteristic Energy Losses in the Curie-Temperature Region of Ni (111). Physical Review Letters 1971;26 1573–1575.

[32] Korsukov, V.E. & Lukyanenko, A.S. The surface relaxation of Al as determined by electron energy loss spectroscopy on plasmons. Zeitschrift für Physik B Condensed Matter 1983;53(2) 143-150.

[33] Rocca, M. Low-energy EELS investigation of surface electronic excitations on metals. Surface Science Reports 1995;22(1–2) 1-71.

[34] Jensen, E.T., Palmer, E.E., Allison, W. & Annett, H.F. Temperature-dependent plasmon frequency and linewidth in a semimetal. Physical Review Letters 1991;66 492–495.

Chapter 8

Stress Measurements in Si and SiGe by Liquid-Immersion Raman Spectroscopy

Daisuke Kosemura, Motohiro Tomita, Koji Usuda and Atsushi Ogura

Additional information is available at the end of the chapter

1. Introduction

Strained Si technology is important for engineering field-effect transistors (FETs) [1,2]. There are two types of the strained Si technologies. One is so-called global strained Si technology. Another is so-called local strained Si technology. The former is the technology of using a strained Si substrate which has a several-dozen-nanometers-thick strained Si layer at the top of the substrate [3-5]. The strained Si layer is obtained by growing Si on SiGe, therefore, large tensile strain with biaxial isotropy can be induced in Si. The isotropic biaxial tensile strain in Si allows for performance improvements for both of n- and p- type FETs. Homogeneous strain distribution can be obtained under the critical thickness of the strained Si layer [6].

In the latter case, the strain is induced only in the desired region, the channel region of FET [7,8]. A SiN film is used as the stressor that can induce tensile or compressive uniaxial stress in Si by changing the deposition conditions of the SiN film [9,10]. The uniaxial tensile strain enhances electron mobility, while the uniaxial compressive strain enhances hole mobility. Various kinds of the local strained Si techniques have so far been suggested by many researchers [11,12]. The combination of the global and local strained Si technologies is considered effective to induce extremely large strain in Si. Fin-type structures have been reported for high-performance FETs [13]. It is considered that the stress relaxation occurs during the fabrication of the fin-shaped strained Si layer. There are many other origins of strain fluctuations, e.g., shallow trench isolation (STI), metal gate electrodes, silicide, interconnections, and its layout. As a result, the stress states in the future generation FETs become complicated. The relationship between the electrical properties of FETs and the strain is also complicated. Therefore, to measure the complicated stress states in Si has great demand in order to improve the FET performance effectively.

Several kinds of strain or stress measurements have been studied, e.g., X-ray diffraction (XRD), transmission electron microscopy (TEM), electron backscattering diffraction (EBSD), and Raman spectroscopy [4,14-16]. Among them, Raman spectroscopy has the advantages such as high sensitive to local strain, submicron spatial resolution, nondestructive measurements, fast measurements, and ease of use. Consequently, Raman spectroscopy has been frequently used by many researchers to measure the strain in Si [3,7-9,17-21]. However, conventional Raman spectroscopy fails to measure the complicated stress states in Si. The reason is as follows. Backscattering geometry from a (001) Si substrate is generally used in Raman measurements of strained Si. In this geometry, only one of three optical phonon modes is Raman active, while two of three modes are Raman inactive. The limitation arises from the extremely high symmetry of the Si crystal. As a result, the weighted average value of the complicated stress state is obtained, that is, it is impossible to perform quantitative measurements of strain by conventional Raman spectroscopy.

Si has three optical phonon modes: one longitudinal optical (LO) phonon mode and two transverse optical (TO) phonon modes, the polarizations of which are parallel and perpendicular to the phonon wave vector, respectively. Recently, the forbidden optical phonon modes, the TO phonon modes, were excited even under the (001) Si backscattering geometry, using a high-numerical aperture (NA) liquid-immersion lens [22-24]. If all of the three optical phonon modes are detectable, the unknown three components of a stress tensor in Si can be obtained in theory [25-35]. The high-NA liquid-immersion Raman spectroscopy has great potential for measuring the complicated stress states in Si with high spatial resolution.

On the other hand, the number of stress tensor components is six. Therefore, the evaluation of nondiagonal stress components, shear stress components, is considered difficult even detecting the TO phonon modes. The shear stress is often generated at the discontinuous region, e.g., around STI and at the edge of a contact etch stop layer. The shear stress often produces dislocations in Si, which cause leakage current during transistor operation [36,37]. The shear stress measurements are desired for failure analysis. The induction of the stress with the nondiagonal components requires the transformation of the Raman tensors. Therefore, Raman spectroscopy is essentially sensitive to the shear stress.

In this study, anisotropic stress states in Si were measured by the high-NA liquid-immersion Raman spectroscopy. Strained SiGe was also measured by the same technique. SiGe has been suggested as the channel material of next generation FETs, because the both mobilities of electrons and holes in SiGe are higher than those in Si. Furthermore, the strain induction in SiGe is considered effective for improving electrical properties of SiGe FETs in the same way as strained Si [38-41]. The nondiagonal stress components, shear stress components, were measured by analyzing the dependence of Raman spectra on the relative polarization direction between sample orientation and electrical fields of incident and scattered light.

2. Experimental procedure

2.1. Excitation of TO phonon modes

The Raman intensity is calculated by the following equation [42];

$$I \propto \sum_j \left| e_s^{\ T} R_j e_i \right|^2 ,$$ (1)

$$R_1 = \begin{pmatrix} 0 & 0 & 0 \\ 0 & 0 & d \\ 0 & d & 0 \end{pmatrix}, R_2 = \begin{pmatrix} 0 & 0 & d \\ 0 & 0 & 0 \\ d & 0 & 0 \end{pmatrix}, R_3 = \begin{pmatrix} 0 & d & 0 \\ d & 0 & 0 \\ 0 & 0 & 0 \end{pmatrix},$$ (2)

where R_j is the Raman tensors of Si [43]. e_i and e_s are the electrical fields of incident and scattered light, respectively. The superscript T denotes transpose. From Eqs. (1) and (2), the TO phonon modes are not excited under the (001) Si backscattering geometry. This is because the component of z polarization of the incident light is reduced to almost zero in the case of the (001) Si backscattering configuration. The z polarization is thus needed to excite the TO phonon modes in Si. It is considered that oblique light relative to the (001) Si surface gives rise to the z polarization.

Fig. 1 shows the experimental set-up for oblique incident light configuration in this study. The glancing angles of the laser against the sample were 30° and 90°, as shown in Fig. 1. Fig. 2 shows the examples of calculations for the Raman intensities in the 90° and 30° configurations, using Eqs. (1) and (2). In the case of the 90° configuration, the (001) Si backscattering configuration, the TO phonon modes are Raman inactive and the LO phonon mode is Raman active, as mentioned above. On the other hand, in the case of the 30° configuration, the oblique incident light configuration, the TO phonon mode is Raman active. This fact arises because the z polarization of the incident light can be obtained in the oblique incident light configuration. In the rough approximations, the Raman intensities are considered to be the same for the 90° and 30° configurations, as shown in Fig. 2.

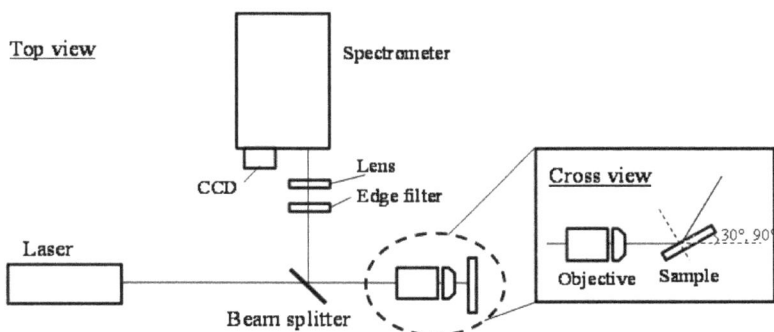

Figure 1. Experimental set-up for oblique incident light configuration

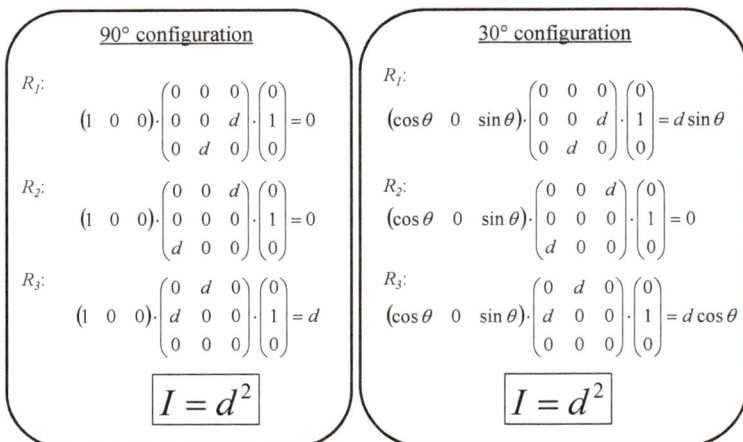

Figure 2. Calculations of Raman intensities for 90° and 30° configurations.

The intensity ratio of the TO phonon mode to the LO phonon mode in the 30° configuration is calculated to be;

$$\frac{I_{TO}}{T_{LO}} = \frac{|d\sin\theta|^2}{|d\cos\theta|^2} \approx 0.046 , \tag{3}$$

where θ is the aperture angle in Si. In the 30° configuration, θ is approximately 12.13° in the case of λ = 532 nm laser because Si has the large refraction index [44]. Therefore, the intensity of the TO phonon mode are much small, compared to that of the LO phonon mode even in the 30° configuration. The detection of the TO phonon modes is basically considered difficult. Moreover, for the oblique incident light configuration, high-resolution measurements cannot be achieved because it is difficult to use the high-NA lens and the beam spot becomes an ellipse. In this study, the high-NA liquid-immersion lens was used in order to obtain the oblique light relative to the (001) Si surface.

An aperture angle θ is calculated by NA = $n\times\sin\theta$ (where n is a refractive index). θ is equal to 44.4° in conventional Raman spectroscopy with the use of NA = 0.7 objective (n = 1.0). On the other hand, θ is equal to 69.0° in high-NA liquid-immersio Raman spectroscopy with the NA = 1.4 liquid-immersio lens (n = 1.5). However, the incident light widely refracts at the interface of the Si surface because Si has the large refractive index as mentioned above. The refractive index of Si for the λ = 364 nm light (where λ is wavelength) is approximately 6.5 [44]. Therefore, θ in Si results in 6.2° in conventional Raman spectroscopy (NA = 0.7). This configuration is almost under the (001) Si backscattering geometry, i.e., the component of the z polarization is reduced to almost zero. This fact causes that the TO phonon modes are Raman inactive in conventional Raman spectroscopy. On the other hand, θ in Si results in 12.4° in high-NA liquid-immersio Raman spectroscopy (NA = 1.4). It is considered that the value of θ in Si is still small to excite the TO phonon modes effectively, although the value is two times larger than that in conventional Raman spectroscopy. It is considered that the use of the UV light has the drawback for the excitation of the TO phonon modes. In the case of visible light (λ = 532 nm), θ in Si are calculated to be 9.8° and 19.8° in conventional and liquid-immersion Raman spectroscopy, respectively. The value for oil-immersion Raman spectroscopy with the use of the visible light is relatively large, therefore, the large component of the z polarization is obtained. Table 1 shows θs as a function of NA. θ_1 and θ_2 are the aperture angles in the medium and Si, respectively.

NA	With use of UV light (λ = 364 nm)			With use of visible light (λ = 532 nm)		
	θ_1	θ_2	TO/LO	θ_1	θ_2	TO/LO
0.7	44.4	6.2	0.06	44.4	9.8	0.25
1.1	57.8	9.7	0.2	57.8	15.5	1.0
1.2	67.4	10.6	0.3	67.4	16.9	1.3
1.4	69.0	12.4	0.5	69.0	19.8	2.0
1.7	-----	-----	-----	70.8	24.4	3.8

Table 1. θs with use of visible and UV light and intensity ratio of TO to LO phonon modes as a function of NA.

It is important to choose the appropriate NA and the wavelength for the excitation of the TO phonon modes. The intensity of the TO phonon mode excited by high-NA liquid-immersion Raman spectroscopy is estimated as follows. The TO phonon modes are excited mainly by the marginal ray of incident light. The Raman intensity can be calculated by the following equation [45]:

$$S_\Omega = A \sum_j \int_{\Omega_s} \int_{\Omega_i} \left(e_i R_j e_s \right)^2 d\Omega_i d\Omega_s , \tag{4}$$

where Ω_i and Ω_s are the solid angles of incident and scattered light, respectively. The intensity ratio of the TO to LO phonon modes is considered to be the intensity ratio of the z component of the marginal ray to the paraxial ray. Fig. 3 shows the intensity ratio of TO to LO phonon modes as a function of NA with the use of visible light. The aperture angle dependence on NA is also shown in Fig. 3.

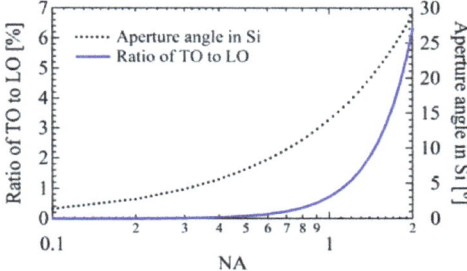

Figure 3. Intensity ratio of TO to LO phonon modes and aperture angle in Si vs. NA

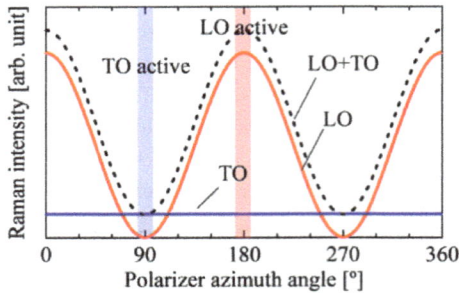

Figure 4. Raman intensities of LO and TO phonon modes vs polarizer azimuth angle

The intensity ratio of the TO to LO phonon modes as well as the aperture angle in Si increases with the increase in NA. Note that drastic increase is confirmed especially beyond NA = 1.0 for the intensity ratio of TO to LO. The value is approximately 2.0% for NA = 1.4 with the use of visible light, while the value is approximately 0.5% with the use of UV light, as shown in Table 1. From the estimations, the intensity of the TO phonon mode is very low, compared to that of the LO phonon mode. Actually, it is important to suppress the intensity of the LO phonon mode for the excitation of the TO phonon modes. This can be accomplished by the Raman polarization selection rules [42].

Fig. 4 shows the Si Raman intensities of the LO and TO phonon modes calculated by Eq. (1) as a function of a polarizer azimuth angle. For the LO phonon mode, the intensity changes in the period of 180°. On the other hand, for the TO phonon mode, the intensity is independent on the polarizer azimuth angle (the value is exaggerated for ease to view). This is because the component of the z polarization obtained by the oblique light remains constant all over the angles. As a result, the measurable Raman intensity profile is the sum of the intensities of the LO and TO phonon modes, which is shown by the dashed line in Fig. 4. From Fig. 4, the LO phonon mode can be detected at the polarizer azimuth angle of 0°, 180°, and 360°. These correspond to the LO active configurations. On the other hand, the TO phonon mode can be detected at the angle of 90° and 270°. These correspond to the TO active configurations. The TO and LO phonon modes can be separately detected by the Raman polarization selection rules. It is possible to evaluate complicated stress states in Si by analyzing multi-optical phonon modes.

2.2. Methodology of measurements for anisotropic biaxial stress states in Si

In this section, the methodology of measurements for anisotropic biaxial stress states in Si by liquid-immersion Raman spectroscopy is shown. In the previous section, it was shown that the z polarization can be created by the oblique light due to the high-NA liquid-immersion lens. Consequently, the TO phonon modes in Si can be excited by the z polarization even under the (001) Si backscattering geometry. The TO phonon modes allow for the measurements of the anisotropic biaxial stress states in Si.

The force constant of a Si crystal changes by the induction of strain. Consequently, the optical-phonon frequencies also change. The difference of the force constant ΔK is represented as a second-rank tensor. The eigenvalues of ΔK after the induction of the strain can be obtained by solving the secular equation [29]. The square roots of the eigenvalues correspond to the changes of the optical-phonon frequencies (the Raman wavenumber shifts). Three equations between the strain (stress) and the Raman wavenumber shifts are obtained because Si has three optical-phonon modes (two TOs and one LO).

Suppose that there is a linear relationship between ΔK and strain [26]. In a coordinate system of x: [100], y: [010], and z: [001], ΔK is represented as the following equation:

$$\Delta K = A\varepsilon , \tag{5}$$

$$A = \begin{pmatrix} p & q & q & 0 & 0 & 0 \\ q & p & q & 0 & 0 & 0 \\ q & q & p & 0 & 0 & 0 \\ 0 & 0 & 0 & 2r & 0 & 0 \\ 0 & 0 & 0 & 0 & 2r & 0 \\ 0 & 0 & 0 & 0 & 0 & 2r \end{pmatrix} , \tag{6}$$

where ε is a strain tensor. A is a fourth-rank tensor whose components are p, q, and r called phonon deformation potentials (PDPs). Generally, transistors are fabricated on (001) Si substrate in the direction of [110] Si. Therefore, the coordinate transformation makes analysis easy [46]. Second-rank and fourth-rank tensors are transformed in the coordinate system of $x' = [110]$, $y' = [-110]$, and $z' = [001]$ by the following equations:

$$T_{ij} = a_{ik}a_{jl}T_{kl} , \tag{7}$$

$$T_{ijkl} = a_{im}a_{jn}a_{ko}a_{lp}T_{mnop} , \tag{8}$$

$$a = \begin{pmatrix} 1/\sqrt{2} & 1/\sqrt{2} & 0 \\ -1/\sqrt{2} & 1/\sqrt{2} & 0 \\ 0 & 0 & 1 \end{pmatrix} , \tag{9}$$

where T and a are a second- or fourth- rank tensor and a transformation matrix, respectively. Hence, Eq. (5) results in:

$$\Delta K' = A'\varepsilon' , \tag{10}$$

where the primes denote the components in the coordinate $x'y'z'$. The secular equation of $\Delta K'$ is below:

$$\begin{vmatrix} \Delta K_{xx}{}' - \lambda & \Delta K_{xy}{}' & \Delta K_{xz}{}' \\ \Delta K_{xy}{}' & \Delta K_{yy}{}' - \lambda & \Delta K_{yz}{}' \\ \Delta K_{xz}{}' & \Delta K_{yz}{}' & \Delta K_{zz}{}' - \lambda \end{vmatrix} = 0 , \tag{11}$$

where λ is the eigenvalues. An anisotropic biaxial stress state is represented as the following second-rank tensor:

$$\sigma' = \begin{pmatrix} \sigma_{xx}{}' & 0 & 0 \\ 0 & \sigma_{yy}{}' & 0 \\ 0 & 0 & 0 \end{pmatrix} , \tag{12}$$

where $\sigma_{xx}{}'$ and $\sigma_{yy}{}'$ are the stress components in the directions of [110] and [–110], respectively. Generally, stress induction changes not only optical-phonon frequencies but also Raman tensors. However, in the case of the stress tensors only with the diagonal components, there are no changes of the Raman tensors. Therefore, the Raman polarization selection rules after the induction of the biaxial stresses σ_{xx} and σ_{yy} remains as those of stress-free Si [46]. The Raman tensors in the coordinate $x'y'z'$ are as follows [46]:

$$R_1{}' = \frac{1}{\sqrt{2}} \begin{pmatrix} 0 & 0 & d \\ 0 & 0 & d \\ d & d & 0 \end{pmatrix} , \; R_2{}' = \frac{1}{\sqrt{2}} \begin{pmatrix} 0 & 0 & d \\ 0 & 0 & -d \\ d & -d & 0 \end{pmatrix} , \; R_3{}' = \begin{pmatrix} d & 0 & 0 \\ 0 & -d & 0 \\ 0 & 0 & 0 \end{pmatrix} \tag{13}$$

Stress tensors are transformed to strain tensors by Hooke's law:

$$\varepsilon' = S'\sigma' , \tag{14}$$

$$S = \begin{pmatrix} S_{11} & S_{12} & S_{12} & 0 & 0 & 0 \\ S_{12} & S_{11} & S_{12} & 0 & 0 & 0 \\ S_{12} & S_{12} & S_{11} & 0 & 0 & 0 \\ 0 & 0 & 0 & S_{44}/4 & 0 & 0 \\ 0 & 0 & 0 & 0 & S_{44}/4 & 0 \\ 0 & 0 & 0 & 0 & 0 & S_{44}/4 \end{pmatrix} , \tag{15}$$

where S expressed by Eq. (15) is the elastic compliance tensor. The components of S, S_{11}, S_{12}, and S_{44} are 7.68×10^{-12}, -2.14×10^{-12}, and 12.7×10^{-12} 1/Pa, respectively [17]. The transformation of the fourth-rank tensor S by Eq. (8) is needed. The strain tensor ε' expressed by Eq. (14) is substituted for Eq. (11) and then the eigenvalues λ_i are calculated. As a result, using Eq. (16), the relationship between the Raman wavenumber shifts $\Delta \omega s$ and the anisotropic biaxial stresses $\sigma_{xx}{}'$ and $\sigma_{yy}{}'$ are obtained as follows [35]:

$$\lambda_i = \omega_i{}^2 - \omega_0{}^2 = (\omega_i + \omega_0)(\omega_i - \omega_0) \approx 2\omega_0(\omega_i - \omega_0) , \; \Delta\omega_i \approx \frac{\lambda_i}{2\omega_0} . \tag{16}$$

$$\Delta\omega_1 = \frac{\lambda_1}{2\omega_0} = \frac{1}{2\omega_0}\left[\frac{1}{2}p(S_{11}+S_{12})+\frac{1}{2}q(S_{11}+3S_{12})+\frac{1}{2}rS_{44}\right]\times\sigma_{xx}{}'$$
$$+\frac{1}{2\omega_0}\left[\frac{1}{2}p(S_{11}+S_{12})+\frac{1}{2}q(S_{11}+3S_{12})-\frac{1}{2}rS_{44}\right]\times\sigma_{yy}{}' \tag{17-1}$$

$$\Delta\omega_2 = \frac{\lambda_2}{2\omega_0} = \frac{1}{2\omega_0}\left[\frac{1}{2}p(S_{11}+S_{12})+\frac{1}{2}q(S_{11}+3S_{12})-\frac{1}{2}rS_{44}\right]\times\sigma_{xx}{}'$$
$$+\frac{1}{2\omega_0}\left[\frac{1}{2}p(S_{11}+S_{12})+\frac{1}{2}q(S_{11}+3S_{12})+\frac{1}{2}rS_{44}\right]\times\sigma_{yy}{}' \tag{17-2}$$

$$\Delta\omega_3 = \frac{\lambda_3}{2\omega_0} = \frac{1}{2\omega_0}\left[pS_{12}+q(S_{11+}S_{12})\right]\times\left(\sigma_{xx}{}'+\sigma_{yy}{}'\right), \tag{17-3}$$

$$\Delta\omega_3 = \frac{\lambda_3}{2\omega_0} = \frac{1}{\omega_0}\left[\frac{S_{12}}{S_{11}+S_{12}}p+q\right]\times\varepsilon_{biaxial} = b\times\varepsilon_{biaxial}, \tag{18}$$

where b is so-called the b coefficient which is used for the evaluation of isotropic biaxial strain $\varepsilon_{biaxial}$ in strained Si substrates using the Raman wavenumber shift of the LO phonon mode $\Delta\omega_3$ [21,47].

Authors (year)	Sample	p/ω_0^2, q/ω_0^2, r/ω_0^2	b cm^{-1}	Citation
Anastassakis et al. (1970)[a]	Si bar	−1.25, −1.87, −0.66	−721	4 (9%)
Chandrasekhar et al. (1978)[b]	Si bar	−1.43, −1.89, −0.59	−696	18 (40%)
Anastassakis et al. (1990)[c]	Si bar	−1.85, −2.31, −0.71	−830	15 (33%)
Nakashima et al. (2006)[d]	Strained Si substrate	----------	−723	1 (2%)
JEITA (2007)[e]	Strained Si substrate	----------	−769	1 (2%)
Others[f-j]	Strained Si substrate	----------	−1040 ~ −715	6 (13%)

[a]Reference 14, [b]Reference 15, [c]Reference 16, [d]Reference 12, [e]Reference 39, [f-j]Reference 46, 47, 71, 78, and 79.

Table 2. Various PDPs suggested so far and statistics of citations.

Various PDPs have so far been suggested by many researchers. The suggested PDPs and the citation count are shown in Table 2. Forty-five papers were confirmed. As shown in Table 2, the values of PDPs are fluctuated. Thirty-seven of forty-five papers, approximately eighty-two percent papers, referred PDPs suggested by the Cardona's group in 1970-1990. Nakashima et al. examined the b coefficient in detail using strained Si substrates by Raman spectroscopy and high-resolution XRD in 2006 [21]. Furthermore, the detailed investigation of the b coefficient was performed in the working group of Japan electronics and information technology industries association (JEITA) in 2007 [48]. Eight organizations attended the working group: three companies for XRD measurements, three companies and one University for Raman measurements, and one company for Rutherford back scattering (RBS) measurements. The b coefficient of -769 cm^{-1} was obtained [16]. Extreme care is needed to choose appropriate PDPs.

In this study, the validity of three sets of PDPs was evaluated by liquid-immersion Raman spectroscopy: first, $p/\omega^2 = -1.25$, $q/\omega^2 = -1.87$, and $r/\omega^2 = -0.66$ reported by Anastassakis et al. in 1970 [49], second, $p/\omega^2 = -1.43$, $q/\omega^2 = -1.89$, and $r/\omega^2 = -0.59$ reported by Chandrasekhar et al. in 1978 [50], and third, $p/\omega^2 = -1.85$, $q/\omega^2 = -2.31$, and $r/\omega^2 = -0.71$ reported by Anastassakis et al. 1990 [51]. The first set of PDPs was obtained from the first investigation. The second set appears to be the most commonly used, and the third set is the most recently reported among the three sets of PDPs.

The relationship between Raman wavenumber shifts $\Delta\omega$s and the anisotropic biaxial stresses σ_{xx}' and σ_{yy}' are obtained by substituting PDPs shown above for Eq. (17). When PDPs reported by Anastassakis et al. in 1970 are used,

$$\Delta\omega_1 = -2.30 \times \sigma_{xx}' - 0.12 \times \sigma_{yy}', \tag{19-1}$$

$$\Delta\omega_2 = -0.12 \times \sigma_{xx}' - 2.30 \times \sigma_{yy}', \tag{19-2}$$

$$\Delta\omega_3 = -2.00 \times \sigma_{xx}' - 2.00 \times \sigma_{yy}'. \tag{19-3}$$

When PDPs reported by Chandrasekhar et al. in 1978 are used,

$$\Delta\omega_1 = -2.31 \times \sigma_{xx}' - 0.37 \times \sigma_{yy}', \tag{20-1}$$

$$\Delta\omega_2 = -0.37 \times \sigma_{xx}' - 2.31 \times \sigma_{yy}', \tag{20-2}$$

$$\Delta\omega_3 = -1.93 \times \sigma_{xx}' - 1.93 \times \sigma_{yy}'. \tag{20-3}$$

When PDPs reported by Anastassakis et al. in 1990 are used,

$$\Delta\omega_1 = -2.88 \times \sigma_{xx}' - 0.54 \times \sigma_{yy}', \tag{21-1}$$

$$\Delta\omega_2 = -0.54 \times \sigma_{xx}' - 2.88 \times \sigma_{yy}', \tag{21-2}$$

$$\Delta\omega_3 = -2.30 \times \sigma_{xx}' - 2.30 \times \sigma_{yy}'. \tag{21-3}$$

2.3. Methodology of measurements for nondiagonal stress components

In the case of the induction of stress with the only diagonal stress components, strain-modified phonon eigenvectors ξ's which are obtained by solving the secular equation expressed by Eq. (11) coincide with the coordinate $x'y'z'$. In this case, the Raman tensors of Si

r∋mains in the same form expressed by Eq. (13). On the other hand, shear stress causes a
∂eviation between the phonon wave vector and ξ_i's, i.e., in the case of the induction of stress
with the nondiagonal stress components, ξ_i's no longer coincide with the coordinate $x'y'z'$
[52]. The difference between ξ_i' and the coordinate $x'y'z'$ requires a change of the Raman
tensors. The new Raman tensors $R_i'^*$ is expressed by:

$$R_i'^* = \left(\xi_i'^* \cdot \xi_1'\right)R_1' + \left(\xi_i'^* \cdot \xi_2'\right)R_2' + \left(\xi_i'^* \cdot \xi_3'\right)R_3', \tag{22}$$

where $\xi_i'^*$ and ξ_i' are the strain-modified eigenvectors for the introduction of stress with the
nondiagonal and only diagonal stress components, respectively. Assuming the stress tensor
shown by Eq. (12), the Raman tensors R_i' changes to $R_i'^*$:

$$R_1'^* = \begin{pmatrix} \times & 0 & \times \\ 0 & \times & \times \\ \times & \times & 0 \end{pmatrix}, \quad R_2'^* = \begin{pmatrix} 0 & 0 & \times \\ 0 & 0 & \times \\ \times & \times & 0 \end{pmatrix}, \quad R_3'^* = \begin{pmatrix} \times & 0 & \times \\ 0 & \times & \times \\ \times & \times & 0 \end{pmatrix}, \tag{23-1)-(23-3}$$

where $\times$ indicates nonzero components (each value is not always the same), some of which
depend on the eigenvalues obtained by solving the secular equation of Eq. (11). $R_2'^*$ has the
same form as the Raman tensor R_2' because of $\tau_{xy} = \tau_{yz} = 0$. Therefore, R_2' corresponds to the
TO phonon mode with the eigenvector in the y' direction. On the other hand, $R_1'^*$ and $R_3'^*$ no
longer correspond to purely transverse and longitudinal modes, respectively, because their
eigenvectors do not coincide with the x' and z' axes, respectively. Consequently, the Raman
intensity is changed by the nondiagonal stress components obeying the Raman polarization
selection rules. The nondiagonal stress components can be evaluated by analyzing the
dependence of the Raman spectra on the relative polarization direction between the sample
orientation and the electrical fields of incident and scattered light. The methodology is
described as follows.

The methodology for evaluating complicated stress states was reported by Ossikovski *et al*
[33]. They employed an experimental configuration that used oblique incident light to
observe the forbidden modes, i.e., the TO phonon modes. In our experiments, the high-NA
liquid-immersion lens was used to observe the TO phonon modes. High spatial resolution
was preserved in liquid-immersion Raman spectroscopy.

First, a stress tensor is considered, and then the strain tensor is calculated by Hooke's law by
Eq. (14). The strain tensor is substituted for the secular equation of Eq. (11). Three phonon
eigenfrequencies of Si have been determined so far. Si is a nonpolar cubic crystal, that is,
there is no difference between the TO phonon modes and LO phonon mode [52]. As a result,
the three determined phonon eigenfrequencies are independent of the phonon wave vector.
Second, the Raman tensors are determined using Eq. (22). Their forms change when the
nondiagonal stress components are nonzero. Subsequently, the Raman intensity of each
phonon mode is calculated by the Raman polarization selection rules given by Eq. (1). Third,
each Raman spectrum is considered to be a Lorentzian function $\Lambda_i(\omega)$ [35]:

$$\Lambda_i(\omega) = \frac{I_i \Gamma^2}{(\omega - \Omega_i)^2 + \Gamma^2}, \tag{24}$$

where ω, Ω_i, and Γ are the Raman shift, the phonon eigenfrequencies , and the half width at half maximum of the spectrum, respectively. It is considered difficult to analyze each spectrum of the TO and LO phonon modes. Therefore the effective phonon eigenfrequency Ω_{eff} is used as a representative value. The effective value is the weighted average of the phonon eigenfrequencies relative to their intensities, as expressed by the following equation [26]:

$$\Omega_{eff} = \sum_i \frac{I_i \Omega_i}{I_T}, \tag{25}$$

where I_T is the total intensity of the three phonon modes. Eq. (25) is valid because the strain-induced splitting between the TO and LO phonon modes is small, compared to Γ. An example of a spectrum with the effective phonon eigenfrequency and the spectra of the TO and LO phonon modes are shown in Fig. 5. A uniaxial stress σ_{xx}' of 1.0 GPa is assumed in the calculation. In Fig. 5, the Raman spectra with the eigenfrequencies of Ω_1, Ω_2, and Ω_3 appear, which correspond to the optical phonon modes with the eigenvectors x', y', and z', respectively. The dashed line shows the Raman spectrum with the weighted average eigenfrequency. The Raman signal of the TO phonon modes with the eigenvectors x' and y' are obtained by the z polarization due to the high-NA lens (the component of z polarization is enlarged for ease to view).

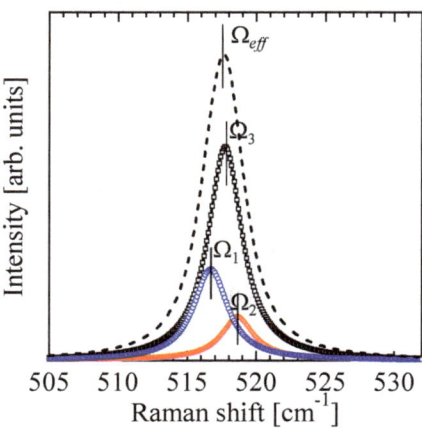

Figure 5. Raman spectrum with effective phonon frwuency and raman spectra with Ω_1, Ω_2, and Ω_3

Finally, the dependence of the Raman spectra on the polarization direction between the sample orientation and the electrical fields of incident and scattered light is obtained as follows. The electrical fields of incident and scattered light are fixed in the y' direction. Regarding e_i, because the high-NA liquid-immersion lens is used, the z polarization can be obtained:

$$e_i = \frac{1}{\sqrt{1+\alpha^2}} \begin{pmatrix} 0 \\ 1 \\ \alpha \end{pmatrix}, \tag{26}$$

where α is the component of the z polarization. For $\alpha = 0$, this is correct in the (001) Si backscattering geometry. α was experimentally determined. Eq. (1) shows that for the Raman intensity, the rotations of the polarization directions of incident and scattered light are equivalent to the rotation of the sample, although the period for the sample rotation is half compared to those for the polarization rotations. In the experiments, the sample was rotated from 0° to 180°, which is represented by the following equations:

$$R_i'^{*}{}_{rot} = TR_i'^{*}T, \tag{27}$$

$$T(\phi) = \begin{pmatrix} \cos\varphi & \sin\varphi & 0 \\ -\sin\varphi & \cos\varphi & 0 \\ 0 & 0 & 1 \end{pmatrix}, \tag{28}$$

where $R_i'^{*}{}_{rot}$ and T are the Raman tensors after rotation by φ and the transformation matrix, respectively.

Fig. 6 shows the dependence of the effective Raman shifts on the sample rotation angle for the various stress states including hydrostatic stress, uniaxial stress, biaxial stress, and stress with the nondiagonal components, which are represented by:

$$\sigma_{hydrostatic} = \begin{pmatrix} 0.33 & 0 & 0 \\ 0 & 0.33 & 0 \\ 0 & 0 & 0.33 \end{pmatrix}, \sigma_{uniaxial} = \begin{pmatrix} 1.0 & 0 & 0 \\ 0 & 0 & 0 \\ 0 & 0 & 0 \end{pmatrix}, \tag{29 and 30}$$

$$\sigma_{biaxial} = \begin{pmatrix} 0.5 & 0 & 0 \\ 0 & 0 & 0 \\ 0 & 0 & 0.5 \end{pmatrix}, \sigma_{shear} = \begin{pmatrix} 0.5 & 0 & 0.5 \\ 0 & 0 & 0 \\ 0.5 & 0 & 0.5 \end{pmatrix}. \tag{31 and 32}$$

The above stress states correspond to the load of 1.0 GPa. A unique profile can be obtained for each stress state, as shown in Fig. 6. The profile for the hydrostatic stress remains constant all over the sample rotation angles because the degeneracy of the long-wavelength optical phonons does not lift under the hydrostatic stress. It should be noted that the profile becomes asymmetric only for the stress state with the nondiagonal components. As a result, the shear stress in Si is considered to be detectable by analyzing the dependence of the effective Raman shifts on the sample rotation angle.

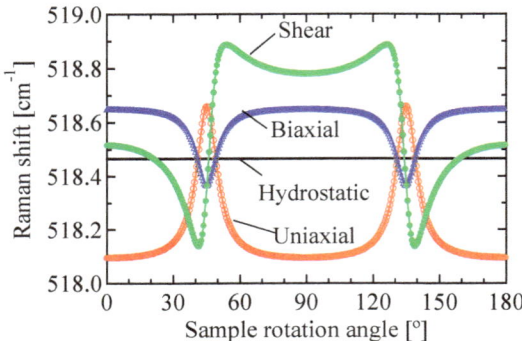

Figure 6. Effective Raman shift dependence on sample rotation angle for hydrostatic, uniaxial, biaxial, and shear stress.

2.4. Samples

(001)-oriented SSOI substrates were used as the samples [53,54]. Fig. 7(a) shows the cross sectional TEM image of the SSOI substrate. The structure of SSOI was strained Si layer/buried oxide (BOX) layer/Si substrate, which is the simplest structure among the strained Si substrates. The low-power consumption operation can be achieved due to the structure of Si on insulator (SOI) [55,56]. The thicknesses of the strained Si layers were 30, 50, and 70 nm. An isotropic biaxial tensile stress state exists in the strained Si layer.

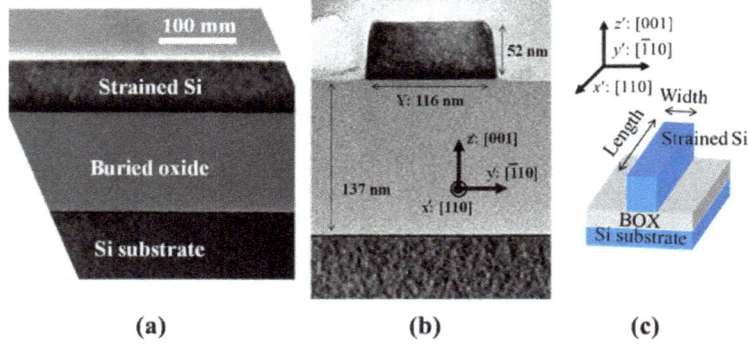

Figure 7. (a) Cross sectional TEM image of SSOI, (b) cross sectional TEM image of SSOI nanostructure, and (c) schematic of SSOI nanostructure

For Si, three long-wavelength optical phonon modes are degenerate at the center of the Brillouin zone. On the other hand, the degeneracy lifts after the induction of stress and the frequency of each mode individually shifts depending on the stress. For the isotropic biaxial tensile stress, the frequency of each mode shifts on the lower-frequency side and splits into singlet and doublet. In the case of (001) Si backscattering geometry, the singlet and doublet correspond to the LO and TO phonon modes, respectively. Fig. 8 shows the optical phonon frequencies for Si and SSOI. Generally, the LO phonon mode which is Raman active under the (001) Si backscattering geometry is measured and the isotropic biaxial stress in the strained Si layer is evaluated using the b coefficient shown in Table 2.

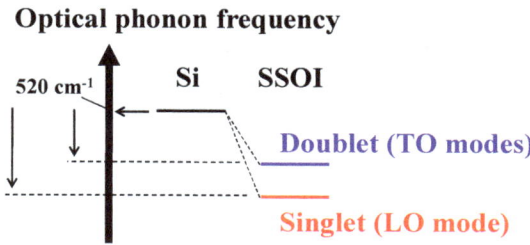

Figure 8. Optical phonon frequencies for Si and SSOI

SSOI nanostructures were fabricated with arbitrary forms by electron beam (EB) lithography and reactive ion etching (RIE). Fig. 7(b) and (c) show the cross sectional TEM image and the schematic of the SSOI nanostructure. The coordinate system in the experiments is also shown in Fig. 7(b) and (c). The SSOI lengths (Ls) were 5.0, 3.0, 2.0, 1.5, 1.0, 0.8, and 0.5 μm. The SSOI widths (Ws) were 1.0, 0.5, 0.2, 0.1, and 0.05 μm. The SSOI nanostructure shapes were anisotropic. Therefore, the stress states are also considered anisotropic. The stress component in the z direction is considered to be zero because of free-standing surface. As a result, the stress tensors in the SSOI nanostructures are considered to be expressed by Eq. (12).

SiGe nanostructures were fabricated as the same manner. SiGe with approximately 30% Ge concentration was epitaxially grown on a Si substrate. The thickness of the SiGe layer was approximately 35 nm. The Ls and Ws of the SiGe nanostructures were the same as those of the SSOI nanostructures. The cross sectional TEM image and the schematic of the SiGe nanostructure are shown in Fig. 9(a) and (b), respectively. As shown in the TEM image, overetching of the Si substrate is confirmed.

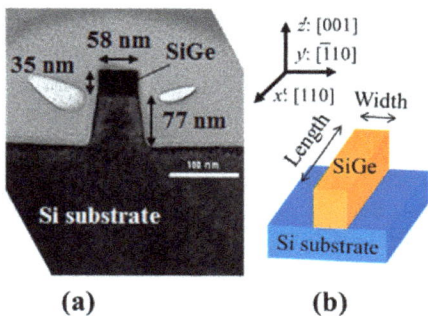

(a)   **(b)**

Figure 9. (a) Cross sectional TEM image of SiGe nanostructure and (b) schematic

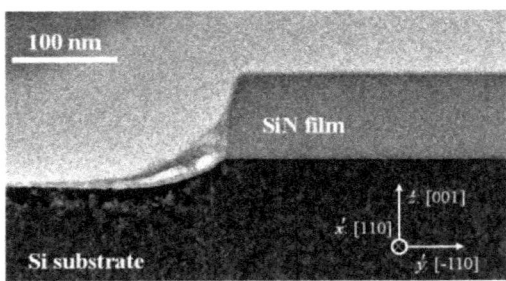

Figure 10. 10 TEM image of SiN film on Si substrate

A 80-nm-thick SiN film was deposited on a Si substrate by low-pressure vapor deposition. The inner stress of the SiN film was compressive due to its high density [10]. The compressive stress of approximately −1.0 GPa was observed by wafer bowing measurements [57]. Subsequently, the SiN film was etched to form an edge structure by EB and RIE. The cross-sectional TEM image of the sample is shown in Fig. 10. The stress distribution in Si around the SiN film edge was reproduced using the edge force model [58]. According to this model, the nondiagonal stress component, i.e., shear stress component, is induced in Si at the edge of the SiN film.

The stress distribution around the edge of the stress film is validated using the following equations of the edge force model [58]:

$$\sigma_{xx} = -\frac{2F_x}{\pi} \cdot \frac{x^3}{(x^2 + z^2)^2}, \tag{33-1}$$

$$\sigma_{zz} = -\frac{2F_x}{\pi} \cdot \frac{xz^2}{(x^2 + z^2)^2}, \tag{33-2}$$

$$\tau_{xz} = -\frac{2F_x}{\pi} \cdot \frac{x^2 z}{(x^2 + z^2)^2}, \tag{33-3}$$

where σ_{xx} and σ_{zz} are the normal stress components in the direction of the x and z axes, respectively. τ_{xz} is the nondiagonal stress component (shear stress component). F_x is the tangential stress at the interface of the stress film and a substrate at the edge of the stress film, which is represented by $f \times t$, where f and t are the inner stress and the film thickness, respectively [58]. Each stress component is a function of x and z, which correspond to the lateral and depth directions of the substrate, respectively. The displacement along the y direction can be ignored because of the geometry. The plane strain assumption gives the stress components σ_{xx}, σ_{yy}, σ_{zz}, and τ_{xz}. Therefore, the stress tensor is represented by:

$$\sigma = \begin{pmatrix} \sigma_{xx} & 0 & \tau_{xz} \\ 0 & \sigma_{yy} & 0 \\ \tau_{xz} & 0 & \sigma_{zz} \end{pmatrix} \tag{34}$$

Fig. 11 shows the stress distribution in the substrate around the edge of the stress film, as calculated by Eq. 33. The inner stress of the film is assumed to be compressive (−1.0 GPa), and the film thickness is 80 nm. The positive and negative values indicate tensile and compressive stresses, respectively. First, large stress is induced around the edge of the stress film. The stress distribution around the edge is steep, especially for the stress components σ_{zz} and τ_{xz}. This fact indicates that high spatial resolution is needed to evaluate the nondiagonal stress component. Second, the opposite stress components σ_{xx} and σ_{zz} are confirmed between the region under the stress film and the space region; tensile stress appears in the region under the stress film, whereas compressive stress appears in the space region.

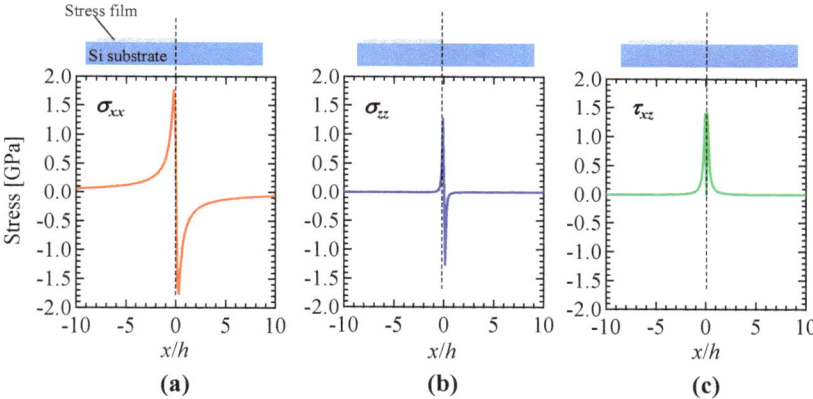

Figure 11. Stress distributions in Si calculated by (a) Eq. (33-1), (b) Eq. (33-2), and (c) Eq. (33-3).

2.5. Experimental configurations

We selectively obtained each optical phonon mode in Si by controlling incident and scattered electrical fields using polarizers and by sample rotation, which was based on the Raman polarization selection rules expressed by Eq. (1). Fig. 12 shows the various polarization configurations in liquid-immersion Raman spectroscopy. In the case of configuration (a), the LO phonon mode is Raman active. As shown in Fig. 12(a), the directions of the incident and scattered electrical fields are parallel to each other. The parallel-polarization configuration is generally applied in conventional Raman spectroscopy. On the other hand, the cross-polarization configuration by rotating the polarizer by 90° for the scattered light shown in Fig. 12(b) results in the fact that the TO phonon modes are Raman active. The TO phonon modes are excited by the z polarization due to the high-NA liquid-immersion lens. In this case, the peak separation of the two TO phonon modes is needed in the analysis. In the case of configuration (c), the sample is rotated by 45° in the parallel-polarization configuration. In this case, one of the two TO phonon modes is Raman active. In the experiments, we applied the configurations (a) and (c) to separately obtain the LO and TO phonon modes.

For the measurements of nondiagonal sress components in Si, the dependence of the Raman spectra from Si at the edge of the SiN film on the relative polarization direction between the sample orientation and the electrical fields of incident and scattered light was analyzed in detail. The experimental polarization configuration in liquid-immersion Raman spectroscopy is shown in Fig. 13. Both of the polarizations of the excitation laser and the scattered light were in the y' direction. The sample was rotated from 0° to 180°, as shown in Fig. 13.

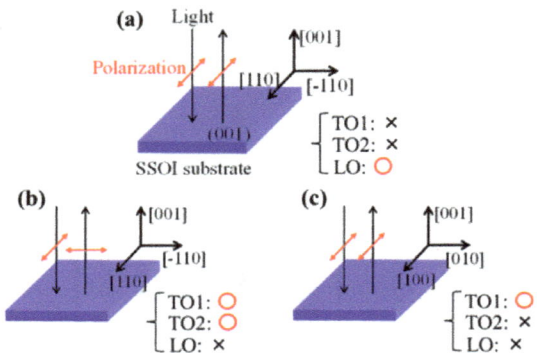

Figure 12. Polarization configurations in oil-immersion Raman spectroscopy: (a) LO active, (b) two TOs active, and (c) one of TOs active configuration

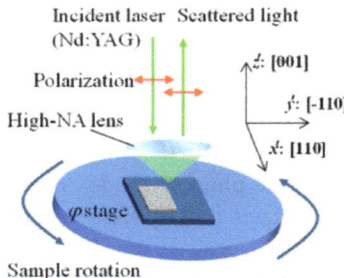

Figure 13. Polarization configuration for measurments of nondiagonal stress components in Si

Second harmonic generation of a neodymium-doped yttrium aluminum garnet (Nd:YAG) laser was used as the excitation source light in liquid-immersion Raman spectroscopy, the optical penetration depth of which is approximately 450 nm into Si [44]. The focal length of the spectroscope and the number of grating grooves were 2,000 mm and 1,800 mm^{-1}, respectively. Therefore, the high-wavenumber resolution of approximately 0.1 cm^{-1} was obtained. The detail explanations of the equipment are shown in Ref. 59. High-NA liquid-immersion lenses were used in this study. An oil-immersion lens with NA of 1.7 was used for the excitation of the TO phonon modes in the SSOI substrate with the 70-nm-thick strained Si layer. The refraction index n of the oil was 1.8. The oil-immersion lens with NA of 1.4 ($n = 1.5$) was used for the measurements of the anisotropic biaxial stress states in the SSOI nanostructures and the strained SiGe nanostructures. A water-immersion lens with NA of 1.2 ($n = 1.3$) was used for the measurements of the nondiagonal stress components. High spatial resolution was achieved owing to the high-NA liquid-immersion lens. The beam spot size was approximately 275, 334 and 390 nm for NA of 1.7, 1.4, and 1.2 liquid-immersion lenses, respectively, according to $0.88 \times \lambda/\mathrm{NA}$ [60]. For the oblique incident light configuration as shown in Fig. 1, NA of the objective was 0.7. The glancing angles were 30° and 90°, as shown above.

2.6. Stress calculation

Stress calculations in the SSOI nanostructures were performed by finite element method (FEM). The results of FEM were compared with the values of the anisotropic biaxial stresses σ_{xx}' and σ_{yy}' obtained by oil-immersion Raman spectroscopy. The virtual biaxial thermal expansion of Si was used and the nodes between the interface of SSOI and BOX were fixed in the FEM calculations. The initial stress value of SSOI before the etching was defined as 1.1 GPa, which was equal to the value obtained by the Raman measurements. The number of meshes was constant for all the SSOI nanostructures: the number of nodes was 13,226 and the number of elements was 11,500. The averaged stress value in the circle area with a diameter of 334 nm corresponding to the beam spot size at the center of the SSOI nanostructure was compared with the measured data. For the depth direction, the stress values throughout the SSOI thickness were averaged because the optical penetration depth of the excitation light was large enough.

3. Results and discussion

3.1. Excitation of TO phonon modes in oil-immersion Raman spectroscopy

Fig. 14 shows the Raman spectra of the LO phonon modes from the SSOI substrate in the oblique incident light configurations with the glancing angles of 30° and 90°. Two peaks are seen in the Raman spectra because the excitation light ($\lambda = 532$ nm) penetrates the strained Si layer, the BOX layer, and reaches the Si substrate. Therefore, the wavenumber on the high-frequency side (defined to be 520 cm^{-1} in this study) originates from the Si substrate, and the wavenumber on the low-frequency side originates from the strained Si layer with the isotropic biaxial tensile stress state. The Raman intensities obtained in the 30° and 90° configurations are almost the same, which is consistent with the calculations shown in Fig. 2. Fig. 15 shows the Raman spectrum from the SSOI substrate in the TO active configuration. The glancing angle was 30°. The fitting curves for the strained Si layer and the Si substrate are also shown in Fig. 15. The explanation about the peak positions of the LO and TO phonon modes for SSOI are shown later.

From the results, the intensity ratio of the TO phonon mode from the strained Si layer obtained in the TO active configuration to the LO phonon mode from the strained Si layer obtained in the LO active configuration was calculated to be approximately 0.04. This value is almost the same as the theoretical value shown by Eq. (3). Using the oblique incident light configuration, the TO phonon modes were excited even for (001) Si. It is possible to completely eliminate the intensity of the LO phonon mode in the oblique incident light configuration in theory. Nevertheless, as shown in Fig. 15, the LO phonon mode was observed in the TO active configuration. This result is considered that there are misalignments of polarization between the incident/scattered light and the orientation of the Si substrate. It is considered that this behavior easily happens because the intensity of the LO phonon mode is much higher than that of the TO phonon mode.

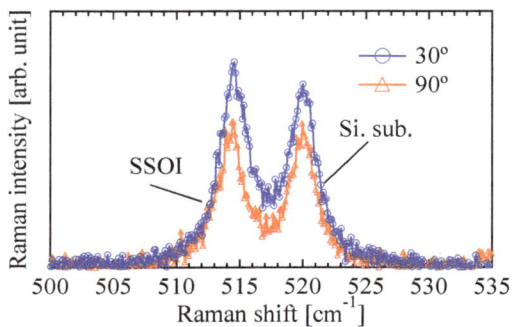

Figure 14. Raman spectra in 30° and 90° configurations

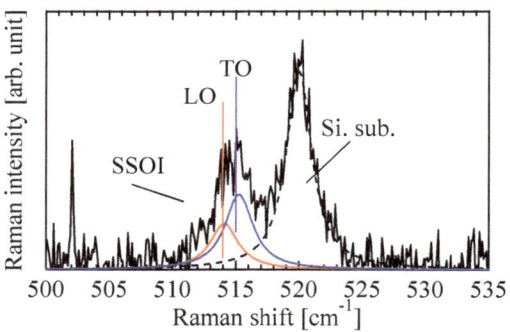

Figure 15. Raman spectrum in TO active configuration with fitting curves

Fig. 16 shows the Raman spectra from SSOI in the LO and TO active conditions, respectively, obtained by oil-immersion Raman spectroscopy. The light-exposure time is 5 s and 300 s for the excitations of the LO and TO phonons, respectively. The Raman intensity shown in Fig. 16 was normalized by the Raman signal from the strained Si layer. It should be noted that the low-frequency peak from strained Si in the LO active condition is lower than that in the TO active condition, although the peaks from the Si substrate in each condition are at the same wavenumber. The difference of the peak positions can be explained by Eq. (17); the LO phonon mode is more affected by biaxial stress than are the TO phonon modes.

Fig. 17 shows the result of fitting each peak. The LO phonon mode is detected irrespective of the TO active condition. This fact arises because the incident light with the polarization in the [010] Si direction generates even in configuration (c) shown in Fig. 12 due to depolarization effects [24]. It is considered difficult to avoid the depolarization effects for the SSOI substrate. On the other hand, it is reported that the contribution of the LO phonon modes can be decreased for the SSOI nanostructures because the depolarization effects relax due to the nanostructure [24]. It is considered that the peak separation of the LO and TO phonon modes is needed for the SSOI substrate to analyze the Raman spectrum obtained in the TO active configuration, while not necessary for the SSOI nanostructure.

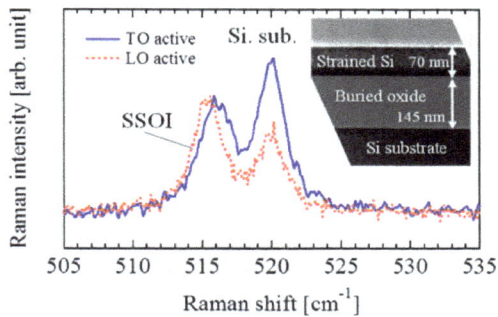

Figure 16. Raman spectra from SSOI in TO and LO active configurations

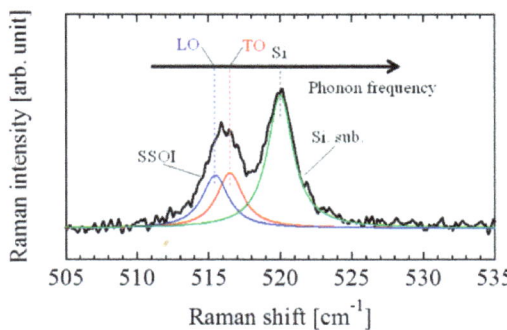

Figure 17. Raman spectrum in TO active configuration with fitting curves

The wavenumber shift of the LO phonon mode for the strained Si layer in the TO active configuration was −4.56 cm^{-1}, which is consistent with the value of −4.60 cm^{-1} for the Raman peak obtained in the LO active configuration. Furthermore, the Raman peak intensity from the Si substrate in the TO active condition is higher than that in the LO active configuration. This behavior indicates that the Raman peak that originates from the LO phonon mode is superimposed onto the Raman peak that originates from the TO phonon mode. We claim that the TO phonon mode was excited by using the high-NA oil-immersion lens.

Fig. 18 shows the Raman spectra from SSOI obtained by conventional Raman spectroscopy with the use of the NA = 0.7 objective. The dashed and solid lines denote the Raman spectra in the LO and TO active configurations, respectively. The light-exposure time of 3600 s for the TO active configuration is 400 times longer than that of 9 s (0.25% of 3600 s) for the LO active configuration. The intensity ratio of the TO to LO phonon modes are anticipated by the calculation shown in Table 2. In Fig. 18, the Raman intensity in each configuration is close to one another. Furthermore, the difference of the peak positions of the strained-Si layer in each configuration is confirmed, similarly to the results in oil-immersion Raman spectroscopy. These results indicate that the TO phonon mode was excited even in conventional Raman spectroscopy. However, the signal to noise ratio of the Raman intensity is bad. Moreover, the extremely long time measurements are necessary. In fact, it is difficult to perform mapping for obtaining biaxial-stress distributions in conventional Raman spectroscopy. We consider that it is important to use the high-NA liquid-immersion lens in order to excite TO phonon mode effectively and obtain biaxial-stress distributions in a realistic time.

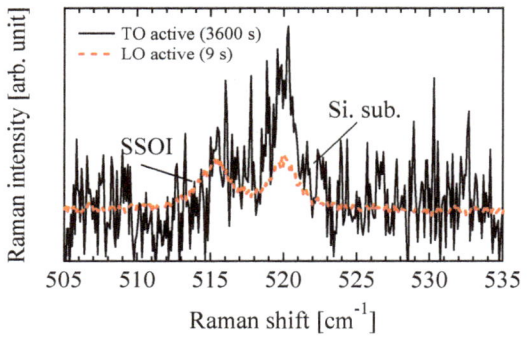

Figure 18. Raman spectra from SSOI with use of dry objective.

3.2. Measurements of anisotropic biaxial stress states in SSOI nanostructures

In-plane XRD measurements were performed to confirm strain in the strained Si layer. The diffraction from Si (220) and (–220) were measured (the results are not shown). As a result, ε_{xx} and ε_{yy} were 7.5×10^{-3} and 7.4×10^{-3}, respectively. These results indicate that the stress state in the strained Si layer is almost isotropic biaxial throughout the wide area which is equivalent to the footprint of the incident X-ray.

PDPs were evaluated by oil-immersion Raman spectroscopy. The calculated biaxial stresses σ_{xx} and σ_{yy} in the strained Si layer are summarized in Fig. 19. In the calculation of the biaxial stresses, PDPs reported in 1970, 1978, and 1990, as mentioned above, were used and Eqs. (19)-(21) were used for the stress calculations. In the oil-immersion Raman measurements, five points were obtained. The solid symbols indicate the average values. From the results, it appears that the biaxial stress values fluctuate in the range of 50–150 MPa, which is attributed to the dislocation conditions in the strained Si layer [61-63]. It should be noted that the apparent anisotropic natures of biaxial stress states were observed in the case of using PDPs reported in 1970 and 1978. The differences in the biaxial stresses are 530 and 170 MPa for PDPs reported in 1970 and 1978, respectively, which are inconsistent with the results of XRD. On the other hand, the isotropic nature was clearly observed in the case of using PDPs reported in 1990. As a result, PDPs of $p/\omega^2 = -1.85$, $q/\omega^2 = -2.31$, and $r/\omega^2 = -0.71$ reported by Anastassakis *et al.* in 1990 are considered the most accurate for evaluating stress in Si among the three sets of PDPs.

Figs. 20(a) and (b) show the Raman spectra from the SSOI nanostructures with $W = 1.0$ and 0.05 μm, respectively, in configuration (c). Ls of the nanostructures were both 5.0 μm. We subtracted the signal of the Si substrate fitting curves from the raw data in order to analyze the spectra from the SSOI nanostructures in detail. From Fig. 20(b), the signal from the SSOI nanostructure even with $W = 50$ nm can be clearly observed. This observation is attributed to the high spatial resolution in oil-immersion Raman spectroscopy.

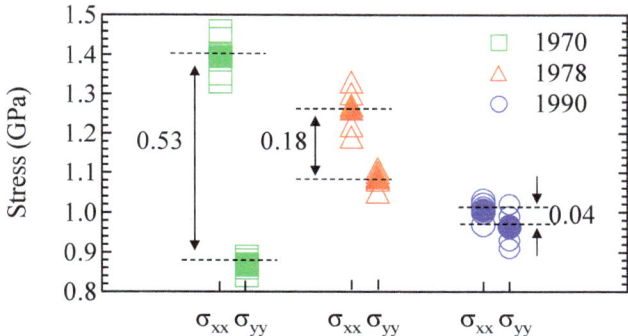

Figure 19. Biaxial stresses σ_{xx} and σ_{yy} in SSOI obtained by oil-immersion Raman spectroscopy using three sets of PDPs.

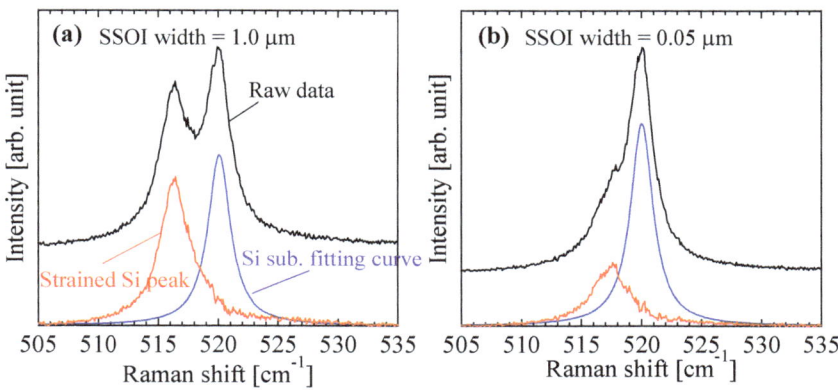

Figure 20. Rama spectra from SSOI nanostructures with Ws of (a) 1.0 and (b) 0.05 μm. *L* is both 5.0 μm.

The normalized Raman spectra as a function of L are shown in Fig. 21. W was fixed to 0.2 μm. The peaks from bulk Si are shown for comparison. The peak positions of strained Si gradually shift toward the high-frequency side with the decrease in L from 5.0 to 0.5 μm, as shown in Fig.21. Using Eq. (21), the anisotropic biaxial stresses σ_{xx}' and σ_{yy}' in the SSOI nanostructures were calculated, as shown in Fig. 22. Figs. 22(a) and (b) show the results for the SSOI nanostructures with the thickness of 50 and 30 nm, respectively. The results of the Raman measurements were compared with those of FEM. The example of the FEM calculations is shown in Fig. 23. Fig. 23 shows the three dimensional distribution of the σ_{xx}' component for the SSOI nanostructure with $L = 1.0$ and $W = 0.2$ μm. From the results of FEM, the stress relaxation is confirmed at the edge of the SSOI nanostructure, while large tensile stress remains in the center of the SSOI nanostructure and at the interface of the strained Si layer and the BOX layer. There is a good correlation between the results of oil-immersion Raman spectroscopy and FEM, as shown in Fig. 22. σ_{xx}' decreases with the decrease in L for the SSOI nanostructures with the thickness of 50 and 30 nm, while σ_{yy}' remains almost constant. Moreover, the values of σ_{yy}' for the SSOI nanostructures with the thickness of 30 nm are larger than those of 50 nm. Therefore, the thin SSOI nanostructures had immunity to the stress relaxation. Using oil-immersion Raman spectroscopy, the evaluation of the anisotropic biaxial stress states was accomplished for the SSOI nanostructures. It is considered that the results obtained in this study have important implications for the SSOI nanostructure fabrication.

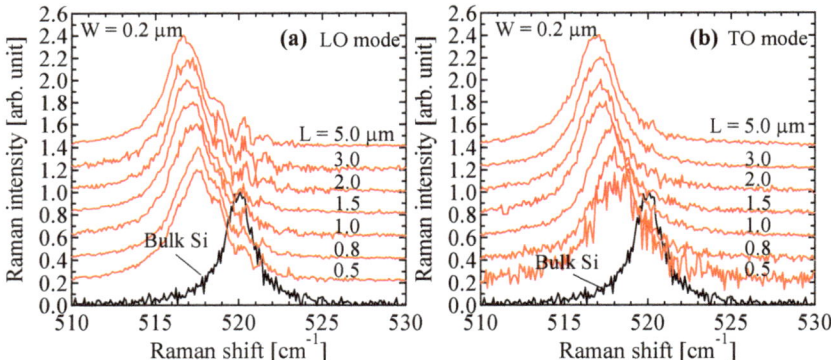

Figure 21. Normalized Raman spectra from SSOI nanostructures as a function of L in (a) LO and (b) TO active configurations.

3.3. Measurements of anisotropic biaxial stress states in strained SiGe nanostructures

In this section, the evaluation of the anisotropic biaxial stress states in the SiGe nanostructures is shown. For the SiGe nanostructures, large compressive stress is induced because the lattice constant of SiGe is larger than that of Si. The stress states in the SiGe nanostructures are considered to be expressed by Eq. (12) similar to the stress states in the

SSOI nanostructures. The crystal structure of SiGe remains diamond type, i.e., the methodology of evaluation for the anisotropic biaxial stress in Si shown above can be directly applied to strained SiGe.

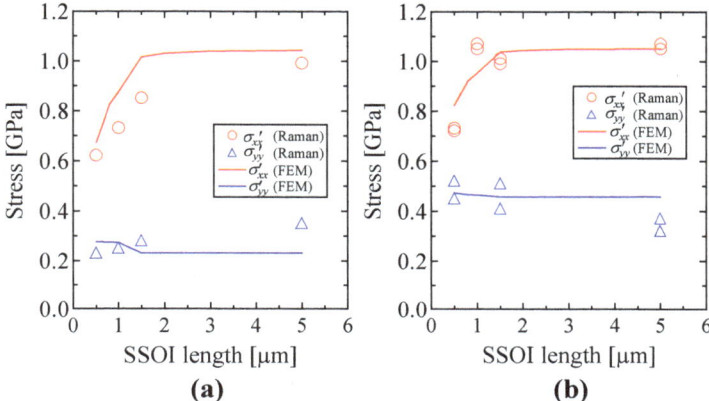

Figure 22. Biaxial stresses σ_{xx}' and σ_{yy}' as a function of L for (a) 50-nm-thick and (b) 30-nm-thick SSOI nanostructures.

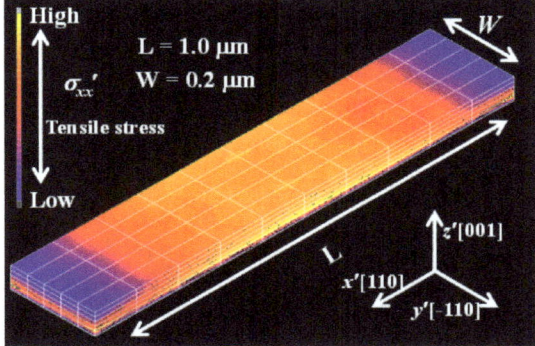

Figure 23. Three dimensional stress distribution obtained by FEM.

Fig. 24 shows the examples of Raman spectra from the SiGe nanostructures. As shown in Fig. 24, the intensity from Si-Si phonon mode in SiGe appeared to be weak, while the intensity of the Si substrate is very strong. This behavior makes the analysis difficult. Nevertheless, the LO and TO phonon modes can be separately obtained. The peak positions of the TO and LO phonon modes for the SiGe nanostructures with L = 5.0 and W = 1.0 μm are clearly different. Moreover, it should be noted that the difference decreases with the decrease in W. For W = 0.1 μm, there is little difference between the peak positions of the TO and LO phonon modes. This behavior indicates that the stress states in the SiGe nanostructures change from a biaxial state to a uniaxial state.

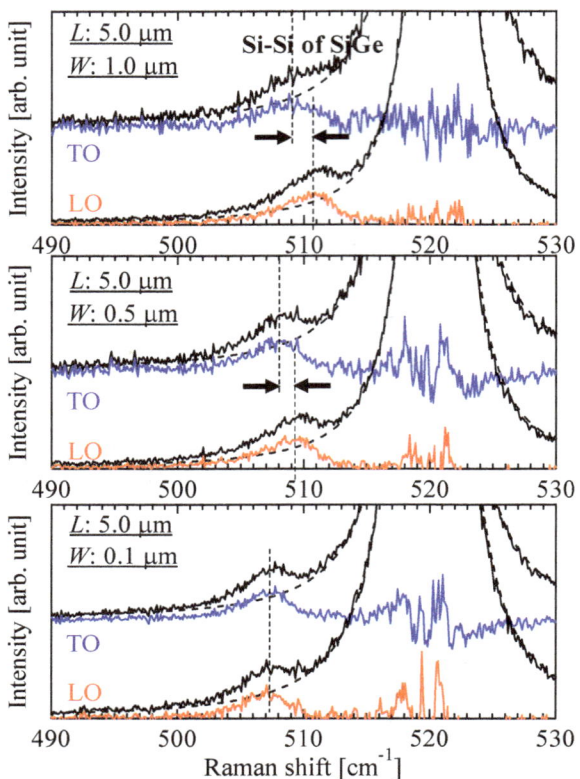

Figure 24. Raman spectra of TO and LO phonon modes from SiGe nanostructures.

Fig. 25 shows the Raman shifts of the TO and LO phonon modes for the SiGe nanostructures with W = 1.0, 0.5, and 0.1 μm as a function of L. First, the large compressive stress exists because the Raman shift of stress-free SiGe with the 30% Ge concentration is approximately 500 cm^{-1} [64,65]. From the results, the clear dependence of the Raman shifts on L and W were observed. It appears that the evaluation of the anisotropic biaxial stress states in strained SiGe is accomplished by oil-immersion Raman spectroscopy similar to evaluating strained Si. However, it is considered that the evaluation of strained SiGe is more difficult than that of Si. There are several unknown parameters for measuring stress in strained SiGe, e.g., PDPs of SiGe, precise Ge concentration, and peak position of stress-free SiGe. Various parameters have so far been suggested [64-70]. These problems are now under investigation.

3.4. Measurements of nondiagonal stress components

First, α (the contribution of z polarization) was determined from the Raman intensity ratio of the TO to LO phonon modes (the results are not shown). From the results, α was calculated to be 0.09 in the water-immersion Raman measurements. Fig. 26 shows the comparison between the measured and calculated data of the Raman shifts dependences on the sample rotation angle. The experimental results were obtained by high-NA water-immersion Raman spectroscopy. The measurement position was the region at the edge of the SiN film (the edge under the SiN film rather than in the space region). Features indicating the induction of stress with the nondiagonal component are clearly observed in the experimental results; the profiles of the Raman shift dependence on the sample rotation angle are asymmetric relative to 45° from 0° to 90° (or relative to 135° from 90° to 180°). As previously shown, this behavior indicates that the shear stress component τ_{xz} is induced in Si at the edge of the SiN film. From Fig. 26, there is a good correlation between the measured and calculated data. On the other hand, disagreements appear around sample rotation angles of 45° and 135°. One possible explanation is that the polarization of the electrical fields of the incident and scattered light is modified by the SiN film. This modification is not included in the calculation.

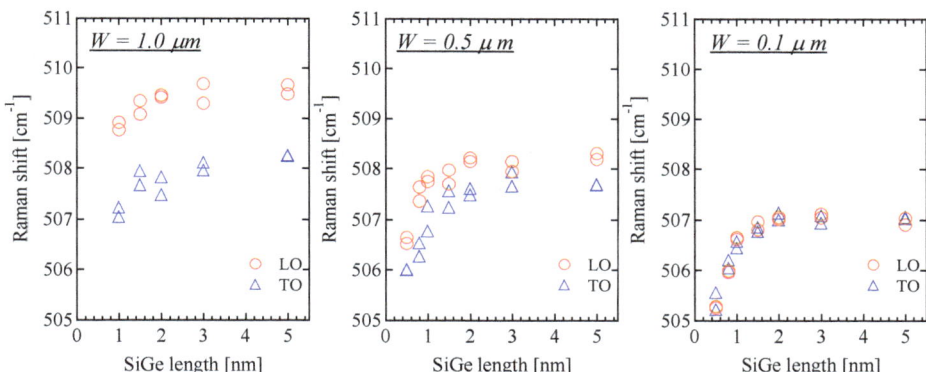

Figure 25. Raman shifts of TO and LO phonon modes for SiGe nanostructures as a function of L.

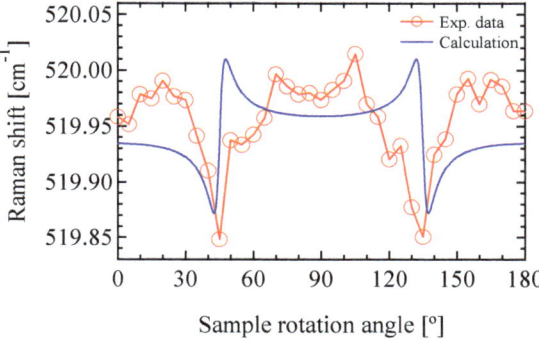

Figure 26. Comparison between measured and calculated data of Raman shift dependences on sample rotation angle.

The Raman shifts appear to be small. This is explained as follows. The stress induced at the edge of the SiN film is localized in width and depth. It is considered that the measured stress is averaged in the region of approximately 390 and 450 nm which are the spot size and optical penetration depth of the laser, respectively. From the water-immersion Raman measurements, τ_{xz}' was approximately 0.1 GPa. This value was consistent with the result obtained by the edge force model after the correction for the spot size and the optical penetration depth of the laser (the results are not shown) [46,58]. The methodology described here has the potential to measure complicated stress states in Si and SiGe even those with nondiagonal stress components.

4. Conclusion

We demonstrated the measurements of the complicated stress states in Si by high-NA liquid-immersion Raman spectroscopy. The z polarization was obtained due to the high-NA liquid-immersion lens, which allows for exciting the forbidden modes, the TO phonon modes, even under the (001) Si backscattering geometry. First, the TO phonon mode of the strained Si layer was observed in oil-immersion Raman spectroscopy. The peak positions of the strained Si layer were clearly separated in the TO and LO active configurations, although the Si substrate peaks remain at the same position in each configuration. This behavior indicates that the biaxial isotropic tensile stress state in the strained Si layer gives rise to the splitting of the optical phonon modes in Si. From the results, the LO phonon mode was more affected by biaxial stress than the TO phonon mode, which was consistent with the result obtained by solving the secular equations. Using the TO phonon mode as well as the LO phonon mode, the anisotropic biaxial stress states in the SSOI nanostructures were measured. As a result, the clear dependences of the biaxial stresses σ_{xx}' and σ_{yy}' on L and thickness were observed. σ_{xx}' decreased with the decrease in L, especially under $L = 1.5$ μm, while σ_{yy}' remains almost constant. The values of σ_{yy}' for the SSOI nanostructures with the thickness of 30 nm are larger than those of 50 nm, which indicates that the thin SSOI nanostructures had immunity to the stress relaxation. The results obtained by oil-immersion Raman spectroscopy were consistent with the FEM calculations. We also measured the anisotropic biaxial stress states in the strained SiGe nanostructures by the same technique. Consequently, the clear dependence of the Raman shifts on L and W were observed similarly to the results for the SSOI nanostructures. Furthermore, the stress with the nondiagonal component, the shear stress component, in Si was measured by water-immersion Raman spectroscopy. As a result, the asymmetric profile was obtained for the dependence of the Raman shifts on the sample rotation angle. This behavior indicates that the shear stress component τ_{xz} is induced in Si at the edge of the SiN film. There is a good correlation between the measured and calculated data. High-NA liquid-immersion Raman spectroscopy enabled us to measure the complicated stress states in strained Si and SiGe with high spatial resolution even those with the nondiagonal stress component.

Author details

Daisuke Kosemura, Motohiro Tomita and Atsushi Ogura
School of Science and Technology, Meiji University, Kawasaki, Japan

Koji Usuda
Green Nanoelectronics Collaborative Research Center, AIST, Tsukuba, Ibaraki, Japan

5. Acknowledgement

The authors thank Ryosuke Shimidzu of PHOTON Design Corporation for the fruitful discussion about the z polarization. The authors thank Dr. Kazuhiko Omote of Rigaku Corporation for his great help in high-resolution XRD measurements. This study was partially supported by the Semiconductor Technology Academic Research Center (STARC), the Japan Society for the Promotion of Science (JSPS) through the "Funding Program for World-Leading Innovative R&D on Science and Technology", "Scientific Research B"and the Japan Science and Technology Agency through the "Adaptable and Seamless Technology transfer Program (A-STEP) through target-driven R&D."

6. References

[1] S. Takagi, J. L. Hoyt, J. Welser, and J. F. Gibborns, J. Appl. Phys. 80, 1567 (1996).

[2] C. S. Smith. Phys. Rev. 94, 42 (1964).

[3] A. Ogura, D. Kosemura, K. Yamasaki, S. Tanaka, Y. Kakemura, A. Kitano, and I. Hirosawa, Solid-State Electronics 51, 219 (2007).

[4] K. Usuda, T. Irisawa, T. Numata, N. Hirashita, and S. Takagi, Semicond. Sci. Technol. 22, s227 (2007).

[5] K. Usuda, T. Mizuno, T. Tezuka, N. Sugiyama, Y. Moriyama, S. Nakaharai, S. Takagi, Appl. Surf. Sci. 224, 113 (2004).

[6] A. Ogura, T. Yoshida, D. Kosemura, Y. Kakemura, M. Takei, H. Saito, T. Shimura, T. Koganezawa, and H. Hirosawa, Solid-State Electronics 52, 1845 (2008).

[7] M. Takei, D. Kosemura, K. Nagata, H. Akamatsu, S. Mayuzumi, S. Yamakawa, H. Wakabayashi, and A. Ogura, J. Appl. Phys. 107, 124507 (2010).

[8] M. Takei, H. Hashiguchi, T. Yamaguchi, D. Kosemura, K. Nagata, and Atsushi Ogura, Jpn. J. Appl. Phys. 51, 04DA04 (2012).

[9] D. Kosemura, Y. Kakemura, T. Yoshida, A. Ogura, M. Kohno, T. Nishita, and T. Nakanishi, Jpn. J. Appl. Phys. 47, 2538 (2008).

[10] A. Ogura, H. Saitoh, D. Kosemura, Y. Kakemura, T. Yoshida, M. Takei, T. Koganezawa, I. Hirosawa, M. Kohno, T. Nishita, and T. Nakanishi, Electrochem. Solid-State Lett. 12, H117 (2009).

[11] S. Mayuzumi, S. Yamakawa, D. Kosemura, M. Takei, Y. Tateshita, H. Wakabayashi, M. Tsukamoto, T. Ono, A. Ogura, and N. Nagashima, IEEE Transactions on Electron Devices 56, 2778 (2009).

[12] T. Yamaguchi, Y. Kawasaki, T. Yamashita, N. Miura, M. Mizuo, J. Tsuchimoto, K. Eikyu, K. Maekawa, M. Fujisawa, and K. Asai, Jpn. J. Appl. Phys. 50, 04DA02 (2011).

[13] W. Xiong, C. R. Cleavelin, P. Kohli, C. Huffman, T. Schulz, K. Schruefer, G. Gebara, K. Mathews, P. Patruno, Y.-M. L. Vaillant, I. Cayrefourcq, M. Kennard, C. Mazure, K. Shin, and T.-J. K. Liu, IEEE Electron Device Lett. 27, 612 (2006).

[14] M. Tomita, D. Kosemura, M. Takei, K. Nagata, H. Akamatsu, and A. Ogura, Jpn. J. Appl. Phys. 50, 010111 (2011).

[15] M. D. Vaudin, Y. B. Gerbig, S. J. Stranick, and R. F. Cook, Appl. Phys. Lett. 93, 193116 (2008).

[16] K. Omote, J. Phys., Condens. Matter. 22, 474004 (2010).

[17] I. D. Wolf, Semicond. Sci. Technol. 11, 139 (1996).

[18] L. Zhu, C. Georgi, M. Hecker, J. Rinderknecht, A. Mai, Y. Ritz, and E. Zschech, J. Appl. Phys. 101, 104305 (2007).

[19] S. C. Jain, B. Dietrich, H. Richter, A. Atkinson, and A. H. Harker, Phys. Rev. B 52, 6247 (1995).

[20] T. Ito, H. Azuma, and S. Noda, Jpn. J. Appl. Phys. 33, 171 (1994).

[21] S. Nakashima, T. Mitani, M. Ninomiya, and K. Matsumoto, J. Appl. Phys. 99, 053512 (2006).

[22] D. Kosemura and A. Ogura, Appl. Phys. Lett. 96, 212106 (2010).

[23] V. Poborchii, T. Tada, and T. Kanayama, Appl. Phys. Lett. 97, 041915 (2010).

[24] A. Tarun, N. Hayazawa, H. Ishitobi, S. Kawata, M. Reiche, and O. Moutanabbir, Nano Lett. 11, 4780 (2011).

[25] E. Bonera, M. Fanciulli, and D. N. Batchelder, J. Appl. Phys. 94, 2729 (2003).

[26] S. Narayanan, S. R. Kalidindi, and L. S. Schadler, J. Appl. Phys. 82, 2595 (1997).

[27] M. Yoshikawa, M. Maegawa, G. Katagiri, and H. Ishida, J. Appl. Phys. 78, 941 (1995) 941.

[28] M. Becker and H. Scheel, J. Appl. Phys. 101, 063531 (2007).

[29] S. J. Harris, A. E. O'Neill, W. Yang, P. Gustafson, J. Boileau, W. H. Weber, B. Majumdar, and S. Ghosh, J. Appl. Phys. 96, 7195 (2004).

[30] D. Kosemura and A. Ogura, Jpn. J. Appl. Phys. 50, 04DA06 (2011).

[31] D. Kosemura, M. Tomita, K. Usuda, and A. Ogura, Jpn. J. Appl. Phys. 51, 02BA03 (2012).

[32] V. Poborchii, T. Tada, K. Usuda, and T. Kanayama, Appl. Phys. Lett. 99, 191911 (2011).

[33] R. Ossikovski, Q. Nguyen, G. Picardi, and J. Schreiber, J. Appl. Phys. 103, 093525 (2008).

[34] T. Tada, V. Poporchii, and T. Kanayama, J. Appl. Phys. 107, 113539 (2010).

[35] G. H. Loechelt, H. G. Cave, and J. Menendez, J. Appl. Phys. 86, 6164 (1999).

[36] P. Kumar, I. Dutta, and M. S. Bakir, J. Electronic Mater. 41, 322 (2011).

[37] M. Feron, Z. Zhang, and Z. Suo, J. Appl. Phys. 102, 023502 (2007).

[38] Y. Moriyama, Y. Kamimuta, Keiji Ikeda, Tsutomu Tezuka, Thin Solid Films 520, 3236 (2012).

[39] K. Ikeda, M. Oda, Y. Kamimuta, Y. Moriyama, and T. Tezuka, Appl. Phys. Exp. 3, 124201 (2010).

[~0] T. Tezuka, E. Toyoda, T. Irisawa, N. Hirashita, Y. Moriyama, N. Sugiyama, K.Usuda, and S. Takagi, Appl. Phys. Lett. 94, 081910 (2009).

[~1] T. Irisawa, T. Numata, T. Tezuka, K. Usuda, N. Hirashita, N. Sugiyama, E. Toyoda, and S. Takagi, IEEE Transactions on Electron Devices 53, 2809 (2006).

[42] E. Anastassakis, J. Appl. Phys. 82, 1582 (1997).

[43] R. Loudon, Adv. Phys. 13, 423 (1964).

[44] D. E. Aspnes and A. Studna, Phys. Rev. B. 27, 985 (1983).

[45] K. Mizoguchi and S. Nakashima, J. Appl. Phys. 65, 2583 (1989).

[46] I. D. Wolf, H. E. Maes, and S .K. Jones, J. Appl. Phys. 79, 7148 (1996).

[47] L. H. Wong, C. C. Wong, J. P. Liu, D. K. Sohn, L. Chan, L. C. Hsia, H. Zang, Z. H. Ni, and Z. X. Shen, Jpn. J. Appl. Phys. 44, 7922 (2005).

[48] Ogura A et al 2007 Report of JEITA Standard Strain MeasurementWorking Group JEITA.

49] E. Anastassakis, A. Pinczuk, E. Burstein, F. H. Pollak, and M. Cardona, Solid State Commun. 8, 133 (1970).

[50] M. Chandrasekhar, J. B. Renucci, and M. Cardona, Phys. Rev. B 17, 1523 (1978).

[51] E. Anastassakis, A. Cantarero, and M. Cardona, Phys. Rev. B 41, 7529 (1990).

[52] E. Anastassakis, J. Appl. Phys. 81, 3046 (1997).

[53] O. Moutanabbir, M. Reiche, A. Ha¨hnel, M. Oehme, and E. Kasper, Appl. Phys. Lett. 97, 053105 (2010).

[54] T. S. Drake, C. Ni Chleirigh, M. L. Lee, A. J. Pitera, E. A. Fitzgerald, D. H. Anjum, J. Li, R. Hull, N. Klymko, and J. L. Hoyt, Appl. Phys. Lett. 83, 875 (2003).

[55] A. Ogura and O. Okabayashi, Thin Solid Films 488, 189 (2005).

[56] A. Ogura, T. Tatsumi, T. Hamajima, and H. Kikuchi, Appl. Phys. Lett. 69, 1367 (1996).

[57] M. Finot, I. A, Blech, S. Suresh, and H. Fujimoto, J. Appl. Phys. 81, 3457 (1997).

[58] S. M. Hu, J. Appl. Phys. 50, 4661 (1979).

[59] A. Ogura, K. Yamasaki, D. Kosemura, S. Tanaka, I, Chiba, and R. Shimidzu, Jpn. J. Appl. Phys. 45, 3007 (2006).

[60] I. D. Wolf, Theoretical and experimental study of the effects of the different optical parameters and lenses on the spatial resolution of the Raman system, STREAM consortium, Doc. No. IST-1999-10341, pp. 1-20.

[61] S. Nakashima, T. Yamamoto, A. Ogura, K. Uejima, and T. Yamamoto, Appl. Phys. Lett. 84, 2533 (2004).

[62] K. Kutsukake, N. Usami, T. Ujihara, K. Fujiwara, G. Sazaki, and K. Nakajima; Appl. Phys. Lett. 85, 1335 (2004).

[63] K. Sawano, S. Koh, Y. Shiraki, N. Usami, and K. Nakagawa, Appl. Phys. Lett. 83, 4339 (2003).

[64] J. C. Tsang, P. M. Mooney, F. Dacol, and J. O. Chu, J. Appl. Phys. 75, 8098 (1994).

[65] M. I. Alonso and K. Winer, Phys. Rev. B 39, 10056 (1989).

[66] D. J. Lockwood and J.-M. Baribeau, Phys. Rev. B 45, 8565 (1992).

[67] M. Holtz, W. M. Duncan, S. Zollner, and R. Liu, J. Appl. Phys. 88, 2523 (2000).

[68] F. Pezzoli, E. Bonera, E. Grilli, M. Guzzi, S. Sanguinetti, D. Chrastina, G. Isella, H. von Känel, E. Wintersberger, J. Stangl, and G. Bauer, J. Appl. Phys. 103, 093521 (2008).

[69] J. P. Dismukes, L. Ekstrom, and R. J. Paff, J. Phys. Chem. 68, 3021 (1964).
[70] F. Cerdeira, A. Pinczuk, J. C. Bean, B. Batlogg, and B. A. Wilson, Appl. Phys. Lett. 45, 1138 (1984).
[71] Y. Hoshi, A. Fukumoto, K. Sawano, I. Cayrefourcq, M. Yoshimi, and Y. Shiraki, Jpn. J. Appl. Phys. 46, 7294 (2007).

Chapter 9

HR-MAS NMR Spectroscopy in Material Science

Todd M. Alam and Janelle E. Jenkins

Additional information is available at the end of the chapter

1. Introduction

In the early to mid-90's, NMR studies were being published that recognized the power of magic angle spinning (MAS) to increase resolution in materials that were not strictly solids by averaging differences in magnetic susceptibility and residual dipolar coupling inherent in these samples. The method of utilizing MAS for non-solid materials to produce liquid-like NMR lines was termed High-Resolution Magic Angle Spinning (HR-MAS). A few of the first HR-MAS examples included investigation of resins for combinatorial chemistry,[1] solvent swollen polystyrene gels,[2] and lipid systems.[3] Then in 1996, Maas *et al.* [4] advanced the field of HR-MAS NMR by adding a magnetic field gradient along the magic angle (see Figure 1). Like high resolution solution NMR, this gradient improved sensitivity and resolution with the ability to more easily select coherence pathways and by reducing indirect dimension (t_1) noise.[4]

There are currently commercially available HR-MAS probes with magic angle gradients from companies like Bruker BioSpin Corporation (Billerica, MA),[5] Agilent Technologies (Santa Clara, CA),[6] JEOL USA, Inc. (Peabody, MA),[7] and Doty Scientific, Inc. (Columbia, SC).[8] In addition to magic angle gradients, many of these probes also have a deuterium (^{2}H) lock channel, allowing improved ease of shimming and long term stability. With the emergence of commercially available probes, HR-MAS NMR has become more popular in the last few years, especially in the biological and biomedical fields. This popularity is mainly due to the heterogeneous nature of tissues and cells that are well suited for HR-MAS. Multiple HR-MAS NMR studies involving different tissue biopsies, like brain, kidney, liver, and muscle tissues for metabonomics studies, as well as identification of abnormal tissues (*i.e.* cancerous tissues) have been reported.[9-11] HR-MAS NMR has also been applied to the characterization of foodstuffs, including the assignment of metabolites in tomatoes and apples, the study of biopolymers in fruit cuticles, quantification of *n*-3 fatty acids content in different fish species, and tracking the chemistry of coffee beans during the roasting process.[12, 13] These types of biological and foodstuff HR-MAS NMR investigations highlight the diverse range of information that can be obtained.

The application of HR-MAS NMR to material science was initially focused almost exclusively on the analysis of solid-phase (*i.e.* utilizing support resins) organic and peptide synthesis, or analysis of combinatorial solid-phase results. In these studies, the material was swollen in appropriate solvents such that the mobility of the attached ligands was increased, allowing high resolution NMR spectra to be obtained. The application of HR-MAS to solid state synthetic chemistry remains an active area of research, but will not be discussed in detail. The readers are encouraged to consult several very extensive reviews in this area.[14-16] In comparison to the numerous HR-MAS NMR studies on biological and solid-phase synthetic chemistry systems, there are fewer examples that focus on the use of this technique to material science. This chapter will review the application of HR-MAS NMR to a wide range of systems, including ceramics, zeolites, liquid crystals, ionic liquids, and surface modified nanoparticles.

1.1. How HR-MAS works

Materials that are crystalline or rigid solids (for example resins, ceramics, *etc.*) exhibit an extremely broad NMR signal due to extensive homo- and hetero-nuclear dipolar coupling, chemical shift anisotropy (CSA), and quadrupolar interactions. A variety of solid state NMR techniques have been developed to improve the resolution and sensitivity including the use of multiple pulse sequences, cross polarization, high power decoupling, multiple quantum NMR and fast MAS NMR. For example, solid state ^1H MAS NMR has found a number of applications in material characterization.[17] Even with these advances, solution-like NMR spectra are rarely realized for solid samples.

For samples that are in the liquid/solid classification, motional averaging will partially reduce or remove many of these broadening interactions. It is also possible to swell or plasticize a material to increase the local mobility, thereby reducing the magnitude of these interactions.[18] Even for liquid environments in a heterogeneous sample, differences in the magnetic susceptibility within the material can drastically reduce the observed resolution. For these types of dynamically averaged or susceptibility broadened systems, MAS even at moderate speeds will produce high resolution NMR spectra: this is the niche of HR-MAS NMR.

This improved resolution due to MAS arises because the Hamiltonians describing the dipolar, CSA and magnetic susceptibility interactions all contain an orientational component that scales as $(3\cos^2\theta-1)$, where θ is the angle between the rotor spinning axis and the magnetic field (Figure 1). When a sample is spun about an axis that is at the "magic angle" ($\theta = 54.7°$) these interactions vanish. A simplistic schematic of a HR-MAS stator is pictured in Figure 1. As mentioned before, the main difference between a standard MAS stator and an HR-MAS stator is the magic angle gradient coil that can produce a gradient along the rotor spinning axis.

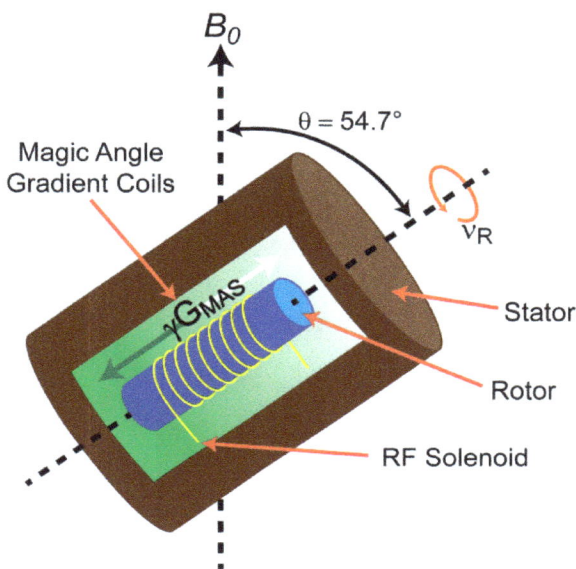

Figure 1. Schematic of a HR-MAS stator with a magic angle gradient along the rotor spinning axis.

For samples where residual dipolar interactions or differences in magnetic susceptibility are small, HR-MAS NMR reduces the observed line widths to be similar to those observed in solution high resolution NMR spectroscopy. For liquids adsorbed into/onto materials MAS speeds between 1 and 2 kHz may result in improved resolution, while for plasticized/swollen materials MAS speeds between 4 and 10 kHz may be necessary to obtain the desired resolution. Care must also be taken to ensure that the residual spinning sidebands (due to incomplete averaging) fall outside the spectral window of interest.

As an example of the enhanced resolution that can be achieved under HR-MAS two different swollen polymer systems are shown in Figure 2. These anion exchange membranes (AEM) operate in methanol solution, and readily adsorb significant amounts of water and methanol from the solution during use. Our group is interested in identifying the different chemical environments that these solvent molecules see within the membrane, as well as directly measuring the diffusion rates for these different species. Under static conditions, the water and methanol are not resolved in the NMR spectrum (Figure 2A, green). This lack of chemical shift resolution precludes any information about the local membrane environment being obtained. The broadening is caused by the magnetic susceptibility differences between the rigid AEM and the methanol solution. In this example, minimal MAS of ~750 Hz (Figure 2A, red) was sufficient to average the magnetic susceptibility allowing individual solvent environments to be easily observed. Increasing the spinning speed to 4 kHz removed the spinning side bands from the spectral region of interest, but did not dramatically increase the resolution of the solvent resonances. In this swollen AEM example, signals from the actual membrane are not observed at 4 kHz (only the swelling solvent is observed) indicating that the dipolar interactions present in the membrane remain much larger than the MAS speed. Since the protons from the membrane are not readily observable, it provides a unique opportunity to further study the solvent behavior in the membrane without interference. Additional discussion of this system is presented in Section 2.1.1. The second example, involves the swelling of a polyButadiene-AcryloNitrile (pBAN) polymer in CDCl₃. Inspection of Figure 2B (green) shows that under static conditions this rather soft material produces a relatively unresolved NMR spectrum. Even after extensive solvent swelling, the NMR spectrum is essentially unchanged (Figure 2B, red). With HR-MAS (4 kHz spinning) the resolution is dramatically enhanced, enabling a more detailed analysis of the pBAN ¹H NMR spectra.

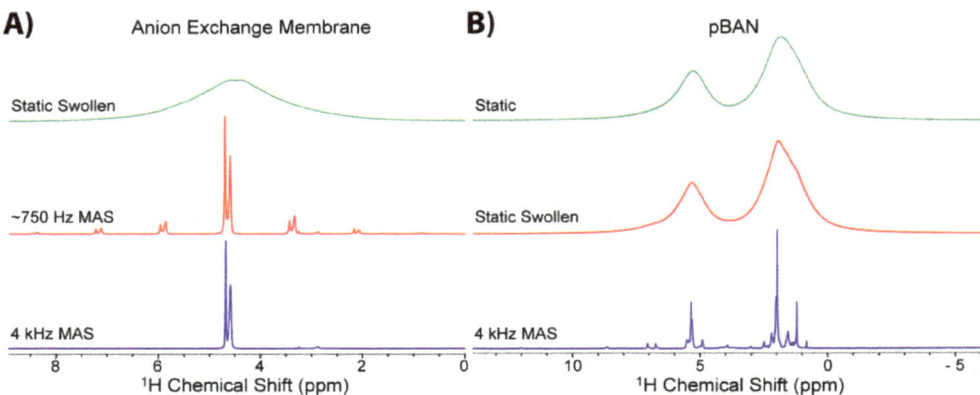

Figure 2. The improved resolution observed using ¹H HR-MAS NMR for the A) methanol swollen anion exchange membrane, and B) the CDCl₃ swollen pBAN (polyButadiene-AcryloNitrile) polymer.

With the dramatic increase in resolution observed utilizing HR-MAS, the arsenal of standard solution NMR techniques can be implemented for HR-MAS experiments, including solvent suppression, gradient-assisted sequences, and multiple-dimensional experiments. These can include INEPT (Insensitive Nuclei Enhanced by Polarization Transfer), COSY (COrrelation SpectroscopY), NOESY (Nuclear Overhauser Effect SpectroscopY), TOCSY (TOtal Correlation SpectroscopY), HETCOR (HETeronuclear CORrelation), HMQC (Heteronuclear Multiple Quantum Coherence), and HMBC (Heteronuclear Multiple Bond Correlation) to name a few. As an example, a gradient-assisted 2D ¹H-¹H HR-MAS COSY NMR spectra for an ionic liquid adsorbed on an aluminum oxide membrane is shown in Figure 3. These types of correlation experiments could not be realized without the resolution afforded by HR-MAS. Additional discussion of this IL material is presented in Section 2.3.1.

It is also important to note that because anisotropic interactions (dipolar, CSA) are still present within these HR-MAS samples (though greatly reduced by molecular motion), it is also possible to apply some solid state NMR techniques during HR-MAS experiments. For example, to measure the residual homonuclear dipole-dipole interaction it is possible to incorporate a dipolar recoupling sequence to re-introduce this interaction under MAS. This allows the incoherent NOE exchange process to be replaced by a coherent polarization transfer process. These types of HR-MAS recoupling experiments have been demonstrated using DQ COSY for solid state synthesis samples,[19] while our group has incorporated a radio frequency dipolar recoupling (RFDR) sequence into the standard NOESY experiment for lipid membranes, including the extension of this dipolar recoupling into a mixing period during ¹H-¹³C Heteronuclear correlation experiments.[20]

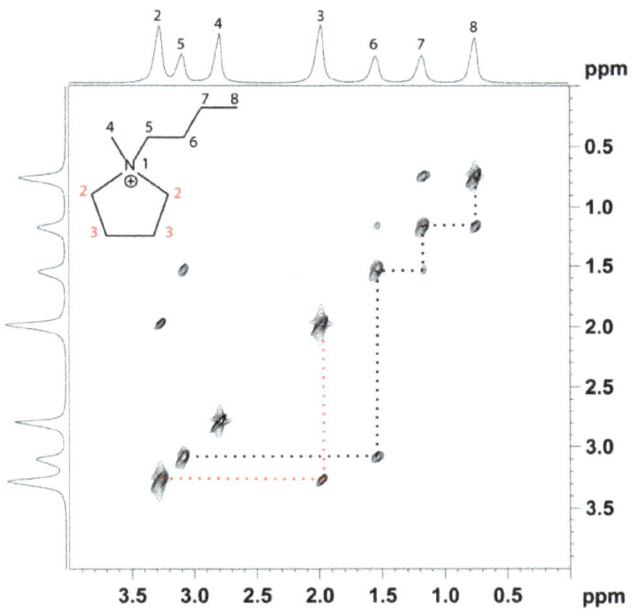

Figure 3. The gradient 2D ¹H HR-MAS NMR COSY spectrum for the ionic liquid [MBPyrr]⁺[TFSI]⁻ adsorbed into an inorganic aluminum oxide membrane. Even though the individual *J* couplings were not resolvable, these types of correlation experiments can still be realized under HR-MAS.

1.2. Experimental protocol and limitations

Many HR-MAS samples contain liquids or liquid-like materials. These samples require special care to prevent dehydration that can occur in a standard rotor with only a rotor cap. Various inserts are commercially available for packing of HR-MAS samples and can be seen in Figure 4. These inserts provide a tight seal to prevent dehydration or loss of solvent during MAS experiments. The inserts used generally depend on the sample size and need of the user. Examples of various insert options from Bruker BioSpin for a standard 4 mm rotor are shown in Figure 4. For ~12 µL samples, a Kel-F® bottom spacer (or a half drilled out rotor) and a top spacer with a seal screw can be utilized (Figure 4B, 4F and 4G). The top spacer insert (Figure 4F) contains a small hole in the top. When placed into the rotor at the properly gauged distance, any additional liquid will protrude through this hole. A Kimwipe® can then be utilized to remove this liquid, followed by using a small Kel-F®

screw to seal the insert (Figure 4B) and prevent dehydration. For ~50 µL volumes, a standard rotor can be utilized with the above top spacer and seal screw (Figure 4B, 4F, 4L). In addition to the spacers, a 30 µL disposable insert is also available (Figure 4I). The disposable Kel-F® inserts use a plug and screw cap to keep the sample well sealed (Figure 4B, 4H). The inserts are efficient if multiple samples need to be run, as samples can be packed into inserts and then easily placed into and removed from rotors, without having to wash the rotors between runs, and without having to own a large number of MAS rotors.

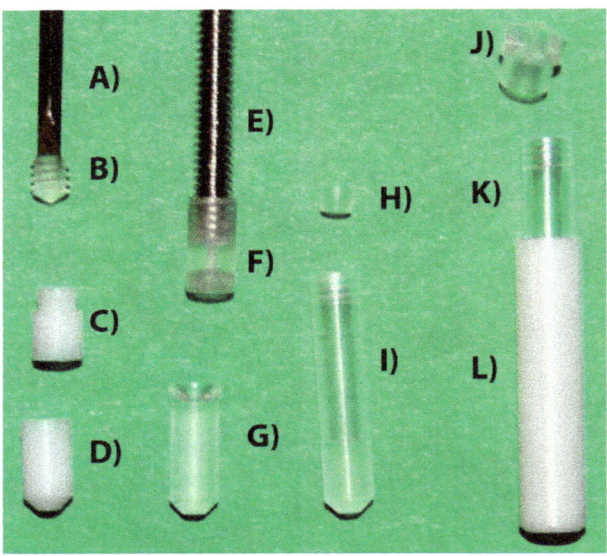

Figure 4. The tools and inserts used for HR-MAS NMR. These include A) the specialized tool for screw cap insertion, B) the sealing screw cap, C) the upper Teflon® insert, D) lower Teflon® insert for 30 µL volume, E) screw for insertion/extraction of top insert, F) top Kel-F® insert, G) bottom Kel-F® insert for 12 µL sample volume, H) plug for disposable insert, I) disposable 30 uL Kel-F® insert, J) 4 mm rotor cap, K) disposable inserted partially in a 4 mm rotor, L) 4 mm zirconia MAS rotor. All these parts are for the Bruker HR-MAS system, and may vary between vendors.

Temperature regulation is important in HR-MAS NMR for both sample purposes, and experimental reasons. The temperature of the bearing gas in many probes can be regulated to maintain a stable sample temperature; however, there is additional frictional heating that comes from spinning the rotor. To compensate for MAS related heating of the sample, the HR-MAS probe temperature should be calibrated at various MAS speeds and changes made to the temperature of the bearing gas to regulate the sample at the desired temperature. Temperature calibration of a HR-MAS probe has been discussed in literature.[21] This study utilized both methanol and glucose as NMR thermometers to determine the impact of MAS on temperature. Temperature changes measured at 2, 4, 6, and 8 kHz MAS where found to be 0.8 ± 0.1, 2.2 ± 0.1, 5.0 ± 0.1, and 7.9 ± 0.2 K in methanol and -0.3 ± 0.1, 1.2 ± 0.2, 3.0 ± 0.2, and 6.1 ± 0.1 K for glucose.[21]

Like all high resolution NMR techniques where resonances can be very narrow, it is important to be able to shim a sample to achieve the best possible resolution. Unlike standard high resolution NMR probes, the HR-MAS probes have the samples spinning at the magic angle, therefore the standard shimming protocols cannot be utilized. Instead, a combination of the standard shims must be used to address inhomogeneity of the B_0 field. By placing the probe so that the rotor is in the (x, z) plane in the laboratory frame, a combination of the laboratory shims can be used to express the Z shims along the magic angle. To obtain the equivalent response of a Z shim in standard shimming protocol, *i.e.* the Z shim along the magic angle (B_Z^{MAS}), the user needs to optimize a combination of the Z and X room temperature shims given by the linear relationship $B_Z^{MAS} = (1/\sqrt{3})B_Z^{LAB} - (\sqrt{2/3})B_X^{LAB}$. Likewise, the equivalent of a Z^2 shim is given by $B_{Z^2}^{MAS} = B_{(X^2-Y^2)}^{LAB} - 2\sqrt{2}B_{XZ}^{LAB}$ and the equivalent Z^3 shim is defined using $B_{Z^3}^{MAS} = -(2/3\sqrt{3})B_{Z^3}^{LAB} - (1/\sqrt{6})B_{XZ^2}^{LAB} + (5/\sqrt{3})B_{(X^2-Y^2)Z}^{LAB} - (5/3\sqrt{6})B_{X^3}^{LAB}$. The higher order Z shims map directly as $B_{Z^4}^{MAS} = -(7/18)B_{Z^4}^{LAB}$ and $B_{Z^5}^{MAS} = -(1/6\sqrt{3})B_{Z^5}^{LAB}$.[22] For the experimental examples discussed in this chapter, the authors typically only adjusted the shims up through $B_{Z^3}^{MAS}$. The Bruker Biospin Manual for High Resolution Magic Angle Spinning Spectroscopy points out that in theory shimming could be performed with just laboratory shims X, XZ,XZ², Z⁴, and Z⁵.[23] In reality, additional shims can be used to compensate for any inefficiency in the shim coil. A previous study has shown that once good shims are obtained, those shims can be utilized for samples with the same detection volume and position, for example the same sample size in the same type of rotor.[24] The shims should be independent of solvent because any susceptibility differences caused by solvent will be averaged out by MAS and does not affect the shims.[24] Piotto *et al.* also demonstrated shimming issues involved in systems containing water. In many cases, improved shims will be detected in an increase in lock signal. Deuterated water (HDO), however, has a strong chemical shift dependence on the temperature. When a sample spins, there can be a temperature gradient across the sample which can cause a portion of the HDO resonances to shift and broaden. If the probe is shimmed again based on the lock level from the water resonance, then any other signal in the spectra will most likely be deshimmed as those signals do not have as large of a chemical shift temperature dependence. With this in mind, Piotto *et al.* emphasize that shimming should be performed based on line shape, not lock level and attention should be taken when working with materials containing water.[24] More recently the MAGIC SHIMMING method has been described which utilizes a conventional homospoil gradient pulse to perform gradient shimming on MAS probes. This techniques does not require a gradient along the magic angle, and will reduce some of the trial and error presently inherent with manual shimming of HR-MAS probes.[25]

2. HR-MAS NMR spectroscopy in materials characterization

In this section a brief review of HR-MAS NMR studies involving the characterization of materials is presented. The majority of studies involved modified surfaces or surface interactions, and demonstrates the characterization need that HR-MAS NMR can fulfill.

2.1. Polymers

For rigid polymer materials the dipolar interactions are very large, leading to broadening of the NMR resonances, and can be characterized using standard solid state MAS NMR techniques. As noted before, these rigid polymers are not readily observed under HR-MAS conditions. For solvent swollen polymers the increased mobility of the polymer segments leads to a semi-solid regime where the moderate spinning speeds of HR-MAS are sufficient to produce high resolution NMR spectra. For example see the improved resolution observed for CDCl3 swollen pBAN (Figure 2B). This ability to obtain spectra for mobile domains or components is the basis for the majority of HR-MAS polymer studies. A variety of different solvents are employed in combinatorial and solid phase synthesis, many of these with multiple ¹H resonances. In addition, there may be reasons to avoid having to dry the sample prior to introducing a deuterated solvent. In these cases, complicated solvent suppression techniques may be required.[26] It has been demonstrated that diffusion-filter HR-MAS spectra (see PFG discussion, Section 3.1) can provide suppression of solvent resonances based on differential diffusion behavior.[27]

One example of using HR-MAS NMR for characterizing polymer materials is the demonstration of the cyclic polyamide receptor threading onto the highly flexible polyethylene glycol (PEG) polymer chain attached to a polystyrene bead. Distinct PEG resonances for the threaded and non-threaded complexes could be easily resolved under HR-MAS conditions, while 2D NOESY spectra showed cross-peaks between the aromatic protons ($\delta \sim$ +8.87 ppm) of the rotaxane and the methylene protons ($\delta \sim$ +3.55 ppm) of the PEG polymer chain.[28] In another study, the complexation of Zn and Ru metalloporphyrins to beads functionalized with pyridyl ligands revealed that supermolecular interactions and changes in the dynamics are directly probed by HR-MAS techniques.[29]

The vulcanization of butadiene rubber (BR) with different curing systems has also been monitored by ¹H and ¹³C HR-MAS NMR. This rubber is a very mobile system, with the increased resolution afforded by HR-MAS allowing the chemical identity of cross links to be determined, and revealing that α-substitution or addition depended on the disulfide cross-linkers employed.[30, 31] Other HR-MAS studies observed the impact of cross-linking in solvent swollen poly(amidoamine) polymers, the nature of water interactions in these same class of polymers, [32, 33] or the cross-linking performance in silicon-containing soybean-oil copolymers.[34] HR-MAS NMR has also been used to study the functionalization of poly hydroxyethyl methacrylate (HEMA) cryogels,[35] cyclomaltoheptaose polymers,[36] the synthesis of hyper-branched bis (hydroxymethyl) propionic acid (bis-MPA) polymers as a function of catalysts,[37] and polymers for nanoparticle stabilization.[38]

HR-MAS NMR has found use in investigation of porous polyalkylvinyl ether polymer particles being used for stationary phases in chromatographic applications. These measurements were performed using the same solvent as HPLC and allowed details about the polymer structure and mobility to be evaluated; these properties are expected to impact the chromatographic process.[39] Interactions between solution molecules and the polymer component of HPLC stationary phases (C18, C30 and PEAA)[40] or the interaction with molecularly imprinted polymers [41] have also been investigated using HR-MAS NMR. These studies used saturation transfer difference pulse sequences to identify those molecules that were involved in binding to the larger stationary phase.

The NMR spectroscopic characterization of polymer degradation is commonly directed towards analysis of the small degradation fragments that are solvent soluble. By using HR-MAS NMR the hydrolytic degradation of the biodegradable photo-initiated cross linked poly(DL-lactide)-dimethacrylate (PDLLA) polymer network was directly monitored. Swelling these polymers in the solvent dimethyl sulfoxide (DMSO) and a combination of 1D and 2D (COSY, NOESY) experiments allowed the different signals in the PDLLA chain to be identified. Degradation in a base was shown to occur through hydrolysis of the ester bonds within the poly(lactide) segment.[42] HR-MAS has also been used to study the role of partial hydrolysis in controlling the composition of hydrophobic polyacrylamide gels.[43]

2.1.1. Example: HR-MAS NMR investigations of anion exchange membranes

The increased resolution achieved with ^{1}H HR-MAS on the solvent swollen anion exchange membrane (AEM) was demonstrated in Figure 2. Due to differences in the magnetic susceptibility between the swelling solution and the polymer membrane, the water and methanol resonances were not resolved under static conditions. Under HR-MAS four distinct NMR resonances were observed for the solvent species; two water resonances and two methanol resonances. Based on line widths and PFG HR-MAS diffusion measurements (Section 3.2.1) the higher ppm shifted water and methanol were assigned to bulk (or free) methanol or water within the membrane, while the lower ppm shifted water and methanol resonances exhibit slower diffusion rates and were assigned to water or methanol associated (bound) within the membrane. The increased resolution obtained under HR-MAS was further exploited to explore the interchange of water and methanol between the different binding environments within the AEM using 2D HR-MAS NMR NOESY exchange experiments (Figure 5). At short mixing times (1 ms) no correlations were observed between any of the resonances. With increasing mixing time (> 10 ms), cross peaks were observed between free water and membrane-associated water, plus cross peaks between free methanol and membrane-associated methanol, resulting from physical exchange of water (red dashed lines) or methanol (red dashed lines) between the free and unbound environments. With long mixing times (> 200 ms), NOE correlation between methanol and water were observed (green dashed line), and result from through space NOE magnetization exchange (not physical spatial exchange). Interestingly, these NOE correlations were only observed between associated water and associated methanol, as well as free water and free methanol, supporting the assignment and indicating that the spatial contact of water and methanol is maintained in these associate membrane environments. Additional exchange studies are ongoing.

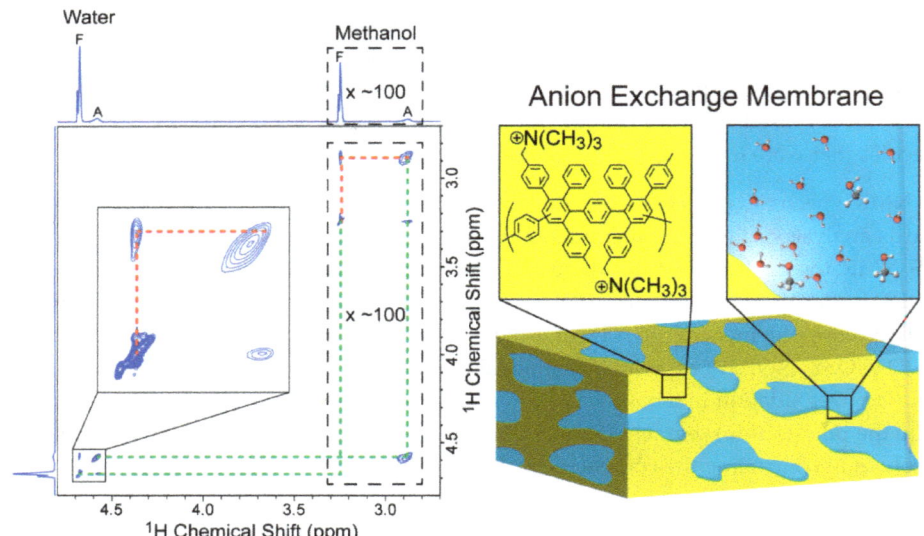

Figure 5. 2D ¹H HR-MAS NMR NOESY spectrum (500 ms mixing time) of an anion exchange membrane swollen in a 1N methanol solution. The increased spectral resolution obtained for the swelling solvent made it possible to detect exchange between free and associated water and between free and associated methanol (red dashed lines), as well as NOE magnetization exchange between free water and methanol and associated water and methanol (green dashed lines).

2.2. Ceramics, zeolites, catalysts surfaces and composites

Due to the rigid or crystalline nature of ceramics, zeolites, catalysts or inorganic/organic composites, solid state MAS NMR spectroscopy remains the dominant NMR characterization method, and can involve a range of nuclei and multi-pulse NMR techniques. In contrast, the limited number of HR-MAS NMR studies is directed towards the investigation of gas/solid adsorption or intercalation within these materials. For example, the degradation of the pesticide diethyl parathion and the chemical warfare agent (CWA) simulant diisopropyl fluorophosphates in functionalized montmorillonite clay using ³¹P HR-MAS NMR has been reported. The HR-MAS removed susceptibility effects present from the clay material, and allowed a direct measure of the decomposition kinetics. The adsorption of organosilanes on the MgCl₂ in Ziegler-Natta catalysts has been investigated using ¹H HR-MAS, and demonstrates differential binding (differential mobility) to the Mg surface with changing surface coverage and as a function of the degree of saturation in the Mg coordination sites.[44]The polymer mobility in the condensed sol-gels produced from organosilanes and phenyl siloxanes was also evaluated.[45]

2.3. Ionic liquids and liquid crystals

HR-MAS NMR should be an ideal tool for the characterization of ionic liquids (ILs) and liquid crystals (LC) due to the high viscosities commonly encountered in ILs, and the thermal- or concentration-induced ordering transitions of LCs. Ionic liquids are interesting compounds, and continue to be used for a wide range of material science applications. ILs are used as liquid electrolytes in energy storage and production devices, solvents for CO_2 capture and green chemistry, solutions for biomass processing, nanostructured synthesis and solvents for catalysis reactions. For many of these applications room temperature ionic liquids (RTIL) are utilized as neat solutions without another solvent. Strong attractions between the IL cation and anion component may lead to the formation of important structural motifs that could be missed or changed with the introduction of a solvent. For NMR characterization, solvent-IL interactions may also mask subtle changes in the chemical shift produced by other molecular interactions of interest. In the case of ILs, it is now recognized that HR-MAS NMR provides a powerful tool for the characterization of these systems.

One of the earliest reports involving HR-MAS NMR of ILs studied silica-immobilized ILs suspended in DMSO,[46] which produced nearly liquid-like resolution. In this same study, they were also able to obtain 2D ^1H-^{13}C HR-MAS NMR HMQC spectra of these attached ILs. Another study explored the use of ILs as a solvent or chemical reaction media. The standard solution NMR of compounds dissolved in these ILs suffered from low resolution, produced by the high viscosities of the ILs, dynamic range issues due to the lack of deuterated ILs, and concerns about chemical shift referencing due to the high magnetic susceptibility of ILs. To demonstrate the capabilities of HR-MAS NMR, Rencurosi and co-workers[47] dissolved *para*-methoxy benzyl acetate and a glucopyranoside in a series of different ILs. Even at moderate spinning speeds there was significant improvement in resolution, and the chemical shift referencing became consistent with that observed in solution NMR using traditional deuterated solvents. In addition, they were able to follow acetylation of *para*-methoxybenzyl alcohol directly in the ILs. The increased resolution inherent in this technique was also important for investigations of CO_2 interactions with imidazolium based ILs using ^{13}C HR-MAS NMR.[48] ^1H HR-MAS NMR has also been used to probe the interaction of the RTIL N-methylimidazolium chloride ([Hmim]$^+$Cl$^-$) with the silica surface of Aerosil. This adsorption was monitored through changes in chemical shift with increased IL loading and Cl$^-$ salt concentration. Based on the magnitude of the chemical shift variations, the interaction was found to preferentially involve the H(2) position of Hmim$^+$, and was a physisorption versus a chemisorption process.[49] In another study, the dynamics of IL's confined in monolithic silica ionogels were followed by ^1H HR-MAS NMR and relaxation experiments, and demonstrated that the IL maintained liquid like behavior with very little reduction in motions even as the pore size diminished from 12 nm to 1.5 nm.[50] Surprisingly, only a single ^1H HR-MAS NMR diffusion study of a non-biological LC material confined in nanopores has been reported.[51]

2.3.1. Example: HR-MAS NMR investigation of ionic liquids on surfaces

While many RTILs are "liquids" at room temperature, they can be highly viscous, making standard solution NMR analysis on the neat ILs challenging (as noted above). In our laboratory, a novel series of quaternary ammonium and cyclic pyrrolidinium RTIL have been studied using different NMR techniques. This includes pulse field gradient (PFG) NMR to measure diffusion and ^{14}N NMR relaxation experiments to determine IL molecular reorientation times.[52] More recently, we have explored using ^{1}H HR-MAS NMR to characterize the interactions between IL and metal oxide surfaces. Figure 6A shows the ^{1}H HR-MAS NMR spectra of the neat ionic liquid, N-methyl-N-(n-butyl) pyrrolidinium bis(trifluoromethanesufonyl) imide ([MBPyrr]$^+$[TFSI]$^-$). The changes in the NMR spectrum for the neat solution under static conditions (top) and under MAS conditions (bottom) demonstrate the simple resolution enhancing properties of HR-MAS (previously discussed in Section 1.1). The static NMR spectrum revealed distorted line shapes reflecting poor magnetic field homogeneity across the sample volume, along with broadening (FWHM ~ 115 Hz) due to differences in local magnetic susceptibility within the sample. The broad line width is commonly encountered when working with highly viscous RTIL that are not dissolved into solvents. Increasing the temperature (within the limitations of the instrumentation) will improve the resolution for these RTIL, but may not be desirable in some instances. Not surprisingly, even slow spinning (~1 kHz) of the neat RTIL provides an immediate improvement in the resolution. For the ([MBPyrr]$^+$[TFSI]$^-$) example, the line width are reduced to ~ 3 Hz (~40 fold reduction), such that small ^{1}H-^{1}H spin-spin J coupling are clearly resolved (Figure 6A, bottom).

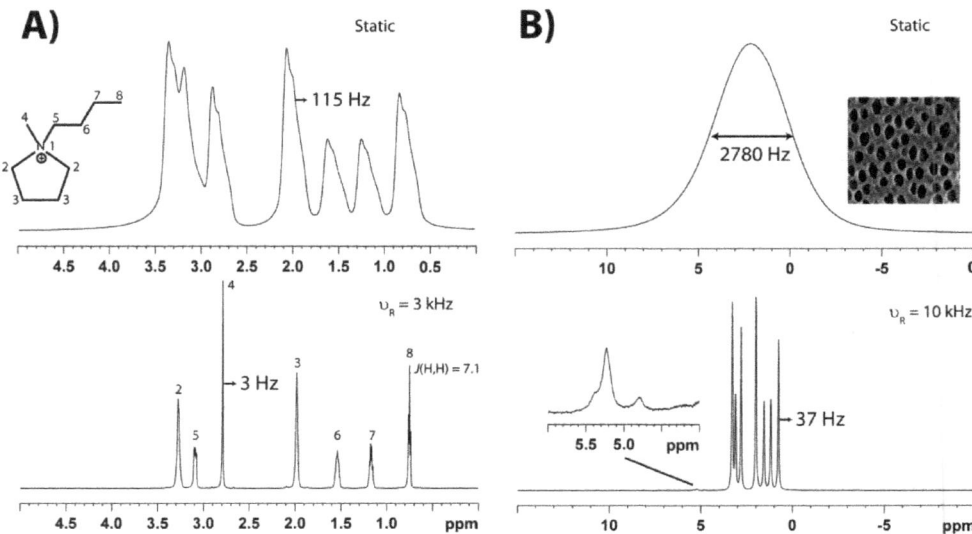

Figure 6. The ^{1}H HR-MAS NMR of the ionic liquid [MBPyrr]$^+$TFSI$^-$ at 298K as a A) neat liquid and B) adsorbed (20 wt%) into an aluminum oxide ANOPORE inorganic membranes (200 nm pore diameter). The spectra in the top portion of each figure were obtained under static conditions, with the spectra in the bottom portions of each figure obtained under MAS conditions.

The loss of resolution is even more dramatic when [MBPyrr]⁺[TFSI]⁻ is adsorbed into the small pores (200 nm) of an inorganic aluminum oxide membrane. The ¹H NMR spectrum of the static sample is now a broad (~2780 Hz) featureless line providing no chemical shift resolution. Under HR-MAS the individual ¹H resonances become resolved, with a line width on the order of 37 Hz. While the individual J couplings are no longer resolved as they were in the neat IL case, the resolved chemical shifts allow one to compare directly with the neat IL. In the adsorbed [MBPyrr]⁺[TFSI]⁻, the n-butyl methyl group protons (H-8) and the ring methylene protons (H-3) show a very small chemical shift change ($\Delta\delta = 0.007$ ppm) due to adsorption with the ANOPORE membrane. In contrast the methyl group (H-4) and the n-butyl methylene protons (H-5) reveal an asymmetric line shape with chemical shift variations ranging from $\Delta\delta = +0.01$ to $+0.02$ ppm. This result suggests that while the adsorption process is weak, it occurs preferentially through the IL N⁺ with the surface. The ability to resolve such small differences under HR-MAS NMR is important in identifying these surface interactions.

Even though the ¹H HR-MAS NMR spectra for the surface adsorbed [MBPyrr]⁺[TFSI]⁻ does not reveal resolvable J coupling, it is still possible to obtain a 2D COSY spectrum for this material, as shown in Figure 3. This correlation experiment immediately confirms the chemical shift assignments for the neat IL sample, and clarifies differences seen for the IL dissolved in a solvent. The 2D ¹H HR-MAS NOESY NMR spectra for the IL [MBPyrr]⁺[TFSI]⁻ adsorbed into the ANAPORE membrane is shown in Figure 7. Interestingly there are numerous cross peaks observed even at moderate mixing times, reminiscent of the liquid-ordered phase observed in lipid bilayers (For example see Ref. [20, 53, 54]). These through-space correlations are stronger than those observed in the neat IL (not shown), and suggest that an increase in the dipolar-dipolar interactions is occurring for the surface adsorbed species. This argues that local motions have become reduced for the adsorbed species. There are also some missing or weak correlations (dashed circle) implying distinct conformations present for the n-butane chain on the adsorbed species. Also of interest is the appearance of new chemical environments for the adsorbed IL. This includes a weak resonance at $\delta = +4.8$ ppm which is assigned to residual water, but shows no correlations with the IL. There are two additional ¹H environments at $\delta = +5.2$ and $+5.3$ ppm, which are attributed to additional surface water species on the aluminum oxide surface. The environment at $\delta = +5.2$ ppm shows a through space NOE correlation with a new resonance near 2 ppm. This small resonance is buried under the shoulder of the dominant methylene (H-2) resonance and has not been presently assigned. These types of correlation experiments demonstrate the capabilities and information that can be obtained from HR-MAS studies of ILs, and suggest future efforts along these lines are warranted.

2.4. Surface modified nanoparticles

HR-MAS NMR studies of surface modified nanoparticles (NP) are closely related to those of the polymer resin studies mentioned in the previous sections, with accurate characterization of the surface-attached ligands being the primary objective. For NPs the solvent plays the role of suspending or dispersing the material, in contrast to the role of swelling that the solvent plays in analysis of resins. For HR-MAS NMR studies of NPs, the optimal solvent provides both high solubility, and good dispersion (prevents aggregation) of the NPs.

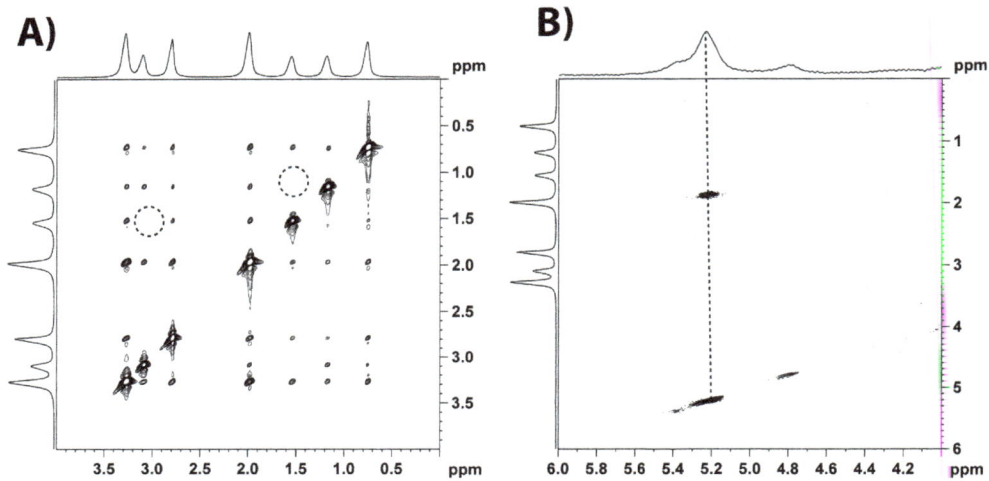

Figure 7. The 2D ^{1}H HR-MAS NOESY NMR correlation spectra for the ionic liquid [MBPyrr]$^+$[TFSI]$^-$ adsorbed into an inorganic aluminum oxide membrane (pore size 200 nm). The spectrum in A) shows numerous through space correlation between the protons of the IL, while B) shows the spectral expansion for the water and surface associate-water species and corresponding NOE correlations.

HR-MAS NMR studies of surface modified NPs include the 1D and 2D ^{1}H NMR investigation of modified gold (Au) NPs using a wide range of high resolution correlation experiments: COSY, TOCSY and HMQC.[55] This revealed that the relative signal sensitivity (intensity) depends on the distance between the detected ^{1}H and the surface of the Au NP. This distance correlation is either a function of reduced local dynamics, or spin-spin T_2 relaxation effects governed by the Au surface. Another study involved the active molecular compounds Aloin A and Aloesin extracted from the leaf of Cape Aloe, and was able to show how these compounds selectively stabilize Au NPs preferentially through the glucose component.[56] HR-MAS NMR has also been used to identify the binding motif of peptides on the Au NP surface.[57] While HR-MAS NMR is typically employed to overcome anisotropy of magnetic susceptibility or residual dipolar interactions, Poito and co-workers[58, 59] have reported an interesting set of experiments that demonstrated HR-MAS can be employed to overcome paramagnetic effects present in iron oxide NPs. Through a careful analysis of these iron oxide systems, the surface structure and ligand binding (chelation) configuration was determined, and in several cases were quite different from the standard single point attachment proposed by others.[58]

^{1}H HR-MAS NMR has also been used to look at the surface modification in polymer based NPs, including the monitoring of multiple synthetic steps employed during the peptide-surface modification of poly(vinylidene fluoride) (PVDF) nanoparticles,[60] or the surface modification of Dendron based NPs.[61] Polystyrene embedded silver clusters produced by a thermolysis reactions have also been characterized using HR-MAS NMR.[62]

2.5. Surface immobilized linkers and catalysts and chiral stationary phases

HR-MAS NMR spectroscopy has also been extensively used in investigating the mobility of linkers and catalysts attached to other materials, including ^{1}H and ^{31}P HR-MAS NMR studies of phosphine, bisphosphinoamine linkers and corresponding metal catalysts.[63-66] These investigations demonstrate very nicely that the surface mobility as probed by HR-MAS is strongly dependent on the swelling solvent employed in the studies.[67] The grafting of organotin catalysts to polystyrene for the tranesterification reaction between ethyl acetate and n-octanol or the ring-opening polymerization of ε-caprolactone has been followed by both 1D and 2D ^{1}H, ^{13}C and ^{119}Sn HR-MAS NMR.[68-70] While the catalytic activity of the attached tin was unchanged following numerous cycles, the NMR revealed there was actually a change in the liquid-solid interface with a reduction in the mobility of the undecyltin trichloride catalyst at the resin surface. For experiments with rapid catalytic turnover rate there was a reduction in performance. This reduction in activity was reversible with solvent extraction, and was confirmed by HR-MAS to involve the change of the Sn environment back to the original state.[69] The use of ^{119}Sn HR-MAS NMR has also been demonstrated to provide a quantitative measure of the tin loading in supporting catalysts.[71] This HR-MAS NMR study was also the first to incorporate the ERETIC (*electronic reference to access in vivo concentration*) method.

HR-MAS NMR techniques have been used to characterize the surfaces of silica and polystyrene particles modified with chiral agents that allow identification of different enantiomers. In one study the NMR was able to probe regiochemistry of the surface chemistry, along with monitoring the stability against decomposition of the polystyrene resins.[72] Another investigation used 2D transfer NOESY NMR experiments to evaluate the stereoselective binding of molecules to the chiral stationary phase based on the negative cross-peaks associated with the transfer NOE effect.[73]

2.6. Soil and humic materials

While standard solid state MAS NMR techniques have become common in the investigations of humic, soil and coal materials, the implementation of HR-MAS techniques for these materials has been more limited. The ability of HR-MAS to identify different motional regimes has proven powerful in the elucidation of the chemistry at water/soil interfaces.[74-79] These NMR studies show that at this interface fatty acids, aliphatic esters and alcohols are the prominent species, and that aromatic functional groups are protected by hydrophobic regions and are not directly accessible to the penetrating water,[74] while surface polymethylene groups may control the sorption properties of organo-clay complexes.[76, 80] These types of studies have been extended to three-dimensional (3D) HMQC-TOCSY to further increase the resolution of the highly overlapping spectra from humic materials.[81]

3. PFG HR-MAS NMR to measure diffusion in materials characterization

Modern HR-MAS probes include a gradient coil that can produce a magnetic gradient along the long axis of the MAS rotor which is set at the magic angle (θ = 54.7°). Standard MAS probes have also been combined with micro-imaging gradient systems in which gradient coils wrapped around the stator were not employed, but instead rely on imaging gradients that were external to the probe. [82] This is not the common configuration, and has been replaced by significant development efforts from the instrumental vendors involving the integration of gradients directly into the HR-MAS probes. With the gradient coil along the magic angle pulsed field gradient (PFG) experiments can be performed under MAS conditions. Fortuitously, enhanced T_2 times are generally observed under MAS allowing PFG experiments with longer diffusion times to be implemented than would have been accessible with static conditions.

During the PFG diffusion experiments, the application of a gradient "tags" a spin with a phase that is related to its spatial position. Figure 8A provides a pictorial representation of the dephasing of spins around the magic angle caused by the magic angle gradient. If the position of the spin does not change during the diffusion period (Δ), this dephasing is refocused and the original signal intensity (S_0) is recovered. If on the other hand the spin changes spatial position (diffuses) during Δ, the dephasing for that spin is not refocused, and the signal intensity decreases. The loss in signal intensity with increasing gradient strength is related to the self-diffusion rate with the classic Stejskal-Tanner equation:

$$\frac{S}{S_0} = e^{-\gamma^2 g^2 \delta^2 \left(\Delta - \frac{\delta}{3}\right)D} \tag{1}$$

where S is the experimental amplitude of the signal, γ is the gyro magnetic ratio, g is the gradient strength, δ is the gradient pulse length, Δ is the diffusion time, and the D is the diffusion constant.[83] By fitting this decay the diffusion constant can be determined. Figure 8B shows an example of the signal intensity loss observed during diffusion experiments for the different two water environments present in swollen AEM with increasing gradient strength.

PFG diffusion HR-MAS NMR experiments can also be used to obtain diffusion-filtered NMR spectra through the separation of different motional regimes present in complex mixtures. This filtering is accomplished by selecting Δ times where the signal intensity (Eqn. 1) for the fast diffusing components has been highly attenuated, while the slow diffusing components have significant signal intensity remaining. Practical aspects of diffusion measurements using HR-MAS have been previously discussed by Viel et al. [84] A significant finding from this study was that sample volume played a role in the reliability of diffusion rates measured. It was shown that small volumes (~12μL) exhibit reproducible diffusion rates, while larger volumes (~50μL, a full 4 mm rotor) produced inconsistent or unreliable data.

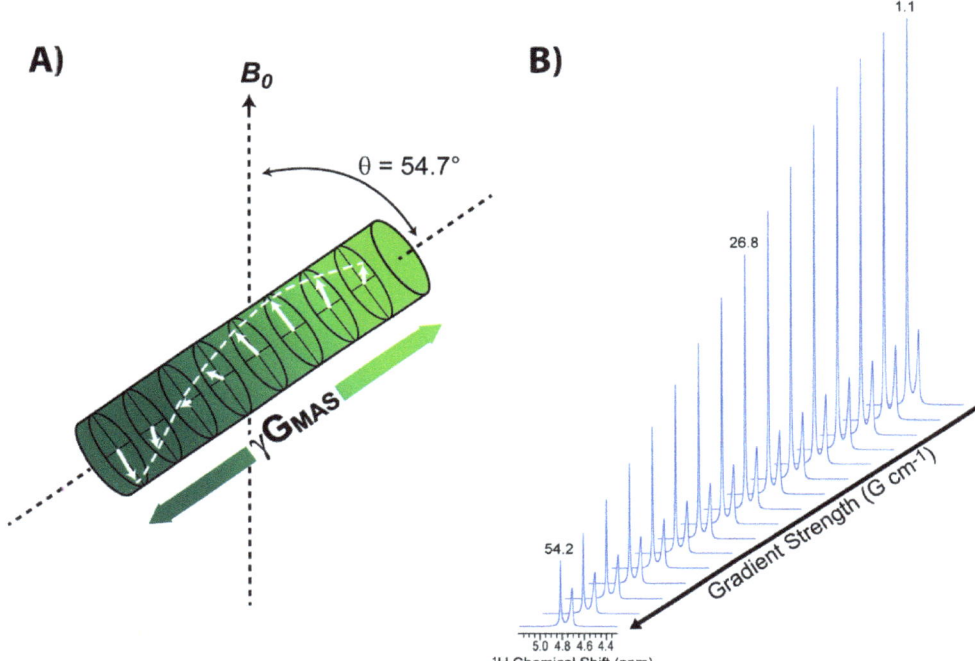

Figure 8. A) Pictorial representation of the gradient produced along the magic angle of the rotor. B) The decay of two different water signals found in a 1N methanol solution of an AEM membrane with increasing gradient strength. Gradient strength values (G/cm) are shown above the stack plot.

Three commonly used PFG diffusion pulse sequences are shown in Figure 9. The basic spin-echo diffusion sequences is depicted in Figure 9A, but is limited by loss of signal intensity due to spin-spin T_2 relaxation during the diffusion period Δ. Two variations of the stimulated echo (STE) sequence are shown in Figure 9B and 9C, and in this case spin-lattice T_1 relaxation is occurring during the diffusion Δ period. For most materials T_1 values are longer than T_1 making these STE sequences the preferred choice for material analysis. It should be noted that all of the gradient pulses in these sequences are trapezoidal shaped to compensate for the inability of instrumentation to generate perfect rectangular gradients. Shaped pulses, like sine or trapezoidal shapes, are used to produce experimentally reproducible gradient pulses. The PFG stimulated echo with dipolar gradients and spoil gradient, depicted in Figure 9B, is beneficial for the use with many HR-MAS samples which exhibit differences in magnetic susceptibility across the sample.[85, 86] The PFG stimulated echo in Figure 9C has an additional delay to the PFG stimulated echo in Figure 9B that is utilized to address eddy currents within the sample.[87]

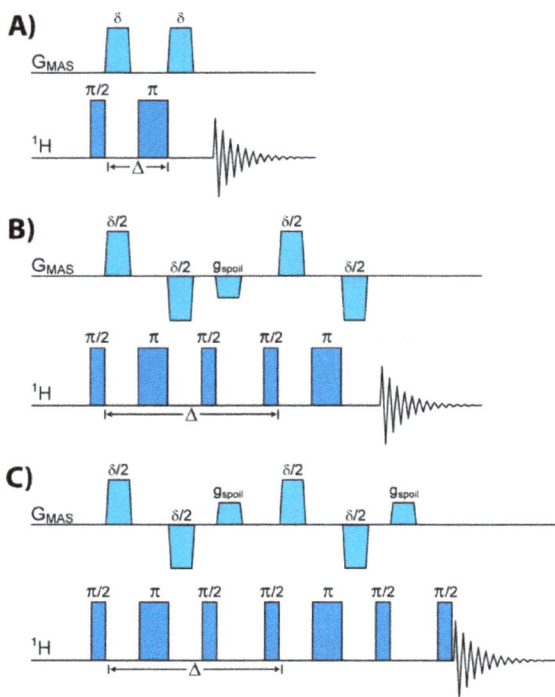

Figure 9. Diffusion pulse sequences. Pulse Field Gradient (PFG) A) Spin-Echo, B) PFG Stimulated Echo with dipolar gradients and spoil gradient based on Cotts *el al.* 13-interval sequence[85], and C) PFG Stimulated Echo with dipolar gradients and spoil gradient with an additional eddy current delay. G_{MAS} indicates that the gradient is applied along the magic angle.

3.1. Diffusion in zeolites, nanoparticles and liquid crystals

As noted for other material studies, heterogeneities in the magnetic susceptibility or restricted molecular motions within zeolite crystallites lead to broadening of the NMR signal, such that resolution of individual species in mixtures becomes difficult. HR-MAS NMR resolves this issue, and has led to the utilization of PFG NMR diffusion experiments on organic mixtures in zeolites. This technique has been used to study the diffusion of *n*-butane in silicalite-1,[88] ethane, water and benzene mixtures adsorbed to the zeolite NaX,[89] acetone-*n*-alkane (C_6 to C_9) mixtures in nanoporous silica,[90] or mixtures of *n*-butane and *iso*-butane adsorbed in MFI zeolite. The increased resolution afforded by these PFG HR-MAS studies on mixtures reveal the obstructive influence of the isopropyl molecules or bulky benzene on the diffusion of other molecular species, [89, 91] and that the creation of acetone-alkane complexes greatly impacts the observed diffusion properties.[90]

PFG HR-MAS NMR has also been used to obtain 2D DOSY (**D**iffusion **O**rdered **S**pectroscop**Y**) spectra of surface modified iron oxide NPs. These results were able to distinguish between NP-bound and free ligands in these materials.[58] These types of PFG

HR-MAS NMR experiments should prove useful in understanding the surface-ligand dynamics present in modified NPs.

Diffusion experiments using HR-MAS NMR has also been found useful in the analysis of transport properties in lipid membranes.[92-94] Due to the anisotropic molecular reorientation of these liquid crystalline (LC) systems, significant dipolar coupling remains, leading to broad lines and short relaxation times. However dipolar coupling can be reduced through the use of MAS. Surprisingly, the use of PFG HR-MAS NMR for non-biological LCs is more limited, with the single investigation of local molecular dynamics of the thermotropic LC 4'-pentyl-4-cyanobiphenyl (5CB) confined in Bioran glasses with pore diameters of 30 nm and 200 nm being reported. By utilizing PFG techniques it was possible to measure the diffusion constants as a function of temperature through the isotropization temperature of the liquid crystal, thus demonstrating that for this case there is only a minor reduction in the diffusion rates with molecular confinement.[51]

3.2. Diffusion in polymers

There is extensive literature on the use of PFG NMR to measure diffusion of polymer solutions and melts, along with PFG diffusion measurements of different species adsorbed into polymers, including gases, water, organic solvents and electrolytes. There have been a limited number of examples where the improved resolution afforded by HR-MAS NMR was coupled with PFG. This work includes the development of diffusion filtered HR-MAS NMR techniques to study the gelation process of super-molecular gels,[95] along with a combination diffusion-filtering and a spin-echo enhanced (T_2 –filtered) experiment on DMF-swollen resins.[96] These techniques allowed the identification of free and surface bound molecules, while eliminating the signal from the immobile bulk resin matrix. In complex mixtures there may be future avenues for HR-MAS to resolve subtle differences in the local chemical environments as demonstrated in the following example.

3.2.1. Example of HR-MAS diffusion in anion exchange membranes

As discussed in Section 2.1.1, four distinct resonances were observed in the 1D ^{1}H HR-MAS NMR spectra of AEM polymers swollen in a 1N methanol solution. In Figure 10A, NMR spectra for three AEM polymers with different ion exchange capacities (IEC) are shown. Two resonances were observed for both water and methanol, and were assigned to free (F) and membrane-associated (A) environments. From this HR-MAS data it is easy to see that there is a correlation between chemical shift of the associated species and the IEC of the membrane, with both the associated water and methanol resonance shifting to lower ppm with decreasing IEC. This decrease in chemical shift is most likely due to a change in hydrogen bonding between the solvent components and the membrane, reflecting how strongly the solvent molecules are associated with the membrane. Recall that the resolution of these individual environments was not observable in the static NMR spectra (Figure 2A). Using ^{1}H HR-MAS PFG diffusion experiments, the self-diffusion constants were obtained for each of these four different environments, which were not accessible from the static data.

Because the ^{1}H signal for the AEM membrane is not readily observable under HR-MAS conditions, the diffusion rates obtained for the resolved solvent resonances were not biased by the polymer membrane. Figure 10B shows the signal decay for the associate methanol environment as a function of gradient strength for the three different IEC levels. The magic angle gradients were used to perform diffusion measurements utilizing a PFG Stimulated Echo with dipolar gradients and spoil gradient with a Δ=100ms (Figure 9B). The signal decay shows that there is a correlation between diffusion rate and IEC, exhibiting a faster diffusion rate with increasing IEC values. A more detailed analysis of this work is forth coming, but this example demonstrates the power of combining HR-MAS and PFG diffusion experiments.

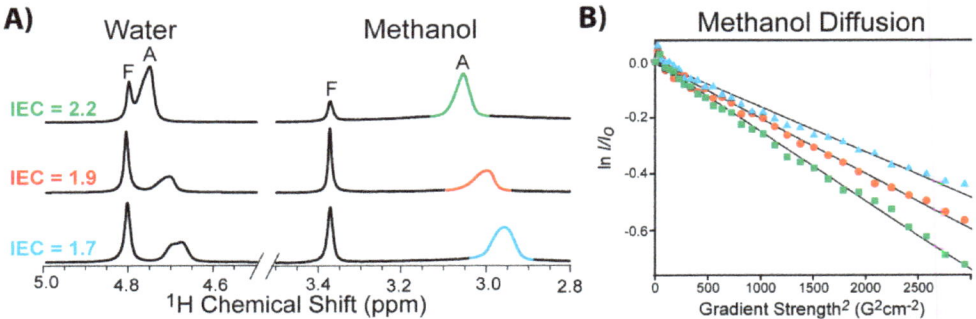

Figure 10. A) ^{1}H HR-MAS NMR spectra with the assigned free [F] and associated [A] water and methanol environments. B) The diffusion rates for the associated methanol in three different anion exchange membranes with varying ion exchange capacity (IEC) values. The colored peaks in the ^{1}H HR-MAS NMR spectra correlate to the colored symbols in the diffusion plot of the associated methanol peak of IEC = 2.2 (■), 1.9 (●), and 1.7 mequiv/g (▲).

4. Conclusions

The application of HR-MAS NMR to the characterization of materials or material interfaces that exist in the semi-solid range has been demonstrated. A wide variety of different material systems have been explored, showing that this technique can provide resolution and dynamic information where standard solution or solid state NMR techniques were unsuccessful. HR-MAS NMR is a powerful tool for the detailed characterization of modified surfaces and surface adsorbed species. This technique also provides a direct probe of differences in local mobility as reflected by line width variations. Through the combination of the enhanced resolution afforded by HR-MAS with pulse field gradient (PFG) capabilities, selective filtering and diffusion measurements of complex heterogeneous materials can also be realized. The ability to resolve and obtain diffusion rates for multiple environments in materials will prove beneficial for understanding the diffusion process in mixed chemical systems. While HR-MAS NMR is considered a mature, relatively routine technique, the application to the materials field is expected to continue being an active area of development. It is hoped that this review will encourage researchers to explore the application of HR-MAS NMR techniques to their different material systems.

Author details

Todd M. Alam* and Janelle E. Jenkins
*Sandia National Laboratories, Department of
Nanostructured and Electronic Materials, Albuquerque, NM, USA

Acknowledgement

Sandia National Laboratories is a multi-program laboratory managed and operated by Sandia Corporation, a wholly owned subsidiary of Lockheed Martin Corporation, for the U.S. Department of Energy's National Nuclear Security Administration under contract DE-AC04-94AL85000. The authors would like to acknowledge Michael Hibbs (Sandia) for providing the anion exchange membranes.

5. References

[1] Sarkar S. K., Garigipati R. S., Adams J. L., Keifer P. A. (1996) An NMR Method To Identify Nondestructively Chemical Compounds Bound to a Single Solid-Phase-Synthesis Bead for Combinitorial Chemistry Applications, *J. Am. Chem. Soc.* 118:2305-2306.

[2] Stöver H. D. H., Frèchet J. M. J. (1991) NMR Characterization of Cross-Linked Polystyrene Gels, *Macromolecules* 24:883-888.

[3] Gross J. D., Costa P. R., Dubacq J.-P., Warschawski D. E., Lirsac P.-N., Devaux P. F., Griffin R. G. (1995) Multidimensional NMR in Lipid Systems. Coherence Transfer Through J Couplings Under MAS, *J. Magn. Reson. Ser. B* 106:187-190.

[4] Maas W. E., Laukien F. H., Cory D. G. (1996) Gradient, High Resolution, Magic Angle Sample Spinning NMR, *J. Am. Chem. Soc.* 118:13085-13086.

[5] Bruker BioSpin, (2012) Biological Tissue Analysis. Available: http://www.bruker-biospin.com/probes_hrmas.html. Accessed 2012 April 10.

[6] Agilent Technologies, (2012) Nano Probes. Available: http://www.chem.agilent.com/en-US/Products/Instruments/magneticresonance/nmr/probes/liquids/nano/pages/default.aspx. Accessed 2012 April 10.

[7] JEOL, (2012) FGMAS. Available: http://www.jeol.cn/?p=1232. Accessed 2012 April 10.

[8] Doty Scientific, (2012) HR-MAS MAG Probe. Available: http://www.dotynmr.com/solids/HRMASMAG.htm. Accessed 2012 April 10.

[9] Lindon J. C., Beckonert O. P., Holmes E., Nicholson J. K. (2009) High-Resolution Magic Angle Spinning NMR Spectroscopy: Application to Biomedical Studies, *Prog. Nucl. Mag. Res. Sp.* 55:79-100.

[10] Zietkowski D., Davidson R. L., Eykyn T. R., De Silva S. S., deSouza N. M., Payne G. S. (2010) Detection of Cancer in Cervical Tissue Biopsies Using Mobile Lipid Resonances

* Corresponding Author

Measured with Diffusion-Weighted ¹H Magnetic Resonance Spectroscopy, *NMR Biomed.* 23:382-390.

[11] Beckonert O., Coen M., Keun H. C., Wang Y., Ebbels T. M. D., Holmes E., Lindon J. C., Nicholson J. K. (2010) High-Resolution Magic-Angle-Spinning NMR Spectroscopy for Metabolic Profiling of Intact Tissues, *Nat. Protoc.* 5:1019-1032.

[12] Valentini M., Ritota M., Cafiero C., Cozzolino S., Leita L., Sequi P. (2011) The HRMAS-NMR Tool in Foodstuff Characterization, *Magn. Reson. Chem.* 49:S121-S125.

[13] Vermathen M., Marzorati M., Baumgartner D., Good C., Vermathen P. (2011) Investigation of Different Apple Cultivars by High Resolution Magic Angle Spinning NMR. A Feasibility Study, *J. Agric. Food Chem.* 59:12784-12793.

[14] Shapiro M. J., Gounarides J. S. (2001) High Resolution MAS-NMR in Combinatorial Chemistry, *Biotechnology Bioengineering (Combinatorial Chemistry)* 71:130-148.

[15] Power W. P. (2003) High Resolution Magic Angle Spinning - Applications to Solid Phase Synthetic Systems and Other Semi-Solids, *Annual Reports NMR Spectroscopy* 51:261-295.

[16] Shapiro M. J., Gounarides J. S. (1999) NMR Methods Utilized in Combinatorial Chemistry Research, *Progress Nuclear Magnetic Resonance* 35:153-200.

[17] Brown S. P. (2012) Applications of High-Resolution ¹H Solid-State NMR, *Solid State Nuclear Magnetic Resonance* 41:1-27.

[18] Schröder H. (2003) High Resolution Magic Angle Spinning NMR for Analyzing Small Molecules Attached to Solid Support, *Comb. Chem. High T. Scr.* 6:741-753.

[19] Thieme K., Zech G., Kunz H., Spiess H. W., Schnell I. (2002) Dipolar Recoupling in NOESY-Type ¹H-¹H NMR Experiments Under HRMAS Conditions, *Organic Letters* 4:1559-1562.

[20] Alam T. M., Holland g. P. (2006) ¹H-¹³C INEPT MAS NMR Correlation Experiments with ¹H-¹H Mediated Magnetization Exchange to Probe Organization in Lipid Biomembranes, *J. Magn. Reson.* 180:210-221.

[21] Nicholls A. W., Mortishire-Smith R. J. (2001) Temperature Calibration of a High-Resolution Magic-Angle Spinning NMR Probe for Analysis of Tissue Samples, *Magn. Reson. Chem.* 39:773-776.

[22] Sodickson A., Cory D. G. (1997) Shimming a High-Resolution MAS Probe, *J. Magn. Reson.* 128:87-91.

[23] Engelke F., Maas W. E. (1997) *High Resolution Magic Angle Spinning Spectroscopy User Manual Version 1.0*: Bruker Instruments, Inc. 50 p.

[24] Piotto M., Elbayed K., Wieruszeski J. M., Lippens G. (2005) Practical Aspects of Shimming a High Resolution Magic Angle Spinning Probe, *J. Magn. Reson.* 173:84-89.

[25] Nishiyama Y., Tsutsumi Y., Utsumi H. (2012) MAGIC SHIMMING: Gradient Shimming with Magic Angle Sample Shimming, *J. Magn. Reson.* 216:197-200.

[26] Smallcombe S. H., Patt S. L., Keifer P. A. (1995) WET Solvent Suppression and Its Applications to LC NMR and High-Resolution NMR Spectroscopy, *J. Magn. Reson., A* 117:295-303.

[27] Warrass R., Wieruszeski J.-M., Lippens G. (1999) Efficient Suppression of Solvent Resonances in HR-MAS of Resin-Supported Molecules, *J. Am. Chem. Soc.* 121:3787-3788.

[28] Ng Y.-F., Meillon J.-C., Ryan T., Dominey A. P., Davis A. P., Sanders J. K. M. (2001) Gel-Phase MAS NMR Spectroscopy of a Polymer-Supported Pseudorotaxane and Rotaxane: Receptor Binding to an "Inert" Polyethylene Glycol Spacer, *Angew. Chem. Int. Ed.* 40:1759-1760.

[29] de Miguel Y. R., Bampos N., de Silva K. M. N., Richards S. A., Sanders J. K. M. (1998) Gel Phase MAS ^{1}H NMR as a Probe for Supramolecular Interactionas the Solid-Liquid Interface, *Chem. Commun.* 2267-2268.

[30] Hulst R., Seyger R. M., van Duynhoven J. P. M., van der Does L., Noordermeer J. W. M., Bantjes A. (1999) Vulcanization of Butadiene Rubber by Means of Cyclic Disulfides. 3. A 2D Solid State HRMAS NMR Study on Accelerated Sulrfur Vulcanizates of BR Rubber, *Macromolecules* 32:7521-7529.

[31] Hulst R., Seyger R. M., Van Duynhoven J. P. M., van der Does L., Noordermeer J. W. M., Bantjes A. (1999) Vulcanization of Butadiene Rubber by Means of Cyclic Disulfides. 2. A 2D Solid State HRMAS NMR Study of Cross-Link Structures in BR Vulcanizates, *Macromolecules* 32:7509-7520.

[32] Calucci L., Forte C., Ranucci E. (2007) Water/Polymer Interactions in a Poly(amidoamine) Hydrogel Studied by NMR Spectroscopy, *Biomacromolecules* 8:2936-2942.

[33] Annunziata R., Tranchini J., Ranucci E., Ferruti P. (2007) Structural Characterisation of Poly(Amidoamine) Networks Via High-Resolution Magic Angle Spinning NMR, *Magn. Reson. Chem.* 45:51-58.

[34] Sacristán M., Ronda J. C., Cádiz V. (2009) Silicon-Containing Soybean-Oil-Based Copolymers. Synthesis and Properties, *Biomacromolecules* 10:2678-2685.

[35] Van Camp W., Dispinar T., Dervaux B., Du Prez F. E., Martins J. C., Frtizinger B. (2009) 'Click' Functionalization of Cryogels Conveniently Verified and Quantified Using High-Resolution MAS NMR Spectroscopy, *Macromolecular Rapid Communications* 30:1328-1333.

[36] Crini G., Bourdonneau M., Martel B., Piotto M., Morcellet M., Richert T., Vebrel J., Torri G., Morin N. (2000) Solid-State NMR Characterization of Cyclomaltoheptoase (β-Cyclodextrin) Polymers Using High-Resolution Magic Angle Spinning with Gradients, *J. Applied Polymer Science* 75:1288-1295.

[37] Komber H., Ziemer A., Voit B. (2002) Etherification as Side Reactions in the Hyperbranched Polycondensationof 2,2-Bis(hydroxymethyl)propionic Acid, *Macromolecules* 35:3514-3519.

[38] Favier I., Gómez M., Muller G., Picurelli D., Nowicki A., Roucoux A., Bou J. (2007) Synthesis of New Functionalized Polymers and their use as Stabilizers of Pd, Pt, and Rh Nanoparticles. Preliminary Catalytic Studies, *J. Applied Polymer Science* 105:2772-2782.

[39] Bachmann S., Hellriegel C., Wegmann J., Händel H., Albert K. (2000) Characterization of Polyalkylvinyl Ether Phases by Solid-State and Suspended-State Nuclear Magnetic Resonance Investigations, *Solid State Nuclear Magnetic Resonance* 17:39-51.

[40] Schauff S., Friebolin V., Grynbaum M. D., Meyer C., Albert K. (2007) Monitoring the Interactions of Tocopherol Homologues with Reversed-Phase Stationary HPLC Phases by ^{1}H Suspended-State Saturation Transfer Difference High-Resolution Magic Angle Spinning NMR Spectroscopy, *Anal. Chem.* 79:8323-8326.

[41] Courtois J., Fischer G., Schauff S., Albert K., Irgum K. (2006) Interactions of Bupivacaine with a Molecularly Imprinted Polymer in a Monolithic Format Studied by NMR, *Anal. Chem.* 78:580-584.

[42] Melchels F. P. W., Velders A. H., Feijen J., Grijpma D. W. (2010) Photo-Cross Linked Poly(DL-Lactide)-Based Networks. Structural Characterization by HR-MAS NMR Spectroscopy and Hydrolytic Degradation Behavior, *Macromolecules* 43:8570-8579.

[43] Feng Y., Billon L., Grassl B., Khoukh A., Francois J. (2002) Hydrophobically Associated Polyacrylamides and their Partially Hydrolyzed Derivatives Prepared by Post-Modification. 1. Synthesis and Characterization, *Polymer* 43:2055-2064.

[44] Busico V., Causà M., Cipullo R., Credendino R., Cutillo F., Friederichs N., Lamanna R., Serge A., Castelli V. V. A. (2008) Periodic DFT and High-Resolution Magic-Angle-Spinning (HR-MAS) ^{1}H NMR Investigation of the Active Surfaces of $MgCl_2$-Supported Ziegler-Natta Catalysts. the $MgCl_2$ Matrix., *J. Phys. Chem. C* 112:1081-1089.

[45] Linder E., Brugger S., Steinbrecher S., Plies E., Mayer H. A. (2001) Investigations on the Mobility of Novel Sol-Gel Processed Inorganic-Organic Hybrid Materials, *J. Mater. Chem.* 11:1393-1401.

[46] Brenna S., Posset T., Furrer J., Blümel J. (2006) ^{14}N NMR and Two-Dimensional Suspension ^{1}H and ^{13}C HRMAS NMR Spectrscopy of Ionic Liquids Immobilized on Silica, *Chem. Eur. J.* 12:2880-2888.

[47] Rencurosi A., Lay L., Russo G., Prosperi D., Poletti L., Caneva E. (2007) HRMAS NMR Analysis in Neat Ionic Liquids: A Powerful Tool to Investigate Complex Organic Molecules and Monitor Chemical Reactions, *Green Chemistry* 9:216-218.

[48] Carvalho P. J., Álvarez V. H., Schröder B., Gil A. M., Marrucho I. M., Aznar M., Santos L. M. N. B. F., Coutinho J. A. P. (2009) Specific Solvation Interaction of CO_2 on Acetate and Trifluoroacetate Imidazolium Based Ionic Liquids at High Pressures, *J. Phys. Chem. B* 113:6803-6812.

[49] Lungwitz R., Spange S. (2008) Structure and Polarity of the Phase Boundry of N-Methylimidazolium Chloride/Silica, *J. Phys. Chem. C* 112:19443-19448.

[50] Le Bideau J., Gaveau P., Bellayer S., Néouze M.-A., Vioux A. (2007) Effect of Confinement on Ionic Liquids Dynamics in Monolithic Silica Ionogels: ^{1}H NMR Study, *Phys. Chem. Chem. Phys.* 9:5419-5422.

[51] Romanova E. E., Grinberg F., Pampel A., Kärger J., Freude D. (2009) Diffusion Studies in Confined Nematic Liquid Crystals by MAS PFG NMR, *J. Magn. Reson.* 196:110-114.

[52] Alam T. M., Dreyer D. R., Bielwaski C. W., Ruoff R. S. (2011) Measuring Molecular Dynamics and Activation Energies for Quanternary Acyclic Ammonium and Cyclic Pyrrolidinium Ionic Liquids using ^{14}N NMR Spectroscopy, *J. Phys. Chem. A* 115:4307-4316.

[53] Huster D., Gawrisch K. (1999) NOESY NMR Crosspeaks Between Lipid Headgroups and Hydrocarbon Chains: Spin Diffusion or Molecular Disorder?, *J. Am. Chem. Soc.* 121:1992-1993.

[54] Huster D., Arnold K., Gawrisch K. (1999) Investigation of Lipid Organization in Biological Membranes by Two-Dimensional Nuclear Overhauser Enhancement Spectroscopy *J. Phys. Chem. B* 103:243-251.

[55] Zhou H., Du F., Li X., Zhang B., Li W., Yan B. (2008) Characterization of Organic Molecules Attached to Gold Nanoparticle Surface Using High Resolution Magic Angle Spinning ^{1}H NMR, *J. Phys. Chem. C* 112:19360-19366.

[56] Krpetic Z., Scarì G., Caneva E., Speranza G., Porta F. (2009) Gold Nanoparticles Prepared Using Cape Aloe Active Compounds, *Langmuir Lett.* 25:7217-7221.

[57] Krpetic Z., Nativo P., Porta F., Brust M. (2009) A Multidentate Peptide for Stabilization of Facile Bioconjugation of Gold Nanoparticles, *Bioconjugate Chem.* 20:619-624.

[58] Polito L., Colombo M., Monti D., Melato S., Caneva E., Prosperi D. (2008) Resolving the Structure of Ligands Bound to the Surface of Superparamagnetic Iron Oxide Nanoparticles by High-Resolution Magic-Angle Spinning NMR Spectroscopy, *J. Am. Chem. Soc.* 130:12712-12724.

[59] Polito L., Monti D., Caneva E., Delnevo E., Russo G., Prosperi D. (2008) One-Step Bioengineering of Magnetic Nanoparticles via a Surface Diazo Transfer/Azide–Alkyne Click Reaction Sequence, *Chem. Commun.* 621-623.

[60] Deshayes S., Maurizot V., Clochard M.-C., Berthelot T., Baudin C., Déléris G. (2010) Synthesis of Specific Nanoparticles for Targeting Tumor Angiogenesis Using Electron-Beam Irradiation, *Radiation Phys. Chem.* 79:208-213.

[61] Costantino L., Gandolfi F., Bossy-Nobs L., Tosi G., Gurny R., Rivasi F., Vandelli M. A., Forni F. (2006) Nanoparticulate Drug Carriers Based on Hybrid poly(D,L-Lactide-*co*-Glycolide)-Dendron Structures, *Biomaterials* 27:4635-4645.

[62] Conte P., Carotenuto G., Piccolo A., Perlo P., Nicolais L. (2007) NMR-Investigations of the Mechanism of Silver Mercaptide Themolysis in Amorphous Polystyrene, *J. Mater. Chem.* 17:201-205.

[63] Blümel J. (2008) Linkers and Catalysis Immobilized on Oxide Supports: New Insights by Solid-State NMR Spectroscopy, *Coordination Chemistry Reviews* 252:2410-2423.

[64] Posset T., Rominger F., Blümel J. (2005) Immobilization of Bisphosphinoamine Linkers on Silica: Identification of Previously Unrecongnized Byproducts vis ^{31}P CP/MAS and Suspension HR-MAS Studies, *Chem. Mater.* 17:586-595.

[65] Guenther J., Reibenspies J., Blümel J. (2011) Synthesis, Immobilization, MAS and HR-MAS NMR of a New Chelate Phosphine Linker System, and Catalysis by Rhodium Adducts Thereof, *Adv. Synth. Catal.* 353:443-460.

[66] Posset T., Guenther J., Pope J., Oeser T., Blümel J. (2011) Immobilized Sonogashira Catalyst Systems: New Insights by Multinuclear HRMAS NMR Studies, *Chem. Commun.* 2011:2059-2061.

[67] Blümel J. (2008) Linkers and Catalysts Immobilized in Oxide Supports: New Insights by Solid-State NMR Spectroscopy, *Coordination Chemistry Reviews* 252:2410-2423.

[68] Pinoie V., Poelmans K., Miltner H. E., Verbruggen I., Biesemans M., Van Assche G., Van Mele B., Martins J. C., Willem R. (2007) A Polystyrene-Supported Tin Trichloride Catalyst with a C11-Spacer. Catalysis Monitoring Using High-Resolution Magic Angle Spinning NMR, *Organometallics* 26:6718-6725.

[69] Poelmans K., Pinoie V., Verbruggen I., Biesemans M., Deshayes G., Duquesne E., Delcourt C., Degée P., Miltner H. E., Dubois P., Willem R. (2008) Undecyltin Trichloride Grafted onto Cross-Linked Polystyrene: An Effecient Catalyst for Ring-Opening Polymerization of ε-Caprolactone, *Organometallics* 27:1841-1849.

[70] Deshayes G., Poelmans K., Verbruggen I., Camacho-Camacho C., Degée P., Pinoie V., Martins J. C., Piotto M., Biesemans M., Willem R., Dubois P. (2005) Polystyrene-Supported Organotin Dichloride as a Recyclable Catalyst in Lactone Ring-Opening Polymerization: Assessment and Catalysis Monitoring by High-Resolution Magic-Angle-Spinning NMR Spectroscopy, *Chem. Eur. J.* 11:4552-4561.

[71] Pinoie V., Biesemans M., Willem R. (2008) Quantitative Tin Loading Determination of Supported Catalysts by ^{119}Sn HRMAS NMR using a Calibrated Internal Signal (ERETIC), *Organometallics* 27:3633-3634.

[72] Porto S., Seco J. M., Espinosa J. F., Quiñoá E., Riguera R. (2008) Resin-Bound Chiral Derivatizing Agents for Assignments of Configuration by NMR Spectroscopy, *J. Organic Chem.* 73:5714-5722.

[73] Hellriegel C., Skogsberg U., Albert K., Lämmerhofer M., Maier N. M., Linder W. (2004) Characterization of a Chiral Stationary Phase by HR-MAS NMR Spectroscopy and Investigation of Enatioselective Interaction with Chiral Ligates by Transferred NOE, *J. Am. Chem. Soc.* 126:3809-3816.

[74] Simpson A. J., Kingery W. L., Sahw D. R., Spraul M., Humpfer E., Dvorstak P. (2001) The Application of ^{1}H HR-MAS NMR Spectroscopy for the Study of Structures and Associations of Organic Components at the Solid-Aqueous Interface of a Whole Soil., *Environ. Sci. Technol.* 35:3321-3325.

[75] Simpson A. J., Simpson M., Smith E., Kelleher B. P. (2007) Microbially Derived Input to Soil Organic Matter; Are Current Estimates Too Low?, *Environ. Sci. Technol.* 41:8070-8076.

[76] Feng X., Simpson A. J., Simpson M. (2006) Investigating the Role of Mineral-Bound Humic Acid in Phenanthrene Sorption, *Environ. Sci. Technol.* 40:3260-3266.

[77] Shirzadi A., Simpson M. J., Kumar R., Baer A. J., Xu Y., Simpson A. J. (2008) Molecular Interactions of Pestisides at the Soil-Water Interface, *Environ. Sci. Technol.* 42:5514-5520.

[78] Combourieu B., Inacio J., Delort A.-M., Forando C. (2001) Differentation of Mobile and Immobile Pesticides on Anionic Clays by ^{1}H HR MAS NMR Spectroscopy, *Chem. Commun.* 2214-2215.

[79] Colnago L. A., Martin-Neto L., Pérez M. G., Daolio C., Ferreira A. G., Camargo O. A., Berton R., Bettiol W. (2003) Application of ^{1}H HR/MAS NMR to Soil Organic Matter Studies, *Ann. Magn. Reson.* 2:116-118.

[80] Simpson A. J., Simpson M., Kingery W. L., Lefebvre B. A., Moser A., Williams A. J., Kvasha M., Kelleher B. P. (2006) The Application of [1]H High-Resolution Magic-Angle

Spinning NMR for the Study of Clay-Organic Associations in Natural and Synthetic Complexes, *Langmuir* 22:4498-4503.

[31] Simpson A. J., Kingery W. L., Hatcher P. G. (2003) The Indentification of Plant Derived Structures in Humic Materials Using Three-Dimensional NMR Spectroscopy, *Environ. Sci. Technol.* 37:337-342.

[82] Pampel A., Zick K., Glauner H., Engelke F. (2004) Studying Lateral Diffusion in Lipid Bilayers by Combining a Magic Angle Spinning NMR Probe with a Microimaging Gradient System, *J. Am. Chem. Soc.* 126:9534-9535.

83] Stejskal E. O., Tanner J. E. (1965) Spin Diffusion Measurements: Spin Echoes in the Presence of Time-Dependent Field Gradient, *J. Chem. Phys.* 42:288-292.

84] Viel S., Ziarelli F., Pagès G., Carrara C., Caldarelli S. (2008) Pulsed Field Gradient Magic Angle Spinning NMR Self-Diffusion Measurements in Liquids, *J. Magn. Reson.* 190:113-123.

[85] Cotts R. M., Hoch M. J. R., Sun T., Marker J. T. (1989) Pulsed Field Gradient Stimulated Echo Methods for Improved NMR Diffusion Measurements in Heterogeneous Systems, *J. Magn. Reson.* 83:252-266.

[86] Tanner J. E. (1970) Use of the Stimulated Echo in NMR Diffusion Studies, *J. Chem. Phys.* 52:2523-2526.

[87] Gibbs S. J., Johnson Jr. C. S. (1991) A PFG NMR Experiment for Accurate Diffusion and Flow Studies in the Presence of Eddy Currents, *J. Magn. Reson.* 93:395-402.

[88] Pampel A., Fernandez M., Freude D., Kärger J. (2005) New Options for Measuring Molecular Diffusion in Zeolites by MAS PFG NMR, *Chem. Phys. Lett.* 407:53-57.

[89] Pampel A., Engelke F., Galvosas P., Krause C., Stallmach F., Michel D., Kärger J. (2006) Selective Multi-Component Diffusion Measurement in Zeolites by Pulsed Field Gradient NMR, *Micropor. Mesopor. Mat.* 90:271-277.

[90] Fernandez M., Pampel A., Takahashi R., Sato S., Freude D., Kärger J. (2008) Revealing Complex Formation in Acetone-*n*-Alkane Mixtures by MAS PFG NMR Diffusion Measurement in Nanoporous Hosts, *Phys. Chem. Chem. Phys.* 10:4165-4171.

[91] Fernandez M., Kärger J., Freude D., Pampel A., van Baten J. M., Krishna R. (2007) Mixture Diffusion in Zeolites Studied by MAS PFG NMR and Molecular Simulation, *Micropor. Mesopor. Mat.* 105:124-131.

[92] Gaede H. C., Gawrisch K. (2003) Lateral Diffusion Rates of Lipid, Water and a Hydrophobic Drug in a Multilamellar Liposome, *Biophys. J.* 85:1734-1740.

[93] Gaede H. C., Gawrisch K. (2004) Multi-Dimensional Pulse Field Gradient Magic Angle Spinning NMR Experiments on Membranes, *Magn. Reson. Chem.* 42: 115-122.

[94] Polozov I. V., Gawrisch K. (2004) Domains in Binary SOPC/POPE Lipid Mixtures Studied by Pulsed Field Gradient ^{1}H MAS NMR, *Biophys. J.* 87:1741-1751.

[95] Iqbal S., Rodríguez-Lansola F., Escuder B., Miravet J. F., Verbruggen I., Willem R. (2010) HRMAS ^{1}H NMR as a Tool for the Study of Supramolecular Gels, *Soft Matter* 6:1875-1878.

[96] Chin J. A., Chen A., Shapiro M. J. (2000) SPEEDY: Spin-Echo Enhanced Difusion Filtered Spectroscopy. A New Tool for High Resolution MAS NMR, *J. Comb. Chem.* 2:293-296.

Chapter 10

From Micro– to Macro–Raman Spectroscopy: Solar Silicon for a Case Study

George Sarau, Arne Bochmann, Renata Lewandowska and Silke Christiansen

Additional information is available at the end of the chapter

1. Introduction

The phenomenon of inelastic scattering of light by matter is referred as Raman spectroscopy named after Sir Chandrasekhara Venkata Raman who first observed it experimentally in 1928 [1]. Because only one photon out of 10^6-10^{12} incident photons is inelastically or Raman scattered, it took some time until lasers with high enough light intensities for efficient Raman excitation and very sensitive detectors for measuring the still low intensity Raman light were developed. Another important step in advancing Raman instrumentation was the efficient rejection of the very intense elastic scattered light, known as Rayleigh light, through a double or triple monochromator or filters [2].

Nowadays, Raman spectroscopy is being successfully applied to both in- and ex-situ analyses of various processes and materials in different states of matter (solid, liquid, gas or plasma). Moreover, Raman spectrometers have become small, portable and easy to use even for nonspecialists. This technique is covering a very broad range of application fields, at scientific and industrial levels, including pharmaceuticals, biology, environment, forensics, geology, art, archaeology, catalysis, corrosion, materials and others. Giving the large amount of specific information related to each of the abovementioned areas, we refer the interested reader to [3,4].

In the field of semiconductors, Raman spectroscopy has shown to be a powerful analytic tool for investigating mechanical stress, crystallographic orientation, doping, composition, phase, and crystallinity of semiconductor materials in bulk, thin film and device form [4-7]. In particular, the use of Raman spectroscopy to study solar silicon materials in form of thin films on glass, wafers, and ribbons, which are then processed to solar cells used as a clean and sustainable energy source has gained new momentum in the context of climate change and energy security. The physics behind Raman scattering in semiconductors or crystals is based on the inelastic interaction of light with lattice vibrations or phonons that are sensitive to internal and external perturbations. A short but relevant theoretical introduction in the case of silicon will be given in Section 2. Back to early 70th, it was Anastassakis et. al. first reporting on the shift of the first-order silicon Raman peak under uniaxial *external* stress [8]. This work triggered the application of Raman spectroscopy in measuring *internal* stresses present in semiconductor materials and structures. Particularly important for the present contribution are the studies on *local* internal stresses in microelectronics devices such as silicon integrated circuits using confocal micro-Raman spectroscopy where the exciting laser light is focused onto the sample's surface through a microscope objective thus enabling investigations on the micrometer scale [4,5].

The first experimental part of this chapter (Sections 3.2 and 4.1) is mainly focused on the application of confocal *micro-Raman spectroscopy* to map the spatial distribution of internal stresses, their magnitude and sign in different solar silicon materials following the existing work in silicon microelectronics. Because internal stresses may decrease mechanical strength increasing the breakage rate and induce recombination active defects when combined with external stresses, their understanding and control will improve both process yields and solar cell efficiencies. In addition to mechanical information, other useful material properties can be obtained from the *same* first-order silicon Raman peak. We will show how internal stresses, defects, doping, and microstructure can be directly correlated with each other on the same map, enabling the basic understanding of their interactions. The micro-Raman measurements are supported and complemented at identical positions by other techniques such as EBSD, EBIC, and defect etching. Such a combination allows the correlation of internal stresses, recombination activity and microstructure on the micrometer scale.

In the second experimental part of this contribution (Sections 3.3 and 4.2), confocal *macro-Raman spectroscopy* is introduced and its application to solar silicon is demonstrated for the first time. Macro-Raman spectroscopy represents the state-of-the-art in fast, large area Raman mapping being initially developed to analyze the chemical homogeneity in pharmaceutical tablets. We will present a statistical analysis using Macro-Raman mapping of solar silicon, which is usually characterized by large spatial properties variations. The combination of the two mapping techniques offers insights into the interplay between solar silicon properties at different length scales. Finally, the potential use of macro-Raman spectroscopy for optimization and in-line quality check in a PV factory will be discussed.

Such detailed Raman studies are not limited to solar silicon materials but they can be performed on all Raman active materials. In this context, it is clear that today Raman spectroscopy is a versatile and mature characterization method, which can be applied both at micro- and macro-scale to learn about the interaction between materials properties and their optimization in relation to individual processing steps.

2. Theoretical background of Raman spectroscopy in silicon

The aim of this section is to provide the basic equations and their interpretation necessary to understand the experimental results shown throughout the present chapter. A rigorous mathematical derivation is extensively documented in many textbooks and papers published during the long history of the Raman effect [4,5,8-14].

In a Raman experiment described from a classical point of view, monochromatic light of frequency ω_i originating from a laser is incident on a crystal in a direction k_i with $E = E_0\exp[i(k_i \cdot r - \omega_i t)]$. The electric field of light will induce an electric moment $P = \varepsilon_0 \chi E$, with the interaction between light and crystal at position r being mediated by lattice vibrations or phonons characterized by a wavevector q_j and a frequency ω_j with $Q_j = A_j\exp[\pm i(q_j \cdot r - \omega_j t)]$. It is the electrical susceptibility χ, which is changed by phonons. This means that the induced electric moment will emit besides the elastic scattered Rayleigh light of ω_i, Raman light of $\omega_i - \omega_j$ and $\omega_i - \omega_j$ resulting from anti-Stokes and Stokes Raman scattering, respectively [5]:

$$P = \varepsilon_0 \chi_0 \cdot E_0 \exp[i(k_i \cdot r - \omega_i t)] + \varepsilon_0 E_0 \left(\frac{\partial \chi}{\partial Q_j}\right)_0 A_j \times \exp[-i(\omega_i \pm \omega_j)t]\exp[i(k_i \pm q_j) \cdot r]. \quad (1)$$

From a quantum mechanical point of view, a photon described by k_i, ω_i produces an electron-hole pair. The electron is excited from the ground state to a higher energy state and interacts with a phonon characterized by q_j, ω_j. As a result of this interaction, the electron gains or losses energy and trough the recombination of the electron-hole pair a photon k_s, ω_s is emitted, where $\omega_s = \omega_i + \omega_j$ and $\omega_s = \omega_i - \omega_j$ for anti-Stokes and Stokes Raman scattering, respectively. In most cases, only the silicon Stokes Raman peak known also as the first-order silicon Raman peak occurring in the absence of internal and external perturbations at $\omega_0 \sim$ 520 cm^{-1} is measured and examined. This peak is referred as the silicon Raman peak throughout the next sections. It corresponds to lower energy scattered photons λ_1 than the incident ones λ_0. The conversion formula from nm to cm^{-1} is written as:

$$\Delta\omega(cm^{-1}) = \left(\frac{1}{\lambda_0(nm)} - \frac{1}{\lambda_1(nm)}\right) \times 10^7. \quad (2)$$

2.1. Orientation evaluation

The Raman scattering efficiency or intensity depends on the polarization direction of the incident (e_i) and backscattered (e_s) light and on the three silicon Raman tensors R'_j which are proportional to $(\partial\chi/\partial Q_j)_0$ (see Equation 1), I_0 is a constant including all fixed experimental parameters [5,10,14]:

$$I(e_i, e_s) \approx I_0 \cdot \sum_{j=1}^{3}\left|e_i \cdot R'_j \cdot e_s\right|^2. \quad (3)$$

Here the polarization directions are defined in the stage coordinate system, while the Raman tensors refer to the crystal coordinate system. The crystal - stage transformation is performed by means of a rotation matrix $T(\alpha,\beta,\gamma)$ applied to the Raman tensors R'_j, where α, β, and γ are the three Euler angles [10]:

$$I(e_i, e_s) \approx I_0 \cdot \sum_{j=1}^{3}\left|e_i \cdot \left(T^{-1}(\alpha,\beta,\gamma) \cdot R'_j \cdot T(\alpha,\beta,\gamma)\right) \cdot e_s\right|^2. \quad (4)$$

By continuously rotating the polarization direction of the incident laser light θ with a $\lambda/2$ plate for two analyzer positions x and y, it is possible to obtain two experimental curves showing the intensity variations of the silicon Raman peak. The data fitting based on Equation (4) results in the numerical evaluation of the three Euler angles which are needed to describe the crystallographic orientation of a particular grain with respect to the stage (reference) coordinate system. Having the grain orientation, the intensity variations of the three optical phonons with polarization settings $I^{X_j}(\theta)$ and $I^{Y_j}(\theta)$ can be simulated separately. These six intensity variations can be transformed into six intensity ratio functions [10]:

$$W_1^{X,Y}(\theta) = \frac{I_1^{X,Y}(\theta)}{I_2^{X,Y}(\theta) + I_3^{X,Y}(\theta)}, W_2^{X,Y}(\theta) = \frac{I_2^{X,Y}(\theta)}{I_1^{X,Y}(\theta) + I_3^{X,Y}(\theta)}, W_3^{X,Y}(\theta) = \frac{I_3^{X,Y}(\theta)}{I_1^{X,Y}(\theta) + I_2^{X,Y}(\theta)} \quad (5)$$

It has been shown that for almost any arbitrary oriented grain, distinct polarization settings (θ, x or y) for which the intensity of one phonon prevails over the intensity sum of the other two phonons can be found [10]. Performing three Raman measurements on the same grain, one for every single-phonon polarization settings, several stress tensor components can be determined as experimentally shown in Section 4.1.1.

2.2. Stress evaluation

In the absence of stress (internal or external), the three Raman optical phonons of silicon are degenerate leading to a single Raman peak at $\omega_0 \sim 520$ cm^{-1}. Large mechanical stresses in the GPa range lift the degeneracy causing frequency shifts of the three optical phonons $\Delta\omega_j = \omega_j - \omega_0$, which appear as separate peaks in the Raman spectrum depending on the direction of the applied stress and measurement conditions [8,12,13]. When the stress level is below 1 GPa, these frequency shifts are too small to be resolved being masked by the natural width of the silicon Raman peak. In such cases, particular polarization settings for the incident and backscattered Raman light can be found that allow the excitation and probing of the three optical phonons almost separately and consequently their frequency shifts can be determined. These settings vary with the orientation of the investigated grain as discussed in the previous section. Next step consists in using the secular equation that relates the frequency shifts to the strain tensor components [5,9,10]:

$$\begin{vmatrix} p \cdot \varepsilon_{xx}' + q \cdot (\varepsilon_{yy}' + \varepsilon_{zz}') - \lambda & 2r \cdot \varepsilon_{xy}' & 2r \cdot \varepsilon_{xz}' \\ 2r \cdot \varepsilon_{xy}' & p \cdot \varepsilon_{yy}' + q \cdot (\varepsilon_{xx}' + \varepsilon_{zz}') - \lambda & 2r \cdot \varepsilon_{yz}' \\ 2r \cdot \varepsilon_{xz}' & 2r \cdot \varepsilon_{yz}' & p \cdot \varepsilon_{zz}' + q \cdot (\varepsilon_{xx}' + \varepsilon_{yy}') - \lambda \end{vmatrix} = 0. \quad (6)$$

Here p, q, and r are material constants so-called phonon deformation potentials being the only three independent components for cubic symmetry crystals such as silicon, ε'_{ij} are the strain tensor components in the crystal coordinate system, while the eigenvalues λ_j (j = 1,2,3) are given by

$$\lambda_j = \omega_j^2 - \omega_0^2 = (\omega_j - \omega_0) \cdot (\omega_j + \omega_0) \approx \Delta\omega_j \cdot 2\omega_0. \quad (7)$$

The stress tensor components are finally obtained from the inverse Hooke's law $\varepsilon'_{ij} = S_{ij} \cdot \sigma'_{ij}$ where S_{ij} represents the elastic compliance tensor whose components are material constants. It is evident from Equation (6,7) that the three frequency shifts $\Delta\omega_j$ are not enough to determine the six independent stress tensor components σ'_{ij}. The probing depth in silicon ranges from a few hundreds of nm to a few µm for visible excitations, and from a few nm to a few tenths of nm for UV excitations. Thus, due to wavelength dependent absorption, only the stress state close to the sample's surface is measured. This implies a predominant planar stress state described by three stress components σ'_{xx}, σ'_{yy}, and τ'_{xy} that can be numerically evaluated using the three frequency shifts $\Delta\omega_j$. The residual stress components in the z direction given by σ'_{zz}, τ'_{xz}, and τ'_{yz} are included in Δ'_z which also serves as a correction parameter [10]:

$$\sigma' = \begin{pmatrix} \sigma'_{xx} & \tau'_{xy} & 0 \\ \tau'_{xy} & \sigma'_{yy} & 0 \\ 0 & 0 & \Delta'_z \end{pmatrix}. \tag{8}$$

The following equation can be used to transform the stress tensor components into average or von Misses stress:

$$\sigma'_{av} = \sqrt{\sigma'^2_x + \sigma'^2_y + \sigma'^2_z - \sigma'_x\sigma'_y - \sigma'_x\sigma'_z - \sigma'_y\sigma'_z + 3(\tau'^2_{xy} + \tau'^2_{xz} + \tau'^2_{yz})} \tag{9}$$

Another more straightforward way to relate measured Raman shifts to stress values is the use of a simple stress model illustrating the stress state in the sample. The classical example in the case of silicon is the presence of uniaxial stress σ along the [100] direction and the measurement of the backscattered Raman signal from the (001) surface [5]. In this configuration, only one stress tensor component is non zero:

$$\Delta\omega_3 (cm^{-1}) = -2 \cdot 10^{-9}\sigma(Pa) \quad or \quad \sigma(MPa) = -500 \times \Delta\omega(cm^{-1}). \tag{10}$$

If biaxial stress in the x-y plane with stress components σ_{xx} and σ_{yy} (or $\sigma_{xx} = \sigma_{yy}$ for isotropic stress) describes the stress distribution in the sample:

$$\Delta\omega_3 (cm^{-1}) = -4 \cdot 10^{-9}\left(\frac{\sigma_{xx} + \sigma_{yy}}{2}\right)(Pa) \quad or \quad \sigma_{xx} = \sigma_{yy}(MPa) = -250 \times \Delta\omega(cm^{-1}). \tag{11}$$

These two formulas (10,11) written in the stage coordinate system are commonly used in the community for a fast and reliable estimation of the *average stress* independently of the crystallographic orientation of grains in multicrystalline silicon. Thus, 1 cm⁻¹ shift of the silicon Raman peak with respect to the stress free value of ~ 520 cm⁻¹ corresponds to a uniaxial stress of 500 MPa or to a biaxial in-plane isotropic stress of 250 MPa. It can be seen that tensile stress shifts the silicon Raman peak to lower frequency, while compressive stress to higher frequency as sketched in Figure 1(a). Experimental examples of the stress state evaluation using both methods described above will be given in Section 4.1.

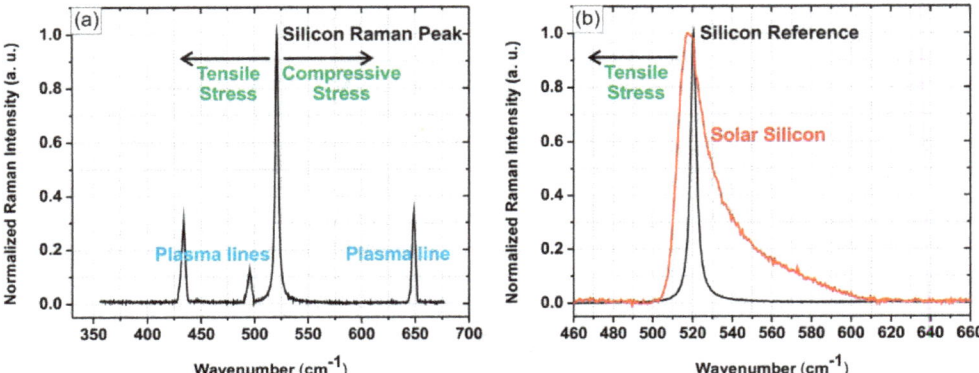

Figure 1. (a) Typical Raman spectrum of a silicon wafer used as reference. In the absence of stress, the three Raman optical phonons of silicon (1 x LO, 2 x TO) are degenerate resulting in a single Raman peak at $\omega_0 \sim 520$ cm^{-1}. The plasma lines originating from an external reference lamp are used to correct the silicon peak position with respect to the thermal drift of the spectrometer grating. Tensile or compressive stresses (internal or external) below 1 GPa may shift the silicon Raman peak to lower or higher frequencies, respectively. (b) Comparison between Raman spectra of stress-free silicon reference and tensile stressed thin film solar silicon on glass. The peak asymmetry caused by high boron doping is clearly visible.

2.3. Doping evaluation

In the case of highly doped silicon, a resonant interaction occurs between the discrete optical phonon states (phonon Raman scattering) and the continuum of electronic states in the valence or conduction bands (electronic Raman scattering) because of electron-phonon coupling. This leads to Fano-type silicon Raman peak asymmetries, which can be observed as tails either on the right side (for p-type doping) or on the left side (for n-type doping) of the otherwise symmetric silicon Raman peak as shown in Figure 1(b) [11,14-17]. The function used to fit the intensity of the silicon Raman peak $I(\omega)$ is given by [11]

$$I(\omega,q,\Gamma,\omega_{max}) = I_0 \frac{[q + 2(\omega - \omega_{max})/\Gamma]^2}{1 + [2(\omega - \omega_{max})/\Gamma]^2}. \tag{12}$$

Here I_0 is a scaling factor, q is so-called symmetry parameter, Γ is the linewidth of the peak, and ω_{max} is the peak position in the presence of Fano interaction. The symmetry parameter q describes the shape of the silicon Raman peak affected by Fano resonances. Large $q > 150$ values correspond to standard doping ($< 10^{16}$ cm^{-3}) resulting in a nearly symmetric peak, while small $q < 50$ values correlate with high doping ($> 10^{18}$ cm^{-3}) and pronounced peak asymmetry. For highly doped silicon, $1/q$ is approximately proportional to the free carrier concentration [11]. Thus, an accurate quantitative evaluation of doping on the micrometer scale is possible, provided a good calibration curve exist.

2.4. Qualitative defect density evaluation

As shown in the previous three sections, Raman spectroscopy can provide detailed information about semiconductor materials, in this contribution solar silicon, including crystal orientation, internal stresses and doping, which can be extracted from the intensity, position and asymmetry of the silicon Raman peak. In addition, the linewidth of the peak relates to the presence of extended crystal defects. In a perfect crystal, the phonon lifetimes are theoretically infinite in the harmonic approximation that neglects third- and higher-order derivatives of the crystalline potential resulting in narrow delta function-like linewidths of the optical phonon Raman spectra [18]. Defects act as anharmonic perturbations leading to finite phonon lifetimes that manifest themselves as a broadening of the peak described by its full-width at half maximum denoted FWHM or Γ. Therefore, the anharmonic lifetimes of phonons are defined as $1/\Gamma$ being evaluated using first principles calculations including both kinematic effects; i.e., the decay of phonons into vibrations of lower frequency and dynamic effects; i.e., the magnitude of the tensor of the third derivates of the crystal potential with respect to atomic displacements that describes the instability of one-phonon states. The calculated Γ values of the Raman-active optical phonons were found to agree well with those determined experimentally in the case of single crystalline semiconductors (defect free) such as diamond, Si, Ge, GaAs, GaP and InP [19,20].

Anharmonic effects and consequently broadening can also be induced by internal or external stresses approaching the stress-induced splitting limit of the peak as discussed in Section 2.2 or by large stress gradients within the probed volume [5]. Moreover, doping and/or impurities can either produce new Raman peaks through their own vibrational modes or alter/broaden the Raman spectrum of the host material through the change in mass and bond length (atomic effects) as well as through the resonant Fano interaction of free carriers (donors or acceptors) with the lattice (electronic effects). Because all these information originates from the silicon Raman peak, one can separate between the effects of stresses, doping and/or impurities and that of defects on the FWHM values. The Raman linewidths were also found to broaden in the case of relatively small grains (in the nm range) due to the phonon confinement effect, that is, the frequency distribution of the scattered light comes from a broader interval in k-space around the Γ-point in the Brillouin-zone since the $\Delta k = 0$ selection rule is partially lifted by the phonon scattering at grain boundaries [21]. Such a broadening does not occur in large grained solar silicon as presented here.

3. Experimental details

3.1. Sample preparation for Raman measurements

Raman spectroscopy investigates materials nondestructively, the appropriate excitation laser power to avoid damage being material dependent, without elaborated sample preparation. In the case of silicon thin film solar cells on glass, no sample preparation is needed because the as-grown material has low roughness. This is different for wafer- and ribbon-based solar cells for which a simple sample preparation procedure is necessary. Their surfaces have to be evened out by mechanical polishing prior to the Raman measurements to avoid artifacts induced by uncontrolled reflections at rough surface facets. The standard polishing procedure applied to small pieces consists in changing gradually from larger to smaller diamond particle sizes with the final polishing step removing most of the previously damage surface layer, thus leaving the samples in a negligible polishing-induced stress state. The cutting into small pieces leads to stress relaxation due to the creation of free surfaces as discussed in Section 4.1.2. After polishing, the samples are Secco-etched [22] for 5 seconds to make the grain boundaries and dislocations visible. This short defect etching step does not affect the Raman scattering or the other measurement techniques used herein.

3.2. Micro-Raman spectroscopy

The incident light needed for Raman excitation is provided by a laser with a main emission with narrow line width. An interferential filter is used to block the other emissions of the laser. After being reflected by an edge or a notch filter, the light is focused onto the sample's surface through a microscope objective, thus giving rise to the term micro-Raman spectroscopy. Depending on the objective (magnification, numerical aperture) as well as on the excitation wavelength, the diameter of the incident laser beam is different. For the 100x objective (numerical aperture 0.9) and 633 nm excitation employed in the micro-Raman measurements presented here, the probing diameter is ~ 1 μm. The laser power density can be quite high, thus a low laser power of ~ 2 mW at the silicon sample's surface should be used. In these conditions, no shift or increase in the FWHM of the silicon Raman peak due to the local heating of the sample by the laser beam were observed.

As already mentioned in Section 2.2, the probing depth is controlled by the material absorption, which is wavelength dependent. In crystalline silicon, an excitation wavelength of 633 nm results in a penetration depth of ~ 3 μm, while 457 nm gives ~ 300 nm. Since we use 633 nm, the entire thickness of the silicon thin films on glass is probed, while only the surface of the ~ 200 – 250 μm thick silicon wafers and ribbons is measured. The backscattered Raman light passes back through an edge or a notch filter which cuts most of the Rayleigh light, it is dispersed using a grating and then detected by a silicon CCD detector. All Raman measurements herein were performed at room temperature in the backscattering configuration using a LabRam HR800 spectrometer from Horiba Jobin Yvon. A schematic picture of our micro-Raman spectrometer is displayed in Figure 2.

In order to draw correct conclusions about materials with varying spatial properties, not only several but many Raman spectra are acquired while moving the microscope stage with the sample in x- and y-directions in steps equal or smaller than the diameter of the laser probing beam as shown in Figure 2. This results in a complete micro-Raman mapping of the investigated areas, which are usually in the range of a few tens of μm². The exposure time is typically up to 1 s per spectrum. These spectra are fitted with a Gauss-Lorentzian function and maps of the shift, FWHM and intensity of the silicon Raman peak corresponding to the spatial distributions of internal stress, defect density and grain orientation are obtained. A Fano-like fitting function as in Equation 12 is used for doping evaluation [11]. To ensure correct interpretation of the Raman data as well as to be able to visualize small mechanical stresses, the effect of the thermal drift of the spectrometer grating on the silicon peak position is corrected [5,14]. This is done by using one of the plasma lines visible in Figure 1(a) emitted by an external lamp located either close to the spectrometer's confocal hole or above the microscope [23,24].

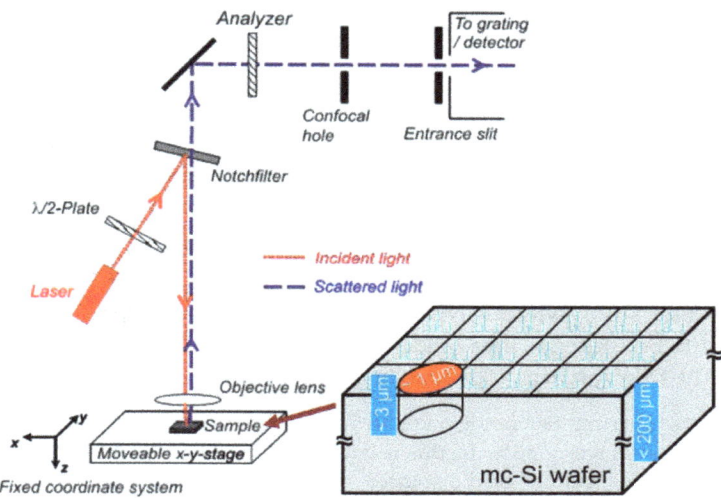

Figure 2. Schematic picture of the used micro-Raman spectrometer. The $\lambda/2$ plate and the analyzer adjust the polarization direction of the incident and backscattered light with respect to the stage (fixed) coordinate system. The polarized micro-Raman procedure enables the evaluation of the crystallographic orientation of arbitrary grains and of stress components as described in Sections 2.1 and 2.2. A graphical representation of micro-Raman mapping obtained by moving the microscope stage with the sample (multicrystalline silicon wafer) in x- and y-directions under a 633 nm exciting laser along with the Raman probing volume are also shown.

3.3. Macro-Raman spectroscopy

Macro-Raman spectroscopy enables fast, large area Raman mapping in the cm² range needed for statistical studies of materials properties and the correlations between them and with processing. In the context of PV, this technique can be used not only for fundamental studies in laboratory scientific research but also for optimization and in-line quality check in a PV factory. Macro-Raman mapping is possible through two add-ons that can be integrated in any existing micro-Raman spectrometer. The two new DuoScan™ (hardware) and SWIFT™ (software) Raman scanning modules developed by HORIBA Jobin Yvon provide significant reduction by orders of magnitude of the measurement times by means of large area probing beam (macro-beam) and high speed detector-stage coordination, respectively. Even faster Raman imaging is possible by combing these two technologies [7].

DuoScan™ Raman imaging technology extends the imaging capabilities of micro-Raman instruments from (sub-) micron to macro-scale mapping. The integration of the DuoScan unit to an existing micro-Raman spectrometer is shown in Figure 3(left). This mode is based on a combination of two orthogonally rotating piezo-mirrors that scan the laser beam across the sample following a user-defined pattern as displayed in Figure 3(middle). The size of the resulting macro-beam is adjustable being limited only by the opening of the used microscope objective. The maximum macro-beam sizes achievable with our 50x (NA 0.80) or 10x (NA 0.30) NIKON microscope objectives are 100 x 100 μm² or 1 x 1 mm². DuoScan allows to integrate the Raman signal over the macro-beam area giving an average spectrum, which contains the same spectral information as that obtained by averaging all micro-Raman spectra for the same area. The gain in acquisition time is evident, macro-Raman being orders of magnitude faster than conventional micro-Raman. For example, if an area of 30 x 30 μm² is entirely probed by macro-Raman in one second, micro-Raman with a spot-size of 1 x 1 μm² needs 900 seconds to cover the same area. The price one has to pay is the loss of lateral resolution.

Furthermore, DuoScan can be used in a step-by-step mode where the mapping takes place without moving the stage with the sample. A minimum step size of 50 nm is reached by deflecting the laser beam, which complements successfully the stepping capability of the stage specified to be ~ 500 nm. This mode applies for Raman imaging of nanoscale objects and features. The DuoScan mapping capabilities are summarized in Figure 3(right).

SWIFT™ Raman imaging technology enables ultra fast mapping without losing lateral resolution and thus image quality. In this mode, the time intervals needed for the stage to accelerate/decelerate as well as for the shutter in front of the detector to open/close for each measurement point are eliminated. Basically, these are dead times, which are not used for the acquisition of the Raman signal. The breakthrough consists in continuously moving the stage with the sample while keeping the shutter open and measuring continuously Raman spectra by means of high speed detector-stage coordination coupled with the high optical throughput of the Raman system. The SWIFT option can also be used for time resolved Raman imaging provided the investigated processes occur on the measurement time scale.

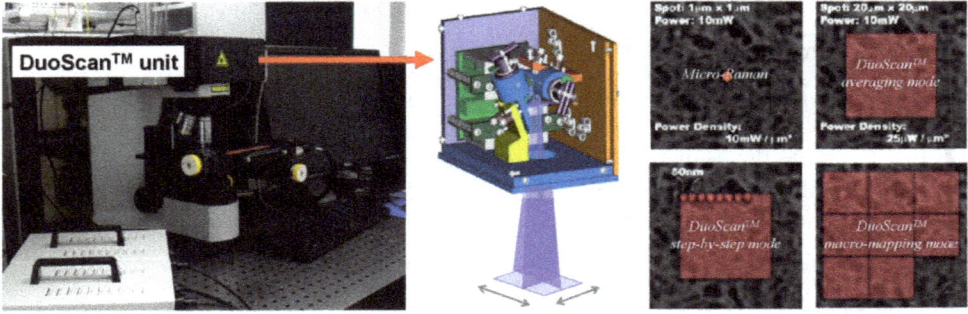

Figure 3. (left) DuoScan unit attached to a micro-Raman spectrometer. (middle) Schematic drawing illustrating the DuoScan working principle. The probing micro-beam is scanned by two orthogonally rotating piezo-mirrors resulting in a macro-beam, thus giving rise to the term macro-Raman spectroscopy. (right) Comparison between standard and DuoScan mapping modes described in the text. The left and middle pictures are taken from HORIBA's official webpage.

3.4. Complementary techniques to Raman spectroscopy for solar silicon studies

The Raman investigations on semiconductor materials represent an important step towards their fundamental understanding and the control of their properties for designing devices with specific functions. As already mentioned, Raman spectroscopy can provide detailed information regarding the spatial distributions of internal stress, defect density, doping, and grain orientation. We will show that even more insight into the interaction between different properties of silicon PV materials can be achieved when Raman measurements are supported and complemented at identical positions by other techniques such as EBSD, EBIC and defect etching.

EBSD measurements are performed using an EDAX system attached to a TESCAN LYRA XMU scanning electron microscope (SEM) to determine the grain orientations and grain boundary types. The crystal orientation is given in the {hkl}<uvw> representation where {hkl} is the crystal plane perpendicular to the sample normal direction (z axis) and <uvw> is the crystal direction aligned with the transverse direction of the sample (y axis). The inverse pole figure (unit triangle) shows the sample normal direction relative to the axes of the measured crystal. The misorientation between adjacent grains is given in the angle/axis notation, that is, the rotation angle about the axis common to both lattices to bring them into coincidence, and in terms of Σ-value which denotes the fraction of atoms in the GB plane coincident in both lattices [25].

EBIC measurements are done with an EVO 40 SEM at 20 keV beam energy both at room temperature and 80K to image most of the electrically active defects. In order to render the inhomogeneities of recombination clearly visible, a color scale is used for the EBIC maps. The maps represent the local EBIC signal normalized by the maximum EBIC signal. The lower the EBIC signal, the higher the recombination activity.

4. Results and discussions

4.1. Micro-Raman measurements

4.1.1. Silicon thin films on glass for solar cells

The unique characterization power of the Raman technique consists in the detailed mechanical and microstructural information that can be extracted from the silicon Raman peak: (1) the peak position map - the distribution, amount and sign of internal stresses, (2) the peak full-with at half maximum (FWHM) map - the distribution and qualitative comparison of defect densities, (3) the peak asymmetry map – the distribution and amount of doping, and (4) the peak intensity map – the grain orientation and the grain boundary pattern [7,10,11].

One representative example is shown in Figure 4 for the laser crystallized silicon seed layers of thin film solar cells on glass (in this example, 110 nm thick silicon seed film). The larger FWHM values in Figure 4(a) indicating a broadening of the Raman spectra are produced by a decrease in the phonon lifetimes, which in turn is mainly due to defects acting as anharmonic perturbations (see Section 2.3). Indeed, the dashed line contours in Figure 4(a) corresponds to low angle GBs indicated by arrows in the EBSD maps shown in Figure 4(e and f). It is well known that low angle GBs consist of dislocation networks/arrays. Their presence at these positions is further supported by the continuously changing crystallographic orientation within the studied grain as indicated by the gradual changing color in the intergranular misorientation gradient map by EBSD in Figure 4(f). Such intergranular misorientation is attributed to geometrically necessary dislocations forming low angle GBs [26].

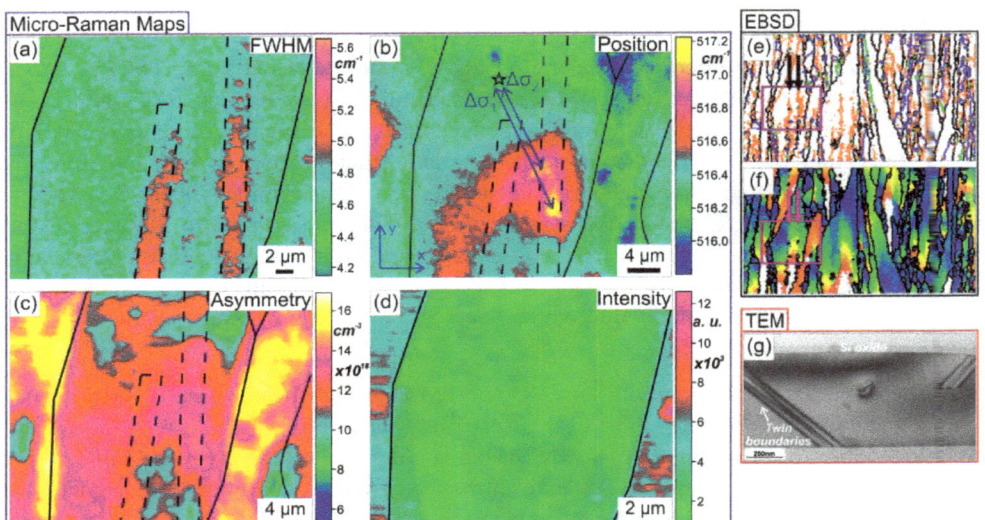

Figure 4. Micro-Raman maps of a laser crystallized silicon seed layer of a thin film solar cell on glass (110 nm thick, nominal boron doping of 2.1×10^{19} cm^{-3}) obtained from the fitting of the first-order Raman spectra of silicon: (a) peak FWHM – defect density map, (b) peak position – internal stress map with two lateral stress gradients $\Delta\sigma_1=227$, $\Delta\sigma_2=197 \pm 12$ MPa (von Mises stresses), (c) peak asymmetry – doping map, (d) peak intensity – grain orientation map. EBSD maps: (e) grain boundary map including high angle GBs (black lines), low angle GBs (orange lines), $\Sigma3$ GBs (blue lines), $\Sigma9$ GBs (green lines), (f) Intergranular misorientation gradient map. The two vertical arrows indicate low angle GBs corresponding to the areas delineating by dashed lines in (a, b and c). (g) TEM cross-section image.

Dislocations are considered to be among the most detrimental type of defect controlling not only the mechanical but also the electrical properties of silicon and other materials. They are produced by the partial or total relaxation of thermally induced stresses during the crystallization and cooling processes as long as plastic deformation is allowed by temperature, this means above the brittle-ductile transition temperature of silicon. Below this temperature, the remaining thermally induced stresses are incorporated as thermally induced residual stresses in the silicon material. Once created, dislocations can move on glide planes leading to further plastic deformation through their multiplication until the lattice friction becomes larger than the effective stress needed for moving the existing dislocations [6].

By comparing the FWHM (defect density) and position (stress) maps displayed in Figure 4(a and b), it can be seen that the two patterns are unlike. Regions of similar FWHM values (similar defect densities) along the two lines of dislocations exhibit different Raman peak positions (different stress levels) showing virtually no influence of internal stresses on peak broadening. The fact that the two dislocation lines are only partly accompanied by stress can be explained by the locally different superposition of stress fields of dislocations and thermally induced residual stresses. Their interaction can lead to local configurations in which internal stresses get cancelled totally, partially or not at all [6,23,27]. Thus, we attribute the stress concentrations in Figure 4(b) to particular combinations of (1) defect configurations/structures, which do not necessary result in higher FWHM values and (2) thermally induced residual stresses. The stress map is corrected for the compressive contributions produced by the Fano effect and by the addition of boron by means of the lattice parameter using Vergard's law [7]. The presence of different types of defects and their nonuniform distribution inside the laser crystallized seed layer are supported by transmission electron microscopy (TEM) investigations shown in Figure 4(g).

Two lateral stress gradients inside the central grain in Figure 4(b) are evaluated in form of stress-tensors following the polarized micro-Raman procedure described in [10]. First, the crystallographic grain orientation is determined using the Raman intensity dependences on the polarization direction of the incident light (θ) and Raman backscattered light (x or y analyzer positions) as explained in Section 2.1 at the point marked by the star within this particular grain. The measured plots are displayed in Figure 5.

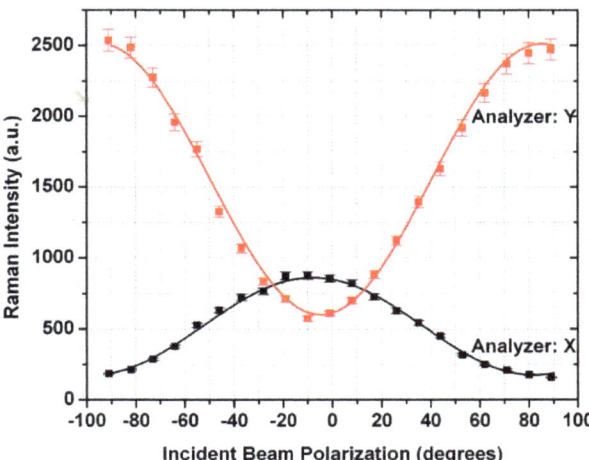

Figure 5. Raman intensity dependences on the polarization direction of the incident light for the X and Y analyzer positions measured at the point marked by a star in Figure 4(b). The error bars account for the ~ 3% intensity variations of the incident laser light. The continuous curves represent fit functions based on Equation 4 used to obtain the three Euler angles: $\alpha = 51° \pm 2°$, $\beta = 27° \pm 2°$, and $\gamma = -2° \pm 2°$ necessary to determine the grain orientation and the stress tensor components.

Their fitting by the Equation 4 gives the following three Euler angles: $\alpha = 51° \pm 2°$, $\beta = 27° \pm 2°$, and $\gamma = -2° \pm 2°$, which in turn provide the following rotation matrix to bring this arbitrary oriented grain in the stage (reference) coordinate system:

$$T(\alpha, \beta, \gamma) = \begin{pmatrix} 0.656 & -0.664 & 0.356 \\ 0.753 & 0.587 & -0.293 \\ -0.014 & 0.461 & 0.887 \end{pmatrix}.$$

Second, since the internal stresses are too small to produce a visible lifting of degeneracy of the three silicon optical phonon frequencies, the polarization settings for the incident and backscattered light for which the intensity of one of the three phonon modes dominates the other two are simulated using the previously determined Euler angles and Equation 5. The simulations of the six intensity ratio functions $W^{XY}_j(\theta)$ are shown in Figure 6. It can be seen that the polarization settings to measure separately the three phonon frequency shits $\Delta\omega_j$ for the given grain are: Phonon 1: 1°, X, Phonon 2: -22°, Y, and Phonon 3: 15°, Y.

Third, three Raman maps of the area in Figure 4(b) are measured for the three polarization settings above. Difference stress-tensors are calculated numerically from the three Raman frequency shifts $\Delta\omega_j$ with respect to the stage (reference) coordinate system as described in Section 2.2:

$$\Delta\sigma_1 = \begin{pmatrix} -184 \pm 10 & -33 \pm 1 & 0 \\ -33 \pm 1 & -192 \pm 10 & 0 \\ 0 & 0 & -151 \pm 10 \end{pmatrix} MPa \, ,$$

$$\Delta\sigma_2 = \begin{pmatrix} -200 \pm 10 & -51 \pm 1 & 0 \\ -51 \pm 1 & -213 \pm 10 & 0 \\ 0 & 0 & -152 \pm 10 \end{pmatrix} MPa.$$

The two lateral stress gradients indicate compressive stresses at these positions with respect to the point marked by the star in Figure 4(b), while the shift towards lower frequencies (<~ 520 cm^{-1}) in the position of the Raman spectra implies tensile stress inside the silicon thin film. Their conversion into average or von Misses stresses using Equation 9 gives $\Delta\sigma_1$=227, $\Delta\sigma_2$=197 $\pm$ 12.5 MPa.

Figure 4(c) shows the asymmetry (doping) map obtained from the symmetry parameter q of the Raman spectra as defined in Section 2.3. For the quantitative doping evaluation, the free carrier concentration vs. q calibration curve in Figure 5 of Reference 11 was used. The free hole concentrations are found to be lower than the nominal boron doping of 2.1x10^{19} presumably due to the incomplete activation of dopants during laser crystallization and cooling [11]. Higher doping is observed both along GBs and inside grains. Regarding the influence of doping/impurities on the FWHM as discussed in Section 2.3, there is no correlation between them as seen by comparing the FWHM map with the asymmetry map displayed in Figure 4(a) and (c), respectively.

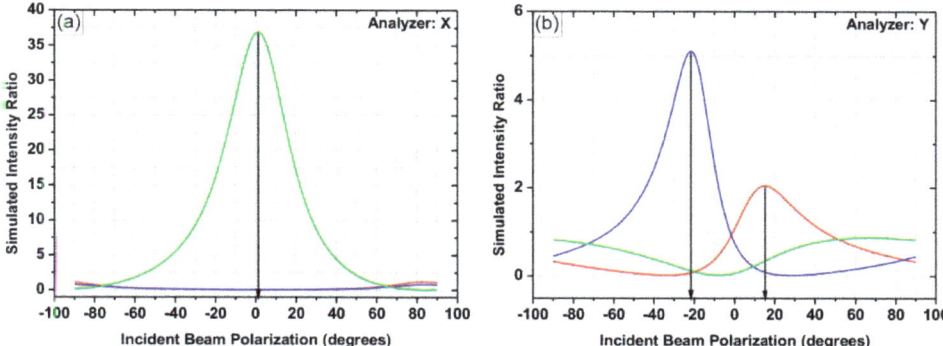

Figure 6. Simulation of the six intensity ratio functions $W^{XY}_i(\theta)$ for the central grain in Figure 4(b). The plot maxima marked by arrows indicate the polarization direction of the incident laser light for the two analyzer positions where the intensity of one of the three silicon phonons dominates over the sum of the other two phonons.

The different Raman scattering efficiencies caused by distinct crystallographic grain orientations and the polarization directions of the incident and backscattered laser light can be used to image the grains and to determine their orientations as shown in Section 2.1. This results in intensity maps such as displayed in Figure 4(d), which enable tracing of GBs represented as solid lines in all Raman maps of Figure 4. Thus, it is possible to relate grains and GBs to defect, stress and doping distributions, all data being provided by the same Raman mapping.

4.1.2. Wafer and ribbon-based silicon solar cells

Next examples illustrate the application of micro-Raman spectroscopy to block-cast and edge-defined film-feed (EFG) multicrystalline silicon materials, two industrial relevant materials with the former having the largest share (> 50%) in the PV market. Internal stresses are the result of the superposition between the thermally induced residual stresses that is the thermally induced stresses at the end of crystallization and cooling processes and the defect-related stresses. By cutting the silicon blocks and EFG tubes into wafers and then into small pieces for micrometer scale investigations, the thermally induced residual stresses are expected to relax to a large extent due to the creation of free surfaces. This is different in the case of silicon thin films on glass, which are measures as-prepared without any cutting. Thus, in the block-cast and EFG samples, the internal stresses produced mainly by defects are measured.

The resolution of micro-Raman can go down to single dislocation characterization as demonstrated in Figure 7(a) in the case of p-type block-cast mc-Si material taken from a PV factory production line. Here, the localized and quite symmetric stress distribution including both compressive (red area) and tensile (blue area) and its decay length resemble the stress field of an edge dislocation, which, in this case, is superimposed on the Σ27a GB [23,27]. The polarized micro-Raman stress measurements give the following difference stress tensors referring to the stage (reference) coordinate system shown in Figure 7(a). They have been evaluated between stressed positions close to the GB and positions at a distance from the GB, which are not affected by the dislocation stress field:

$$\Delta\sigma_1 = \begin{pmatrix} -40 \pm 10 & -14 \pm 3 & 0 \\ -14 \pm 3 & -38 \pm 10 & 0 \\ 0 & 0 & -25 \pm 10 \end{pmatrix} MPa,$$

$$\Delta\sigma_2 = \begin{pmatrix} 33 \pm 10 & -7 \pm 1 & 0 \\ -7 \pm 1 & 31 \pm 10 & 0 \\ 0 & 0 & 34 \pm 10 \end{pmatrix} MPa.$$

As shown in the previous example, it is worth to combine at the same position micro-Raman with other techniques not only to support the interpretation of the Raman results but also to get new insights into other material properties and their interplay. The EBIC images in Figure 7(b and c) show lower signal corresponding to reduced minority carrier lifetime of 79% at 300K and 63% at 80K in the region of the GB trajectory change where the edge dislocation is located as well as a signal variation along the Σ27a GB. By comparing the stress and EBIC images, it can be seen that the stressed area close to the change in the Σ27a GB trajectory denoted K and the stress-free area above and below it show similar EBIC signals, and thus similar recombination activities.

Figure 7(e) shows a Raman stress map of the same Σ27a GB at a distance of several millimeters from the position displayed in Figure 7(a). The compressive (red area) and tensile (blue area) stresses are more extended along the GB, less symmetric, and change positions with respect to the GB as compared with the stress map in Figure 7(a). This stress distribution is attributed to the stress field of an array of edge dislocations superimposing the GB. The band-like less compressed region on the right-hand side of the Σ27a GB can be explained by the presence of dislocations (edge, screw and/or mixed), in the grain and close to the GB, which have locally rearranged during crystal growth and cooling to reduce the strain energy and thus, the stresses in this region [23,27]. The following stress-tensor gradients referring to the stage (reference) coordinate system shown in Figure 7(e) have been determined by polarized micro-Raman:

$$\Delta\sigma_3 = \begin{pmatrix} 29 \pm 10 & -7 \pm 1 & 0 \\ -7 \pm 1 & 28 \pm 10 & 0 \\ 0 & 0 & 36 \pm 10 \end{pmatrix} MPa,$$

$$\Delta\sigma_4 = \begin{pmatrix} -34 \pm 10 & -3 \pm 1 & 0 \\ -3 \pm 1 & -37 \pm 10 & 0 \\ 0 & 0 & -37 \pm 10 \end{pmatrix} MPa.$$

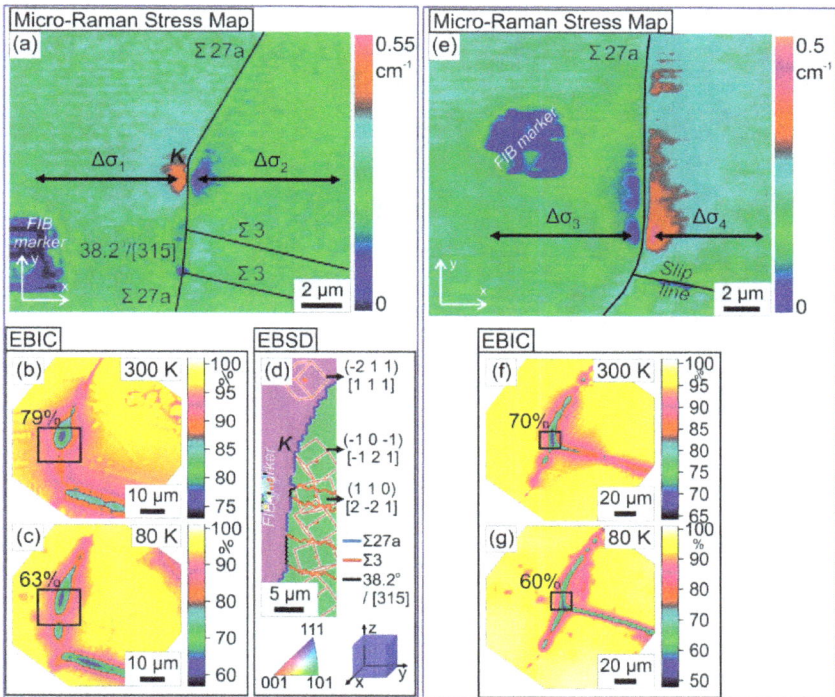

Figure 7. Micro-Raman, EBIC and EBSD studies of block-cast solar silicon at two positions along the same Σ27a GB. The Raman stress distributions are attributed to a single edge dislocation (a) and to an array of edge dislocations (e) superimposing the GB. The regions enclosed by rectangles in the EBIC images (b, c) and (f, g) correspond to the Raman mapped areas in (a) and (e), where the numbers indicate the maximum EBIC signal. The lower the EBIC signal, the higher the recombination activity. The focused ion beam (FIB) markers in (a, e) allows exact spatial correlation between different measurement techniques. (d) EBSD map showing the grain orientations and GB types along with the orientation triangle and the sample reference frame.

Like in the previous case the stressed and stress-free areas around the GB in Figure 7(e) are located in a region of similar (lower) EBIC signal of 70% at 300K and 60% at 80K as indicated in Figure 7(f) and (g), respectively and the recombination activity is inhomogeneous along the Σ27a GB. These two representative examples demonstrate the presence of spatial variations in mechanical and electrical properties of block-cast solar silicon on the micrometer scale.

Similar spatial properties variations are observed in the p-type EFG mc-Si material taken also from a PV factory production line. In order to illustrate the correlation between internal stresses, defect structure and electrical activity in the EFG material, we show here three positions along the same GB that contain representative examples of this correlation. Here, the internal stresses are evaluated using Equation 11 without employing the polarized micro-Raman procedure as in the case of silicon thin films on glass and block-cast mc-Si.

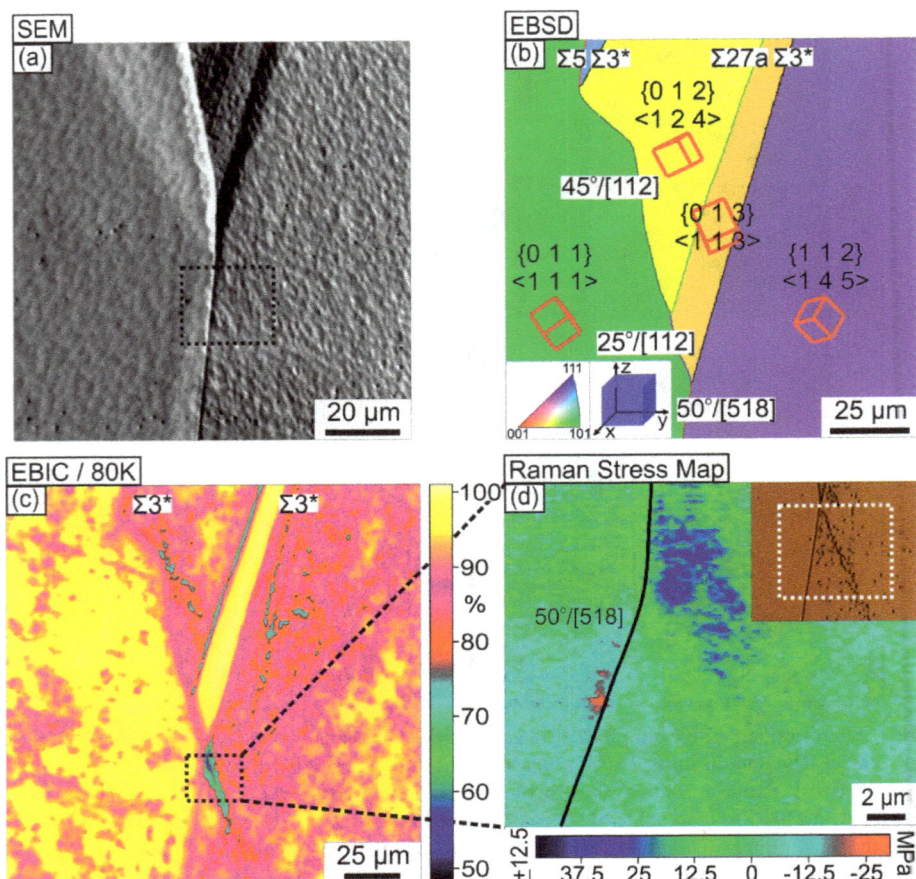

Figure 8. Position 1 (a) SEM image of the as-grown EFG wafer before mechanical polishing. (b) EBSD map showing the grain orientations and GB types along with the orientation triangle and the sample reference frame. (c) EBIC image taken at 80K where the inhomogeneous recombination activity inside grains and at GBs is mainly attributed to dislocations decorated with metallic impurities. (d) Not all dislocations visible in the defect etching image shown in the inset or measured by EBIC are accompanied by internal stresses as probed by micro-Raman. The dashed rectangle in the inset represents the Raman mapped area. At this position, the lowest EBIC current corresponds to the largest (tensile) stress.

The EBSD, EBIC, and micro-Raman measurements at the first position are displayed in Figure 8. The Raman stress map shows concentrated tensile (in blue) and compressive (in red) stresses close to a large-angle random GB described by a misorientation angle/axis of 50°/[518]. Except these areas, nearly no stresses are found neither along the GB nor inside the two adjacent grains of {011}<111> and {112}<145> orientations. By comparing the stress map with the corresponding EBIC map enclosed by the rectangle in Figure 8(c), it can be seen that not all recombination active dislocations visible at 80K are accompanied by stresses.

That is because dislocations interact with each other and tend to locally rearrange in configurations of minimum strain energy that can result in stresses or virtually no stresses. The EBIC image in Figure 8(c) shows an inhomogeneous electrical activity along different types of GBs as well as inside grains of different crystallographic orientations indicated in Figure 8(b). The recombination-active Σ3 GBs {60°/[111]} in Figures 8-10 are marked with an asterisk to distinguished them from the recombination-free Σ3 GBs in Figure 9.

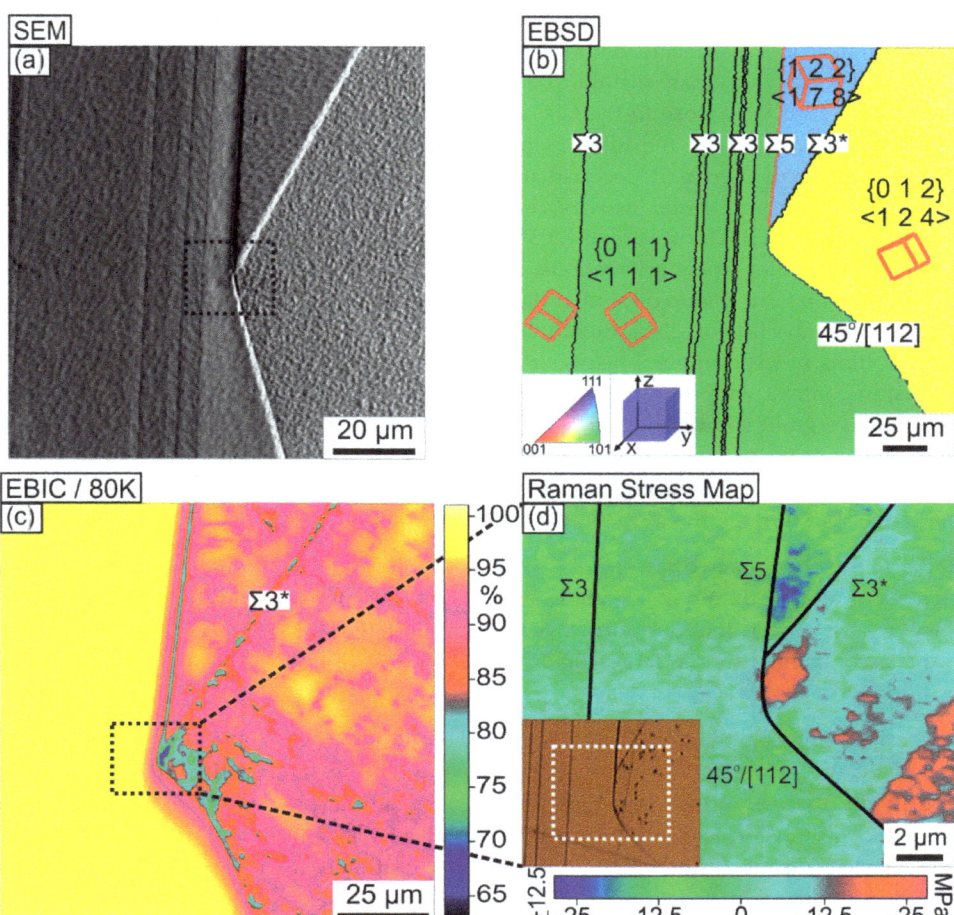

Figure 9. Position 2 (a) SEM image of the as-grown EFG wafer before mechanical polishing. (b) EBSD map. (c) EBIC image where the same left-hand side grain like in Figure 8 shows at this position no electrical activity. The Σ3 GBs are either recombination-free (Σ3) or recombination-active (Σ3*), while being virtually stress-free. (d) The defect etching image in the inset indicates that the presence of dislocation etch pits on Σ3* GBs leads to electrical activity provided the dislocations are decorated with metallic impurities. Here, the highest recombination activity corresponds to the largest (compressive) stress.

The EBSD, EBIC, and micro-Raman results obtained at the second position are shown in Figure 9. Like in the previous case, we did not find a one-to-one correspondence between electrically active dislocations and stresses, both exhibiting inhomogeneous spatial and magnitude distributions. These findings are similar to those on block-cast mc-Si displayed in Figure 7. Different at this position is the presence of tensile (in blue) and compressive (in red) stresses concentrated close to a GB triple point where a Σ5 GB {36.86°/[100]}, a Σ3 GB, and a large-angle random GB {45°/[112]} meet. It is worth noting that GBs of the same type, here Σ3 GBs, can be either recombination-free (Σ3) or recombination active (Σ3*), while being both nearly stress-free. Essentially, independent of the GB type, such large differences in electrical activity originate mainly from the absence or presence of recombination-active dislocations on or very close to the GB. This point is confirmed by comparing the defect-etched optical image with the EBIC map: the Σ3 GBs decorated by dislocation etch pits (denoted Σ3*) show increased electrical activity, while the Σ3 GBs without dislocation etch pits show no recombination activity. It can be seen that despite the same GB type assignment by the EBSD software, the Σ3* and Σ3 GBs are formed between adjacent grains of different crystallographic orientations. This fact suggests distinct kinematic conditions at these Σ3 GBs that can lead to dissimilar thermally induced stress levels and as a result to the generation or absence of dislocations. On the other hand, Raman measures only those configurations of dislocations (including the recombination-free dislocations) that lead to stresses. In contrast with the first (Figure 8) and third (Figure 10) positions, the {011}<111> left-hand side grain shows no reduction of the EBIC signal at the second position despite quite similar grain geometries at these three positions. This indicates that the thermally induced stresses present during the EFG growth relaxed not through the generation of dislocations but through the formation of twins found at the second position by EBSD.

Similar to the previous two cases, we observe non-uniform distributions of electrical activity and stresses along GBs and inside grains at the third position as displayed in Figure 10(c, d). However, we choose this position to show that the largest recombination activity is not always accompanied by the largest internal stresses as in the case of the first and second positions.

The local variations in the sign and values of the dislocation-related stresses as well as in the strength of the recombination activity are attributed to the cumulative effect of metallic impurity decoration, intrinsic structure, type, density, and distribution of dislocations inside grains and on GBs. This non-uniform distribution of dislocations originates from locally different mechanisms of nucleation, motion, multiplication and annihilation of dislocations controlled by the grain structure including the orientation, size and geometry of the grains, the kinematic constraints at GBs, and temperature. The presence of impurities is confirmed by EBIC measurements which show quite strong reduction of the EBIC signal at room temperature up to 70-80% which further reduces with decreasing temperature up to 50-65% at 80K. Such EBIC behavior corresponding to increasing recombination can be explained by the interaction of shallow levels related to the strain fields of dislocations with deep levels due to metallic impurity decoration and/or intrinsic core defects at dislocations. It is known that impurity accumulation in silicon can be enhanced due to the presence of stresses

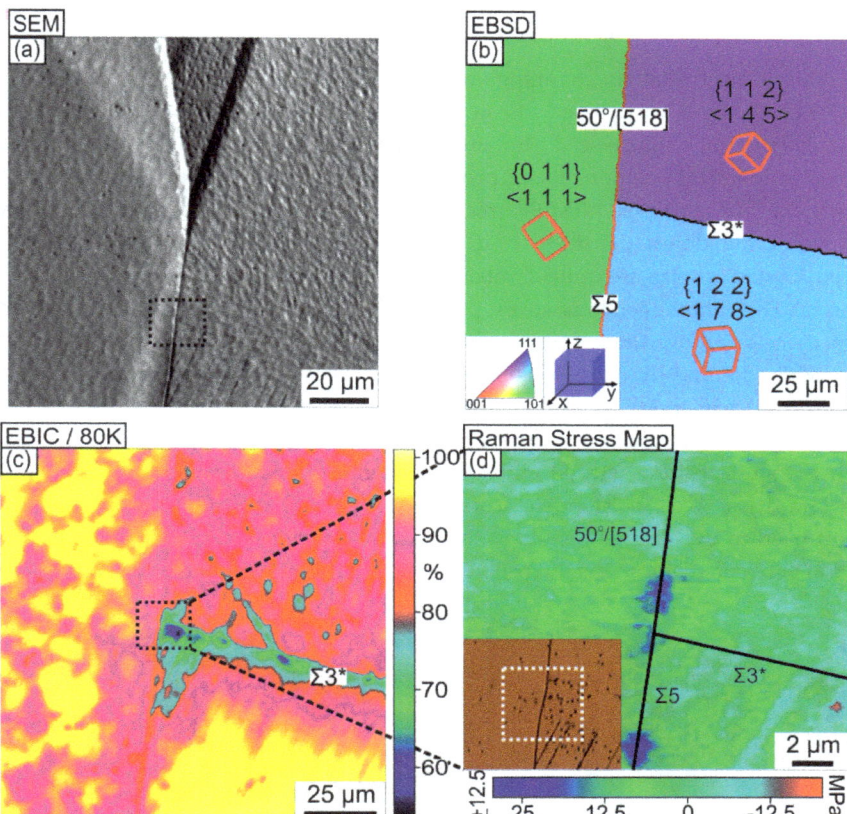

Figure 10. Position 3 (a) SEM image of the as-grown EFG wafer before mechanical polishing. (b) EBSD map. (c) EBIC image where the same left-hand side grain like in Figure 8 and 9 shows recombination activity. (d) The lowest EBIC current is not accompanied by stress at this position.

(thermally induced residual stresses and/or defect-related stresses) at temperatures where both impurities and dislocations are mobile. This can explain the increased electrical activity at regions of higher dislocation densities as at the first and second positions where dislocations are spatially distributed in such a way that their stress fields cancel partially or not at all so that an overall long-range stress field from these dislocations is measured by micro-Raman. The dislocations arranged in configurations in which their stress fields cancel totally (or below the detection limit of our Raman spectrometer of ± 12.5 MPa) are only visible by EBIC (when recombination-active) but not by micro-Raman, as at the third position. Point-by-point correlation of the micro-Raman and EBIC measurements indicates that internal stresses of several tens of MPa do not influence the minority carrier recombination in block-cast and EFG mc-Si. Comparably high stresses of up to 1.2 GPa are necessary in silicon in order to influence its electrical properties such as enhanced carrier mobility in the transistor channel through band structure modification and effective mass reduction [6,23].

4.2. Macro-Raman measurements

Representative macro-Raman mappings acquired using the DuoScan option described in Section 3.3 on the laser crystallized silicon seed layers of thin film solar cells on glass (in this example, 290 nm thick silicon seed film) are displayed in Figure 11. These measurements are performed at identical positions using probing macro-beams of 30 x 30 μm² and 100 x 100 μm² with the 50x and 10x NIKON microscope objectives. The distribution of internal stresses (a, c) and defect densities (b, d) obtained from the position and FWHM of the measured Raman spectra are quite similar when measuring with different DuoScan macro-beam sizes. The inhomogeneous stress patterns in (a, c) are the result of the interaction between defects through their own intrinsic stress fields and thermally induced residual stresses, while the line shape regions in (b, d) correlate with the laser traces where higher defect densities corresponding to larger FWHM values develop predominantly at adjacent laser scan lines where irradiated areas overlap. It can be seen that there is no correlation between the shift/position (stress) and FWHM (defect density) maps both at macro-scale (Figure 11(a-d)) as well as at micro-scale (Figure 4(a and b)). This further supports the argument used to explain the results in the previous sections, namely the locally different interaction between dislocations themselves and with thermally induced residual stresses.

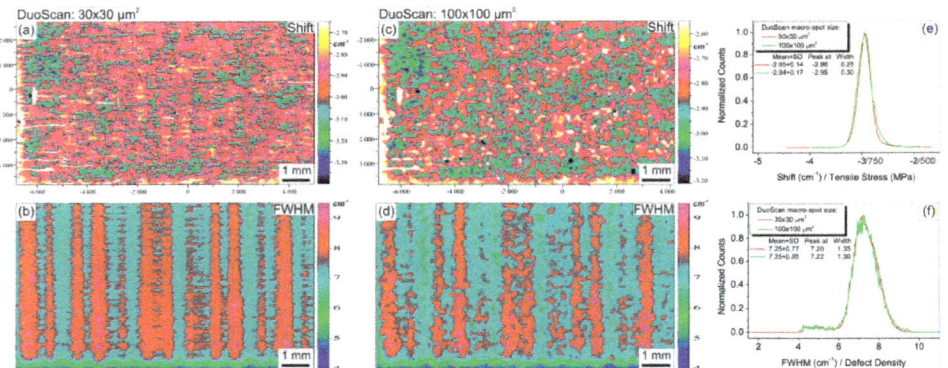

Figure 11. DuoScan Raman maps of the same area using probing macro-beams of 30 x 30 μm² (a, b) and 100 x 100 μm² (c, d), where the sharpness of the features decreases due the loss of lateral resolution. (a, c) The Raman peak position shifts with respect to a stress-free silicon reference are negative indicating the presence of tensile stresses inside the 290 nm thick laser crystallized silicon thin film on glass. (b, d) The FWHM maps show areas of different crystal quality related to different defect densities, which correlate with the laser traces as visible from the line shape character of the FWHM distributions. (e, f) Statistical evaluation using histograms for the two macro-beam sizes demonstrating that DuoScan can be used for large scale mappings without losing average information.

As expected, the sharpness of the features decreases with increasing the size of the probing macro-beam due to the loss of lateral resolution. However, the spectral information is not altered by integrating the Raman signal over the 30 x 30 μm² or 100 x 100 μm² macro-beam areas. Indeed, the normalized histograms in Figure 11(e, f) calculated from the shift and

FWHM DuoScan maps show a good overlap demonstrating that DuoScan can be used for large-scale mapping of PV and other Raman active materials without losing average spectral information. The presence of thermally induced residual stresses is reflected in the position of the shift/stress histogram in Figure 11(e). Smaller thermally induced residual stresses corresponding to shift/stress histograms closer to or at zero prevent cracking or peeling off the silicon thin film solar cells or even substrate bending minimizing the breakage risk and processing/handling difficulties. When small, they also impede under external mechanical and thermal loads the occurrence of new stress-induced defects, which are commonly recombination active. The qualitative estimation of defects is apparent in the position of the FWHM/defect density histogram in Figure 11(f) where FWHM values closer to ~ 3 cm^{-1} corresponding to defect-free silicon indicate lower defect densities in the silicon thin film solar cells. Thus, macro-Raman can be used to evaluate statistically the materials properties and to see clearly the changes originating from different preparation conditions and processing.

5. Conclusions

The characterization power of the Raman technique at micro- and macro-scale in the case of multicrystalline solar silicon materials is demonstrated. Raman investigations at length scales ranging from μm^2 to cm^2 are possible through two new developed scanning modules, DuoScan™ (hardware) and SWIFT™ (software), which can be integrated in any standard micro-Raman spectrometer. The statistical evaluation of the large area Raman maps measured by macro-Raman spectroscopy shows that macro-scale Raman mapping integrates data over the macro-beam area giving an average spectrum that contains the full spectral information at the cost of decreasing lateral resolution. Moreover, Macro-Raman enables significant reduction by orders of magnitude of the acquisition time: if an area of 30 x 30 μm^2 is entirely probed by macro-Raman in one second, micro-Raman with a spot-size of 1 x 1 μm^2 needs 900 seconds to cover the same area. Deeper insights into the interplay between internal stresses, defects, doping, microstructure, and recombination activity with practical impact on the mechanical stability and conversion efficiency of solar cells have been obtained by combining Raman, EBSD, EBIC, TEM, and defect etching techniques. By tuning the crystallization process, the interaction between dislocations driven by the strain energy minimization can be used to reduce internal stresses resulting in mechanically stronger wafers and cells and to prevent metallic impurity precipitation at dislocations that should lead to improved energy conversion efficiencies.

Author details

George Sarau and Silke Christiansen
Max Planck Institute for the Science of Light, Erlangen, Germany
Institute of Photonic Technology, Jena, Germany

Arne Bochmann
Institute of Photonic Technology, Jena, Germany

Renata Lewandowska
Horiba Scientific, Villeneuve d'Ascq, France

Acknowledgement

This work was financially supported by (1) the German Federal Ministry for the Environment, Nature Conservation and Nuclear Safety and all the industry partners within the research cluster "SolarFocus", (2) the Max-Planck Society within the project "Nanostress", and (3) the European Commission within the FP7-Energy priority project "High-EF". Within the "High-EF" project, Horiba Scientific developed the DuoScan hardware and the SWIFT software that permit fast, large area Raman analyses. The first demonstrator was deployed to the Max Planck Institute for the Science of Light for further development of this technique with respect to applications in photovoltaics. The authors are thankful for this great research opportunity. We would also like to thank A. Gawlik of IPHT for preparing the laser crystallized samples used in this study as well as to M. Holla and W. Seifert of Joint Lab IHP/BTU for their help with the EBIC measurements and fruitful discussions. The content of this publication is the responsibility of the authors.

6. References

[1] Raman C.V, Krishnan K.S (1928) A New Type of Secondary Radiation. Nature 121: 501-502.

[2] Ferraro J.R, Nakamoto K, Brown C.W (2003) Introductory Raman Spectroscopy. Academic Press.

[3] Smith E, Dent G (2005) Modern Raman Spectroscopy - A Practical Approach. John Wiley & Sons.

[4] Pelletier M.J (Ed.) (1999) Analytical Applications of Raman Spectroscopy. Blackwell Publishing.

[5] Wolf I (1996) Micro-Raman Spectroscopy to Study Local Mechanical Stress in Silicon Integrated Circuits. Semicond. Sci. Technol. 11: 139-154.

[6] Sarau G, Christiansen S, Holla M, Seifert W (2011) Correlating Internal Stresses, Electrical Activity and Defect Structure on the Micrometer Scale in EFG Silicon Ribbons. Sol. Energy Mater. Sol. Cells 95: 2264–2271.

[7] Sarau G, Christiansen S, Lewandowska R, Roussel B (2010) Future of Raman in PV Development. Proc. 35th IEEE PVSC: 001770-001775.

[8] Anastassakis E, Pinczuk A, Burstein E, Pollak F.H, Cardona M (1970) Effect of Static Uniaxial Stress on the Raman Spectrum of Silicon. Solid State Commun. 8: 133-138.

[9] Wolf I, Maes H.E, Jones S.K (1996) Stress measurements in silicon devices through Raman spectroscopy: Bridging the gap between theory and experiment. J. Appl. Phys. 79: 7148-7156.

[10] Becker M, Scheel H, Christiansen S, Strunk H.P (2007) Grain orientation, texture, and internal stress optically evaluated by micro-Raman spectroscopy. J. Appl. Phys. 101: 063531 (1-10).

[11] Becker M, Gösele U, Hofmann A, Christiansen S (2009) Highly p-doped Regions in Silicon Solar Cells Quantitatively Analyzed by Small Angle Beveling and micro-Raman Spectroscopy. J. Appl. Phys. 106: 074515 (1-9).

[12] Anastassakis E (1999) Strain Characterization of Polycrystalline Diamond and Silicon Systems. J. Appl. Phys. 86: 249-258.

[13] Puech P, Pinel S, Jasinevicius R.G, Pizani P.S (2000) Mapping the three-dimensional strain field around a microindentation on silicon using polishing and Raman spectroscopy. J. Appl. Phys. 88: 4582-4585.

[14] Hanbücken M, Müller P, Wehrspohn R.B (Eds.) (2011) Mechanical Stress on the Nanoscale: Simulation, Material Systems and Characterization Techniques. Wiley-VCH.

[15] Fano U (1961) Effects of Configuration Interaction on Intensities and Phase Shifts. Phys. Rev. 124: 1866- 1878.

[16] Nickel N.H, Lengsfeld P, Sieber I (2000) Raman Spectroscopy of Heavily Doped Polycrystalline Silicon Thin Film. Phys. Rev. B 61: 15558-15561.

[17] Lengsfeld P, Brehme S, Brendel K, Genzel Ch, Nickel N.H (2003) Raman Spectroscopy of Heavily Doped Polycrystalline and Microcrystalline Silicon. phys. stat. sol. (b) 235: 170–178.

[18] Weber W.H, Merlin R (Eds.) (2000) Raman Scattering in Materials Science. Springer. pp. 56-64.

[19] Debernardi A, Baroni S, Molinari E (1995) Anharmonic Phonon Lifetimes in Semiconductors from Density-Functional Perturbation Theory, Phys. Rev. Lett. 75: 1819–1822.

[20] Debernardi A (1998) Phonon linewidth in III-V semiconductors from density-functional perturbation theory, Phys. Rev. B 57: 12847.

[21] Iqbal Z, Veprek S, Webb A.P, Capezzuto P (1981) Raman Scattering from Small Particle Size Polycrystalline Silicon, Solid State Commun. 37: 993-996.

[22] Secco D' Aragona F (1972) Dislocation Etch for (100) Planes in Silicon, J. Electrochem. Soc. 119: 948-951.

[23] Sarau G, Becker M, Christiansen S, Holla M, Seifert W (2009) Micro-Raman Mapping of Residual Stresses at Grain Boundaries in Multicrystalline Block-Cast Silicon Solar Cell Material: Their Relation to the Grain Boundary Microstructure and Recombination Activity, Proc. 24th European Photovoltaic Solar Energy Conference: 969-973.

[24] Sarau G, Becker M, Berger A, Schneider J, Christiansen S (2007) Stress Distribution in Polycrystalline Silicon Thin Film Solar Cells on Glass Measured by Micro-Raman Spectroscopy, Mater. Res. Soc. Symp. Proc. 1024E: 1024-A07-04.

[25] Grimmer H, Bollmann W, Warrington D.H (1974) Coincidence-Site Lattices and Complete Pattern-Shift in Cubic Crystals, Acta Cryst. A30: 197-207.

[26] Niederberger Ch, Michler J, Jacot A (2008) Origin of Intragranular Crystallographic Misorientations in Hot-Dip Al–Zn–Si Coating, Acta Materialia 56: 4002–4011.

[27] Hull D, Bacon D.J (2001) Introduction to Dislocations. Butterworth-Heinemann. Chapters 4 and 9.

Permissions

The contributors of this book come from diverse backgrounds, making this book a truly international effort. This book will bring forth new frontiers with its revolutionizing research information and detailed analysis of the nascent developments around the world.

We would like to thank Muhammad Akhyar Farrukh, for lending his expertise to make the book truly unique. He has played a crucial role in the development of this book. Without his invaluable contribution this book wouldn't have been possible. He has made vital efforts to compile up to date information on the varied aspects of this subject to make this book a valuable addition to the collection of many professionals and students.

This book was conceptualized with the vision of imparting up-to-date information and advanced data in this field. To ensure the same, a matchless editorial board was set up. Every individual on the board went through rigorous rounds of assessment to prove their worth. After which they invested a large part of their time researching and compiling the most relevant data for our readers. Conferences and sessions were held from time to time between the editorial board and the contributing authors to present the data in the most comprehensible form. The editorial team has worked tirelessly to provide valuable and valid information to help people across the globe.

Every chapter published in this book has been scrutinized by our experts. Their significance has been extensively debated. The topics covered herein carry significant findings which will fuel the growth of the discipline. They may even be implemented as practical applications or may be referred to as a beginning point for another development. Chapters in this book were first published by InTech; hereby published with permission under the Creative Commons Attribution License or equivalent.

The editorial board has been involved in producing this book since its inception. They have spent rigorous hours researching and exploring the diverse topics which have resulted in the successful publishing of this book. They have passed on their knowledge of decades through this book. To expedite this challenging task, the publisher supported the team at every step. A small team of assistant editors was also appointed to further simplify the editing procedure and attain best results for the readers.

Our editorial team has been hand-picked from every corner of the world. Their multi-ethnicity adds dynamic inputs to the discussions which result in innovative outcomes. These outcomes are then further discussed with the researchers and contributors who give their valuable feedback and opinion regarding the same. The feedback is then collaborated with the researches and they are edited in a comprehensive manner to aid the understanding of the subject.

Apart from the editorial board, the designing team has also invested a significant amount of their time in understanding the subject and creating the most relevant covers. They scrutinized every image to scout for the most suitable representation of the subject and create an appropriate cover for the book.

The publishing team has been involved in this book since its early stages. They were actively engaged in every process, be it collecting the data, connecting with the contributors or procuring relevant information. The team has been an ardent support to the editorial, designing and production team. Their endless efforts to recruit the best for this project, has resulted in the accomplishment of this book. They are a veteran in the field of academics and their pool of knowledge is as vast as their experience in printing. Their expertise and guidance has proved useful at every step. Their uncompromising quality standards have made this book an exceptional effort. Their encouragement from time to time has been an inspiration for everyone.

The publisher and the editorial board hope that this book will prove to be a valuable piece of knowledge for researchers, students, practitioners and scholars across the globe.

List of Contributors

Claudia Maria Simonescu
Department of Analytical Chemistry and Environmental Engineering, Faculty of Applied Chemistry and Materials Science, „Politehnica" University of Bucharest, Romania

S.Lakshmi Reddy
Dept. of Physics, S.V.D.College, Kadapa, India

Tamio Endo
Dept. of Electrical and Electronics Engineering, Graduate School of Engineering, Mie University, Mie, Japan

G. Siva Reddy
Dept. of Chemistry, Sri Venkateswara University, Tirupati, India

Wieslawa Urbaniak-Domagala
Technical University of Lodz, Department of Material and Commodity Sciences and Textile Metrology, Poland

Taesam Kim and Chhiu-Tsu Lin
Northern Illinois University, Illinois, USA

Lidia Martínez and Elisa Román
ICMM-CSIC, Dept. Surfaces and Coatings, Madrid, Spain

Roman Nevshupa
CISDEM-CSIC, Spain

Jesús Rodarte Dávila, Jenaro C. Paz Gutierrez and Ricardo Perez Blanco
Department of Electrical and Computer Engineering, Juarez City Autonomous University, North Charro Avenue, Juarez City, Chihuahua, México

Vitaliy Tinkov
Department of the Surface Atomic Structure and Dynamic, Institute for Metal Physics of NAS of Ukraine, Ukraine

Daisuke Kosemura, Motohiro Tomita and Atsushi Ogura
School of Science and Technology, Meiji University, Kawasaki, Japan

Koji Usuda
Green Nanoelectronics Collaborative Research Center, AIST, Tsukuba, Ibaraki, Japan

Todd M. Alam and Janelle E. Jenkins
Sandia National Laboratories, Department of Nanostructured and Electronic Materials, Albuquerque, NM, USA

George Sarau and Silke Christiansen
Max Planck Institute for the Science of Light, Erlangen, Germany
Institute of Photonic Technology, Jena, Germany

Arne Bochmann
Institute of Photonic Technology, Jena, Germany

www.ingramcontent.com/pod-product-compliance
Lightning Source LLC
Chambersburg PA
CBHW070733190326
41458CB00004B/1145